Scientific Computation

Series Editor

Gang Zhang, Yangtze River Delta Research Institute, Beijing Institute of Technology, Jiaxing, Zhejiang, China

This series publishes original research monographs on topics described by the term "computational science". This includes theoretical aspects such as numerical analysis and mathematical modeling. The focus is on important recent developments at the interface of mathematical analysis and computational algorithms,as well as, the application of these approaches throughout the disciplines of engineering, physics, chemistry, earth sciences, biology, social sciences, finance and economics.

Zhengjun Liu · Yutong Li
Editors

Intelligent Analysis of Optical Images

Springer

Editors
Zhengjun Liu
School of Physics
Harbin Institute of Technology
Harbin, Heilongjiang, China

Yutong Li
School of Physics
Harbin Institute of Technology
Harbin, Heilongjiang, China

ISSN 1434-8322 ISSN 2198-2589 (electronic)
Scientific Computation
ISBN 978-981-95-3719-8 ISBN 978-981-95-3720-4 (eBook)
https://doi.org/10.1007/978-981-95-3720-4

This Springer imprint is published by the registered company Springer Nature Singapore Pte Ltd.
The registered company address is: 152 Beach Road, #21-01/04 Gateway East, Singapore 189721, Singapore

Preface

This book mainly introduces intelligent analysis methods for optical images. Focus on introducing analytical methods and programming knowledge. Automatic image processing and intelligent analysis are the trend of future applications. It is also an important analysis of machine vision. Combined with the optical imaging process, computing technology can realize operations such as extracting, modifying and organizing information. Studying these intelligent analysis methods is inseparable from science and technology such as optics, mathematics, computing, and artificial intelligence.

Collecting and organizing relevant technologies for intelligent analysis has important application significance in fields such as autonomous driving, computational vision, artificial intelligence, and multimodal information processing. Accurate, automatic and rapid acquisition of image information is an urgent application requirement. Image processing technology has been paid attention and studied for a long time. The difference between image analysis and image processing is difficult. Given the wealth of image information, it becomes even more necessary to extract the needed information from the image. It can save time and calculation for subsequent applications. For optical images, sorting out intelligent analysis methods is a relatively urgent task. It has great significance and social benefits for application and research.

Intelligent analysis methods are very important for image acquisition and image applications. It plays a connecting role as a bridge. Combined with deep learning technology, it can promote more comprehensive and in-depth research and help relevant information and technologies serve application tasks. This book contains current representative research results and believes that there will be more exciting research results in the future. This book introduces various methods for academic issues of intelligent analysis technology, brings together new theories and new technologies in optical image analysis, and provides new technical support for the field of image analysis.

This book provides readers with a comprehensive overview of the principles, methodology and latest developments in optical image analysis. This book is a valuable reference for engineers, scientific researchers, graduate students and senior undergraduates majoring in optics, imaging, signal, media processing service, etc. With the technical classification of optical image analysis, we divided this book into eight chapters:

Chapter 1–"Basic Programming Methods".
Chapter 2—"Target Recognition".
Chapter 3—"Autofocusing".
Chapter 4—"Optical Speckle Analysis".
Chapter 5—"Moving Images for Velocity Measurement".
Chapter 6—"Image Stitching".
Chapter 7—"Super-Resolution Imaging".
Chapter 8—"Shape Analysis".

Harbin, China

Zhengjun Liu
Yutong Li

Acknowledgements The editors would like to take this opportunity to express their sincere gratitude to the authors of the chapters for extending their wholehearted support in sharing some of their latest results. Without their significant contribution, this book volume could not have fulfilled its mission. All the authors are indebted to the editorial staff at Springer, for their hard working and selfish attitude on helping our book plan.

Contents

Contributors

Xiaoying Chen School of Physics, Harbin Institute of Technology, Harbin, China

Longhui Fu School of Physics, Harbin Institute of Technology, Harbin, China

Guancheng Huang School of Physics, Harbin Institute of Technology, Harbin, China

Yutong Li School of Physics, Harbin Institute of Technology, Harbin, Heilongjiang, China

Ziyang Li School of Physics, Harbin Institute of Technology, Harbin, China

Zhengjun Liu School of Physics, Harbin Institute of Technology, Harbin, Heilongjiang, China

Zhe Wang School of Physics, Harbin Institute of Technology, Harbin, China

Zhengyu Wu School of Physics, Harbin Institute of Technology, Harbin, Heilongjiang, China

Yulin Xue School of Physics, Harbin Institute of Technology, Harbin, China

Chapter 1
Basic Programming Methods

Zhengjun Liu and Yutong Li

Abstract This chapter introduces basic programming methods. This chapter includes image acquisition, image processing methods, program structure, combination of numbers and shapes, imaging and recognition, optimization methods, image analysis methods, image preprocessing methods, and image transformation methods. Learning these materials will help readers build computing programs for intelligent image analysis.

1.1 Image Acquisition

Images are mostly collected by optical systems. It is stored in a computer or other storage device in discrete and digital form, and it can be processed and analyzed by programs, transmitted over the Internet, etc.

1.1.1 Image Formation

The main forms of image acquisition include point-by-point scanning, line-by-line scanning, and full-area recording, as Fig. 1.1. Point-by-point scanning includes three scan interval methods. Overlap scans collect a large amount of data, such as stacked scan imaging [1]. The non-interval scanning method is a common method. Sparse scanning can collect less data, and combined with compressed sensing technology can supplement more information, realizing an imaging mode with less measurement and more calculation. The line scanning method obtains vector data one time, and

Z. Liu (✉) · Y. Li
School of Physics, Harbin Institute of Technology, Harbin, China
e-mail: zjliu@hit.edu.cn

Y. Li
e-mail: yutong_li@hit.edu.cn

Z. Liu and Y. Li (eds.), *Intelligent Analysis of Optical Images*, Scientific Computation,
https://doi.org/10.1007/978-981-95-3720-4_1

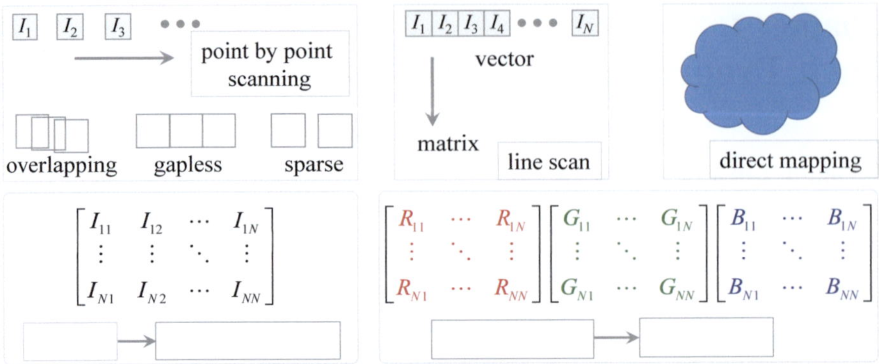

Fig. 1.1 Image acquisition and presentation

splicing them together to obtain a matrix to form an image. The direct mapping method uses an area array detector to record an image.

Image processing or analysis algorithms often used gray level images as test objects. Some scientific computing languages, such as MATLAB or Python, can be used to convert a grayscale image into a matrix. Color images are generally represented using three matrices, which correspond to three primary colors.

Color images can also be represented using indexed images. In order to solve the problem that color image calculation and storage consume a large amount of space, the data format of the index image is similar to a dictionary, and each pixel value corresponds to a color. It is generally used in scenes with simple color composition. With the emergence of some image compression methods and image formats, such as jpg images, it is not commonly used to use indexed images to represent color images.

Here is an image format for animation storage, which is gif images. It saves many images together and can be easily embedded into PowerPoint software. Playing gif file can demonstrate the dynamic process.

Application commands can save the matrix as an image. The image data format is mostly 8-bit integer data. For image display, the data format is sufficient in terms of visual observation requirements. Users can save in image formats such as jpg, tiff, and emf, as needed.

1.1.2 Image Recording

Objects are recorded in imaging through the optical system, as Fig. 1.2. Imaging is a process of information loss, and its inverse process is difficult to perform perfectly, and in some cases it is even a pathological model. The imaging limit resolution is determined by the wavelength and numerical aperture.

The imaging model can be described by the following convolution formula

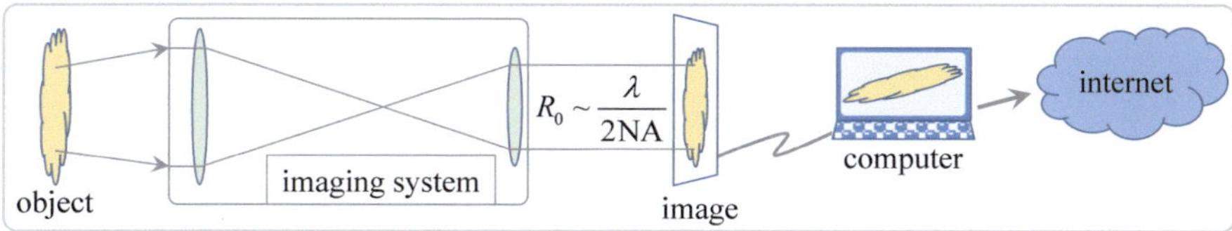

Fig. 1.2 Imaging model

$$I(x, y) = f(x, y) \otimes h(x, y) + n(x, y) \tag{1.1}$$

where, $I(x, y)$ is imaging result. $f(x, y)$ is characterization light field information of objects. $h(x, y)$ is point spread function of imaging system. The function is determined by the properties of the imaging system and is independent of the system input and output. Its Fourier transform is called the optical transfer function. $n(x, y)$ is noise function.

If there is relative motion in the shooting system, then the imaging result is related to motion blur. The manufacturing error of the optical elements of the imaging system is reflected in the expression of the point spread function. There is atmospheric turbulence influence or underwater imaging during the imaging process, and the response correction term needs to be considered in the point spread function. Defocus conditions can cause blurred images.

1.1.3 Imaging Error and Correction

Optical experimental systems are famous for being difficult to adjust because they require high concentric and coaxial optical elements. With the rise of computational optical imaging technology, optical systems are combined with technology, and some mechanical adjustments and manual operations can be completed by digital compensation, or even automatically and efficiently.

Figure 1.3 shows the pose errors of three optical elements. Inclined illumination will cause the imaging light signal to weaken, and actually more high-frequency information will enter the imaging system. This situation is also used for super-resolution imaging [2]. The eccentricity of the optical element is equivalent to the translation of the center of the lens. The recording surface is tilted, and at a small angle approximation, it is equivalent to increasing the linear phase, which will cause distortion of the recorded image.

The correction method can be judged based on the acquired image conditions. Tilt lighting will cause spectrum shift, and the tilt angle can be determined based on the spectrum shift. If the lens is eccentric, the amount of translation can be judged based on the image translation. The tilt of the recording surface will cause pattern deformation, and deformation correction can be performed based on the

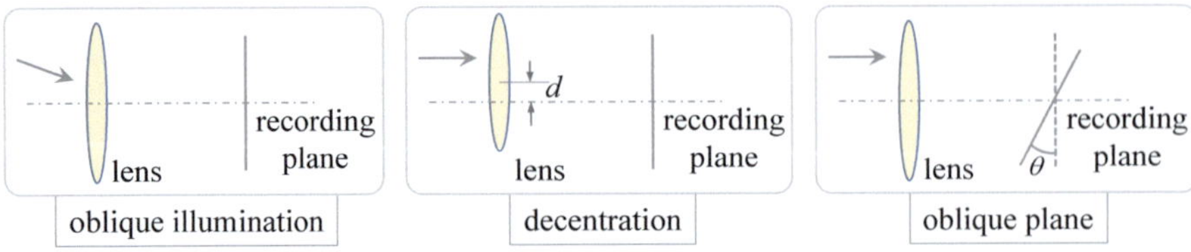

Fig. 1.3 Image degradation caused by several pose errors

affine transformation parameters. For pattern translation, correlation operations can be applied to determine the amount of translation.

1.2 Image Processing Methods

Digital images stored in computers can be processed by professional software or computing code to obtain the required image form. Researchers prefer to design algorithms to process digital images. The content of the book is limited and fixed. Readers who want to get good image processing skills training should complete more programming training on practical problems, so that they can improve their ability to solve problems.

1.2.1 Binarization

The binarization makes image data simple. Binary data is suitable for processing by some logical operations, such as morphological operations. The key to binarization processing is to determine the threshold v_t. It can be determined by an average or an adaptive algorithm, as the graph in Fig. 1.4. The data of the image is constrained to two values by a threshold.

The outline of the optical image becomes blurred due to diffraction, and the approximate range of the target can be determined after binarization. Binarization patterns are also used to localize a target range, called a mask pattern. The binary map

Fig. 1.4 Image binarization

$P_{\text{mean}} = \frac{1}{9}\begin{bmatrix} 1 & 1 & 1 \\ 1 & 1 & 1 \\ 1 & 1 & 1 \end{bmatrix}$	$S_x = \begin{bmatrix} -1 & 0 & 1 \\ -2 & 0 & 2 \\ -1 & 0 & 1 \end{bmatrix}$ $S_y = \begin{bmatrix} -1 & -2 & -1 \\ 0 & 0 & 0 \\ 1 & 2 & 1 \end{bmatrix}$	$L = \begin{bmatrix} -1 & -1 & -1 \\ -1 & 8 & -1 \\ -1 & -1 & -1 \end{bmatrix}$
mean operator smoothing filtering	Sobel operator image edge	Laplace operator high-pass filtering

Fig. 1.5 Filtering operators

limits the calculation amount of optical image processing to some image elements, reducing the calculation amount and improving algorithm execution speed.

1.2.2 *Image Filtering*

Image filtering is a big concept, and many image processing operations can be regarded as filtering. An average filter can be constructed using an all-one matrix. If the filter matrix has difference coefficients, such as the Sobel operator, they can be used to extract the edges of the image. The Laplacian operator calculates the difference between the central pixel and the eight neighborhood pixels, which can be used to highlight high-frequency information of the image. The calculation formulas for these operators are shown in Fig. 1.5. In the neighborhood of the center pixel, the median value is calculated as the filtering result. This method is called median filtering. Filtering using these operators is called spatial filtering. The two-dimensional inner product of these filtering operators and graphic elements is traversed to complete the filtering.

Frequency domain filtering adjusts the Fourier spectrum of the image matrix, which can be converted to the frequency domain using fast Fourier transform algorithm to achieve filtering purposes. Reduce the spectral amplitude data based on the spectral area where noise may occur. Given the randomness of noise, noise reduction is an endless research topic. The noise-free state of actual images is difficult to achieve.

1.2.3 *Image Moments*

$(p+q)$-order image moments is defined as

$$m_{pq} = \sum_{i=1}^{M} \sum_{j=1}^{N} i^p j^q I(i, j), \quad p, q = 0, 1, 2, \ldots \tag{1.2}$$

The central moment is expressed as

$$\mu_{pq} = \sum_{i=1}^{M} \sum_{j=1}^{N} (i - \bar{i})^p (j - \bar{j})^q I(i, j) \tag{1.3}$$

where

$$\bar{i} = m_{10} \big/ m_{00}, \; \bar{j} = m_{01} \big/ m_{00} \tag{1.4}$$

The normalized central moment is given as

$$\eta_{pq} = \mu_{pq} \big/ \mu_{00}^{k}, \; k = \frac{p+q}{2}, \; p + q = 2, 3, \ldots \tag{1.5}$$

From the above definition, we can determine the seven invariant moments that are insensitive to translation, scaling, and rotation. They are defined as

$$\varphi_1 = \eta_{02} + \eta_{20} \tag{1.6}$$

$$\varphi_2 = (\eta_{02} - \eta_{20})^2 + 4\eta_{11}^2 \tag{1.7}$$

$$\varphi_3 = (\eta_{30} - 3\eta_{12})^2 + (\eta_{03} - 3\eta_{21})^2 \tag{1.8}$$

$$\varphi_4 = (\eta_{30} + \eta_{12})^2 + (\eta_{03} + \eta_{21})^2 \tag{1.9}$$

$$\begin{aligned} \varphi_5 = {} & (\eta_{30} - 3\eta_{12})(\eta_{30} + \eta_{12})\left[(\eta_{30} + \eta_{12})^2 - 3(\eta_{21} + \eta_{03})^2\right] \\ & + (3\eta_{21} - \eta_{03})(\eta_{21} + \eta_{03})\left[3(\eta_{30} + \eta_{12})^2 - (\eta_{21} + \eta_{03})^2\right] \end{aligned} \tag{1.10}$$

$$\varphi_6 = (\eta_{20} - \eta_{02})\left[(\eta_{30} + \eta_{12})^2 - (\eta_{21} + \eta_{03})^2\right] + 4\eta_{11}(\eta_{30} + \eta_{12})(\eta_{21} + \eta_{03}) \tag{1.11}$$

$$\begin{aligned} \varphi_7 = {} & (3\eta_{21} - \eta_{03})(\eta_{30} + \eta_{12})\left[(\eta_{30} + \eta_{12})^2 - 3(\eta_{21} + \eta_{03})^2\right] \\ & + (3\eta_{21} - \eta_{30})(\eta_{21} + \eta_{03})\left[3(\eta_{30} + \eta_{12})^2 - (\eta_{21} + \eta_{03})^2\right] \end{aligned} \tag{1.12}$$

In the case of zooming and panning, applying these seven moments can provide very important reference data for the target to be measured, and the moment values can be used to filter and determine the target in the image.

1.3 Program Structure

Program structure is the skeleton of program design. Programmers can draw draft flowcharts or determine program calculations in their heads. On the premise that the calculation results are correct, optimize the program to improve execution speed.

1.3.1 Structure Design

Ordinary program structure is a form of executing statements in sequence. All statements are calculated in order to get the final result. Note that the code is executed in the order of assignment first and call later. To enhance readability, it is best to put related calculation statements together.

Loop structures are generally used for repetitive calculation tasks, and some traversal calculations can be calculated using this structure. When loop calculations are terminated through conditional statements, you need to pay attention to the number of loop calculations within expectations to avoid long-term invalid execution of the program.

Recursive structure is a very concise form of program. It requires a calculation formula similar to a sequence of numbers. For example, factorial operations can be calculated using a recursive structure. This program structure can be used for the drawing of some fractal shapes, such as the Koch curve, and its self-similarity is the basis for the design of recursive program structures.

1.3.2 Program Debugging

Writing a program is inseparable from program debugging. Textbooks and open source code are the basis for program reference code. In order to obtain new computing functions, users need to modify existing reference programs.

One type of program error is a syntax error. These errors can be prompted by the programming language, and users can correct the code based on the prompts. It is easy to discover and correct.

Modifying other program errors is time-consuming and laborious. The program can run, but it cannot get the expected results. Programmers need to check statement by statement.

Program optimization can increase execution speed. Programs can be accelerated through parallel calculations. Similar to the MATLAB language, you can calculate tasks through vectorized programs and perform parallel calculations of many matrix elements on vectors or matrices.

1.3.3 Toolbox

A toolbox is a collection of functionally relevant and frequently used code. It can be provided by software developers. Users can also make their own toolbox. It can consist of files such as data, code, and help documents. The new code can call data and program code in the toolbox.

Commonly used functions will be included in the toolbox. Programmers try to design programs based on toolboxes, which can save programming time.

1.4 Number-Shape

Image processing is more information processing based on data. Shape can be regarded as a feature expressed by an image. Geometric distortion issues sometimes occur when taking images.

1.4.1 Affine Transform

Affine transformation is the application of a 2×2 matrix transform as

$$\begin{bmatrix} x' \\ y' \end{bmatrix} = M \begin{bmatrix} x \\ y \end{bmatrix} = \begin{bmatrix} a & b \\ c & d \end{bmatrix} \begin{bmatrix} x \\ y \end{bmatrix} \tag{1.13}$$

It can also be extended to three-dimensional coordinate transformation. When the identity matrix M is used, the coordinates do not change before and after.

Figure 1.6 shows several results for affine transforms. When the matrix M is sample, the square mesh is converted to the shape of a parallelogram. If we add random small amounts to each element of the matrix, such as the two patterns behind Fig. 1.6, the square will become a random quadrangle. Here $R(x, y)$ is a random number corresponding to the coordinates of the grid point, and its value is between 0 and 1.

Affine transform is simpler and more interesting. Its inverse process corresponds to shape correction, also known as camera calibration. For ordinary affine transform, that is, transform into a parallelogram, we choose the other two points to form the following equation

$$\begin{bmatrix} x_1' & x_2' \\ y_1' & y_2' \end{bmatrix} = M \begin{bmatrix} x_1 & x_2 \\ y_1 & y_2 \end{bmatrix} \tag{1.14}$$

The calculation of matrix M is as

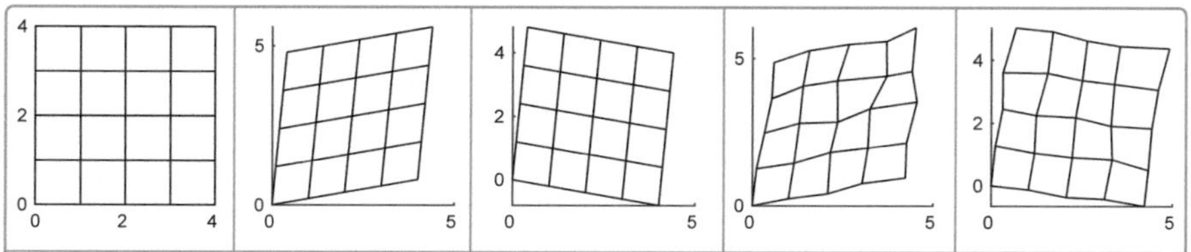

$M_0 = \begin{bmatrix} 1 & 0 \\ 0 & 1 \end{bmatrix}$	$M_1 = \begin{bmatrix} 0.1 & 1 \\ 1.2 & 0.2 \end{bmatrix}$	$M_2 = \begin{bmatrix} 0.1 & 1 \\ 1.2 & -0.2 \end{bmatrix}$	$M_1 + \frac{R(x,y)}{12.5}$	$M_2 + \frac{R(x,y)}{12.5}$

Fig. 1.6 Affine transform

$$M = \begin{bmatrix} x_1' \ x_2' \\ y_1' \ y_2' \end{bmatrix} \begin{bmatrix} x_1 \ x_2 \\ y_1 \ y_2 \end{bmatrix}^{-1} \tag{1.15}$$

Equation (1.15) holds if the inverse matrix exists, i.e. $\det([x_1, x_2; y_1, y_2]) \neq 0$. If it doesn't exist, we need to change the coordinates. Figure 1.7 shows the relationship between the determinant values corresponding to the coordinates of two points. If the determinant values corresponding to the coordinates of two points are equal to 0, the red dotted line is used for the connection line; otherwise, the black solid line is used for connection. Table 1.1 gives a set of data that can be used to verify the inversion of Eq. (1.15), and the corresponding determinant value is equal to 3.

When there are random affine coefficients, the restoration of this affine relationship becomes more complex. Complex deformations can be modeled as the result of affine matrices. Affine matrix elements are position-dependent distribution functions. The difficulty of the problem is that when determining the affine matrix, it is difficult to find coordinate data that requires two points because the amount of geometric distortion at each point is different. It becomes an ill-conditioned equation problem.

In the topography measurement method, fringe projection can be used to obtain the topography changes of the sample surface. The deformation of an optical imaging system can be equivalent to the object of topography measurement. Deformation variables are reflected in phase changes. Figure 1.8 shows the corresponding calculation process. In the input plane of the imaging system, the phase function is characterized as a fringe pattern.

A pattern display device is used to use these stripes as input images to the imaging system. These deformed fringes can be collected at the output, and the phase distribution can be solved from these deformed fringes. To understand the calculation phase, the initial phase φ_0 can be changed. Measured output pattern is I_n. An example is as follows

$$(\varphi_0 = 0) \to I_0,\ (\varphi_1 = \pi/2) \to I_1,\ (\varphi_2 = \pi) \to I_2,\ (\varphi_3 = 3\pi/2) \to I_3 \tag{1.16}$$

$$\phi' = \arg\left\{\frac{1}{2}(I_2 - I_0) + \frac{i}{2}(I_3 - I_1)\right\} \tag{1.17}$$

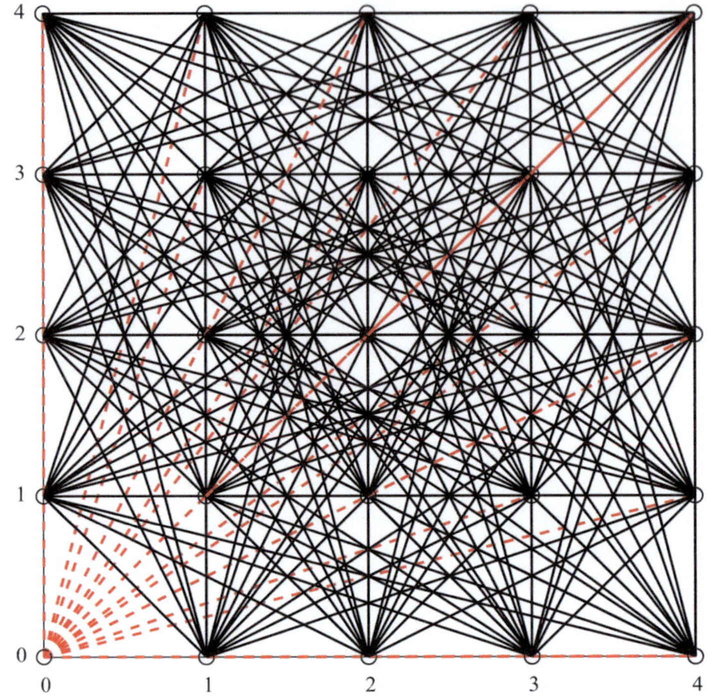

Fig. 1.7 Relation between determinant values

Table 1.1 Coordinate points

x_1	x_2	y_1	y_2	x'_1	x'_2	y'_1	y'_2
1	1	0	3	0.1	3.1	1.2	1.8

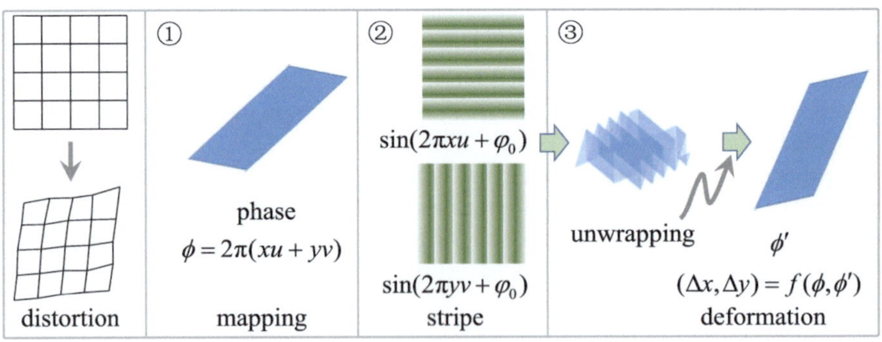

Fig. 1.8 Determining shape variable using fringe projection

Equation (1.17) uses a four-step phase shift method to calculate the phase distribution, which can remove the DC component and suppress the influence of noise. Three-step phase shift and two-step phase shift methods can also be used for calculation and measurement methods.

The phase result obtained in Eq. (1.17) requires unwrapping to obtain the actual phase distribution ϕ', in which it includes the interference of deformation factors on the phase. According to the phase transformation, the positional changes with the same phase value can be retrieved, and the distribution of distortion variables in the orthogonal direction $(\Delta x,\ \Delta y)$ can be calculated. Image deformation can be restored by interpolation method. Interpolation is difficult to perfectly restore the exact shape of the original figure because it is not a reversible operation.

1.4.2 Path

Designing paths is very interesting and requires some knowledge of geometry. Especially in some ray tracing methods, it is very important. We simulate the situation where a point light source emits a beam of light in an ellipse and reflects multiple times in the ellipse. In order to design the program, several calculation methods need to be explained here.

Elliptic equation is defined as

$$\frac{x^2}{a^2}+\frac{y^2}{b^2}=1 \tag{1.18}$$

Let the coordinates of the light source be (x_0, y_0). It is inside the ellipse. The angle at which the light source emits a ray is α. The ray equation is

$$y = kx + b_c \tag{1.19}$$

where, $k = \tan\alpha$. Considering that the light source is on this straight line, the following relation exists

$$y_0 = kx_0 + b_c \tag{1.20}$$

Removing b_c, we have

$$y = kx + y_0 - kx_0 \tag{1.21}$$

Substitute into the elliptic equation and get

$$\frac{x^2}{a^2}+\frac{[kx+y_0-kx_0]^2}{b^2}=1 \tag{1.22}$$

$$\frac{x^2}{a^2}+\frac{k^2x^2}{b^2}+\frac{2kx(y_0-kx_0)}{b^2}+\frac{[y_0-kx_0]^2}{b^2}=1 \tag{1.23}$$

$$\left(\frac{1}{a^2}+\frac{k^2}{b^2}\right)x^2+\frac{2kx(y_0-kx_0)}{b^2}+\frac{[y_0-kx_0]^2}{b^2}-1=0 \tag{1.24}$$

The discriminant is

$$\Delta=\frac{[2k(y_0-kx_0)]^2}{b^4}-4\left(\frac{1}{a^2}+\frac{k^2}{b^2}\right)\frac{[y_0-kx_0]^2-b^2}{b^2} \tag{1.25}$$

The two solutions of the equation are obtained as

$$x_{c1,c2}=-\frac{1}{2}\left(\frac{2k(y_0-kx_0)}{b^2}\pm\sqrt{\Delta}\right)\left(\frac{1}{a^2}+\frac{k^2}{b^2}\right)^{-1} \tag{1.26}$$

There are four possible solutions to the corresponding ordinate of the intersection of a straight line and an ellipse

$$y_{c1,1,2}=\pm b\sqrt{1-x_{c1}^2/a^2},\quad y_{c2,1,2}=\pm b\sqrt{1-x_{c2}^2/a^2} \tag{1.27}$$

Three points can be excluded based on the direction of light travel to obtain the intersection coordinates (x_c, y_c).

When $\alpha = 0,\ \pi/2,\ \pi,\ 3\pi/2$, the light travels in the horizontal or vertical direction. At this time, the intersection coordinates can be calculated according to Eq. (1.27) or Eq. (1.18). The unique intersection point needs to be determined according to the direction of the light travel.

Another problem that needs to be addressed is the reflection of light when it encounters an ellipse, considering the reflection calculation at the intersection. Normal direction on the inside of the ellipse ϕ_n can be calculated by

$$\phi_n=\arg(x_c/a+\mathrm{i}\cdot y_c/b)+\pi \tag{1.28}$$

where, the function arg is to calculate the phase of a complex number.

The reflection is expressed as

$$\alpha_r=2\phi_n-(\alpha+\pi) \tag{1.29}$$

where, α_r is direction of reflected light. In calculation, $a = 4$, $b = 3$, $\alpha = 0.66 \times \pi$, $x_0 = 1$, $y_0 = 2$. The result of multiple reflections of light is shown in Fig. 1.9.

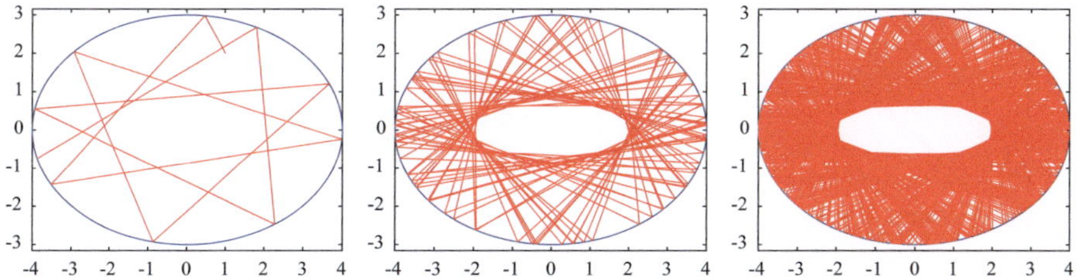

Fig. 1.9 The trajectory of multiple reflections of light

In this example, we mainly calculate the intersection of a straight line and an ellipse. The distribution of normal methods on the inside of the ellipse is easy to calculate. Because the coordinate calculation formula of the intersection point is calculated based on the formula, its error will be very small. In other cases, intersection coordinate calculation and reflection processing will become complicated.

1.4.3 Convex Polygon

Consider calculating a circumscribed convex polygon [3] that surrounds some points, as Fig. 1.10. Randomly generate several point coordinates and plot them. By calculating the minimum value of the abscissa, select the leftmost point of these points as the first determined polygon vertex. Select a point as the vertex of the polygon, such as (x_2, y_2). Given a function $f(x, y)$, and $f(x, y) = 0$ corresponds to the equation of the straight line through these two points. The flow chart in Fig. 1.10 shows the process of finding the vertices of a polygon. It is completed by double loop calculation.

If the product of the maximum and minimum values of the function $f(x, y)$ corresponding to these point coordinates is non-negative, then indicating that these points are on one side of the current line, this constraint comes from the definition of a convex polygon. At this time, the coordinates of the current point are recorded as vertex coordinates, and the determined vertex for the next search is updated as the current vertex.

If the distance from the current vertex to the first vertex is equal to 0 or very small, and the number of polygon vertices obtained is greater than 3, the calculation is terminated and the polygon vertex coordinates are output.

As an extension, the three points (x_1, y_1, z_1), (x_2, y_2, z_2) and (x_3, y_3, z_3), that are not collinear, the determined plane equation can be written into a determinant as

$$\begin{vmatrix} x - x_1 & y - y_1 & z - z_1 \\ x_2 - x_1 & y_2 - y_1 & z_2 - z_1 \\ x_3 - x_1 & y_3 - y_1 & z_3 - z_1 \end{vmatrix} = 0 \tag{1.30}$$

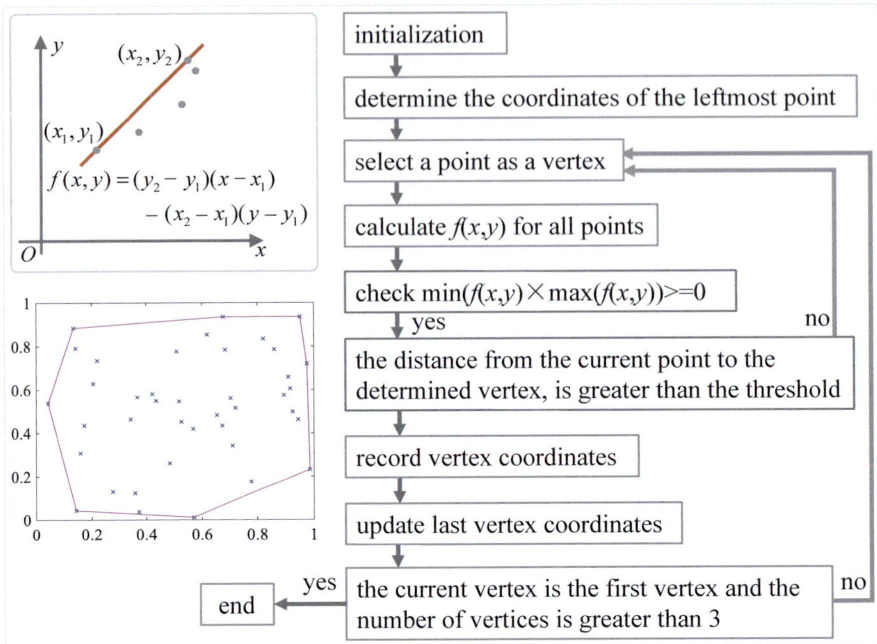

Fig. 1.10 Determination of vertices of circumscribed convex polygon

The calculation formula of Eq. (1.30) is long and is important for writing statements in programming. Based on the value of the determinant on the left side of the equation, the relative positions of the points in the plane can be judged. Its dimensionality reduced form

$$\begin{vmatrix} x - x_1 & y - y_1 \\ x_2 - x_1 & y_2 - y_1 \end{vmatrix} = 0 \tag{1.31}$$

It is related with $f(x, y) = 0$.

Here is a supplement to the formula for the distance from a point to the plane. For a plane equation $Ax + By + Cz + D = 0$, the distance d_2 from a point (x_0, y_0, z_0) to the plane is

$$d_3 = \frac{|Ax_0 + By_0 + Cz_0 + D|}{\sqrt{A^2 + B^2 + C^2}} \tag{1.32}$$

Let the left determinant of Eq. (1.30) be written as $M_2 = f_3(x, y, z)$. Here we consider to calculate the coefficients A, B, C, and D as

$$D = f_3(0, 0, 0),\ A = f_3(1, 0, 0) - D,\ B = f_3(0, 1, 0) - D,\ C = f_3(0, 0, 1) - D \tag{1.33}$$

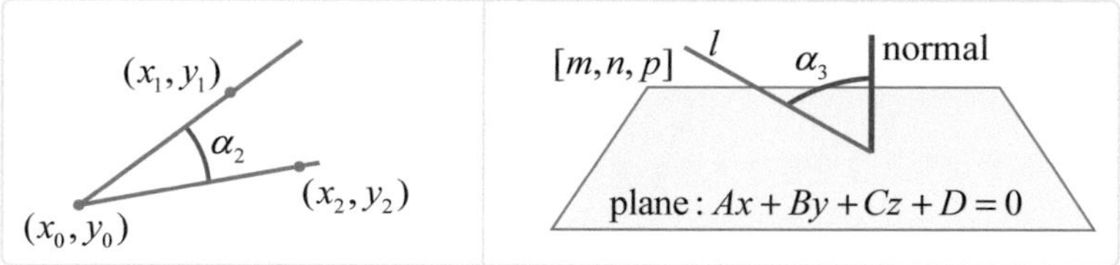

Fig. 1.11 Model for calculating angles

For 2D case, the calculation of distance d_2 is listed as follows

$$\begin{cases} Ax + By + C = 0 \\ d_2 = \dfrac{|Ax + By + C|}{\sqrt{A^2 + B^2}} \\ C = f_2(0, 0),\ A = f_2(1, 0) - C,\ B = f_2(0, 1) - C \end{cases} \tag{1.34}$$

where, $M_2 = f_2(x, y)$ corresponds to the determinant form of the straight line equation in two-dimensional cases. Applying these formulas to calculate distances makes writing code simple and easy.

The calculation from coordinates to angles is shown in Fig. 1.11. For two-dimensional cases, the angle α_2, can be obtained using complex numbers. Its value is

$$\alpha_2 = \arg([x_1 - x_0] + \mathrm{i} \cdot [y_1 - y_0]) - \arg([x_2 - x_0] + \mathrm{i} \cdot [y_2 - y_0]) \tag{1.35}$$

The normal vector of the plane is $[A, B, C]$. The angle formula can be obtained from the inner product of vectors, as

$$\begin{aligned} \cos\alpha_3 &= \langle [A, B, C], [m, n, p] \rangle \\ &= \frac{|Am + Bn + Cp|}{\sqrt{A^2 + B^2 + C^2} \cdot \sqrt{m^2 + n^2 + p^2}} \end{aligned} \tag{1.36}$$

where, $[m, n, p]$ is the direction vector for the line l.

1.4.4 Circle Fitting

For some points distributed near the circumference, as Fig. 1.12, we can use the fitting method to determine the parameters of the circle, namely the center and radius.

Let the coordinates of these points be (x_1, y_1), (x_2, y_2), …, (x_n, y_n), the equation for the circle is written as

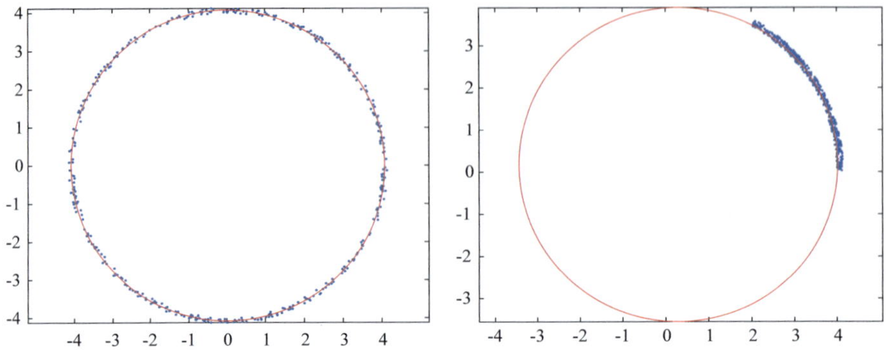

Fig. 1.12 Results of circle fitting

$$(x - x_c)^2 + (y - y_c)^2 = r^2 \tag{1.37}$$

The following formula can be obtained

$$2xx_c + 2yy_c - x_c^2 - y_c^2 + r^2 = x^2 + y^2 \tag{1.38}$$

Let $d = x_c^2 + y_c^2 - R^2$, we have

$$\begin{bmatrix} x_1^2 + y_1^2 \\ x_2^2 + y_2^2 \\ \vdots \\ x_n^2 + y_n^2 \end{bmatrix} = \begin{bmatrix} 2x_1 & 2y_1 & -1 \\ 2x_2 & 2y_2 & -1 \\ \vdots & \vdots & \vdots \\ 2x_n & 2y_n & -1 \end{bmatrix} \begin{bmatrix} x_c \\ y_c \\ d \end{bmatrix} \tag{1.39}$$

Defining A and B

$$B = \begin{bmatrix} x_1^2 + y_1^2 \\ x_2^2 + y_2^2 \\ \vdots \\ x_n^2 + y_n^2 \end{bmatrix}, A = \begin{bmatrix} 2x_1 & 2y_1 & -1 \\ 2x_2 & 2y_2 & -1 \\ \vdots & \vdots & \vdots \\ 2x_n & 2y_n & -1 \end{bmatrix} \tag{1.40}$$

According to the least squares method, we have

$$\begin{bmatrix} x_c \\ y_c \\ d \end{bmatrix} = \left(A^{\mathrm{T}}A\right)^{-1}A^{\mathrm{T}}B \tag{1.41}$$

The radius of the circle is $r = \sqrt{x_c^2 + y_c^2 - d}$. Circle fitting and arc fitting can be performed by applying Eq. (1.41), as Fig. 1.12.

The calculation method of ellipse fitting is supplemented. Let the elliptic equation $f(x, y) = 0$, as

$$f(x, y) = x^2 + ay^2 + bx + cy + dxy + q = 0 \tag{1.42}$$

$$f(x, y) = x^2 + ay^2 + bx + cy + dxy + q = E \tag{1.43}$$

where E represents the deviation of a point from an ellipse. A cumulative error function F_e is defined as

$$\begin{aligned} F_e &= \sum_{\forall i} E_i^2 \\ E_i &= f(x_i, y_i),\ i = 1, 2, ..., n \end{aligned} \tag{1.44}$$

To get the minimum of F_e, the partial derivative is calculated

$$\frac{\partial F_e}{\partial a} = 0,\ \frac{\partial F_e}{\partial b} = 0,\ \frac{\partial F_e}{\partial c} = 0,\ \frac{\partial F_e}{\partial d} = 0,\ \frac{\partial F_e}{\partial q} = 0 \tag{1.45}$$

Consider all points (x_i, y_i), we have

$$\frac{\partial F_e}{\partial a} = \sum_{\forall i} 2E_i \frac{\partial E_i}{\partial a} = \sum_{\forall i} 2\left(x_i^2 + ay_i^2 + bx_i + cy_i + dx_iy_i + q\right)y_i^2 = 0 \tag{1.46}$$

and

$$\sum_{\forall i} y_i^2 \left[y_i^2\ x_i\ y_i\ x_iy_i\ 1\right]\left[a\ b\ c\ d\ q\right]^{\mathrm{T}} = -\sum_{\forall i} y_i^2 x_i^2 \tag{1.47}$$

$$\sum_{\forall i} x_i \left[y_i^2\ x_i\ y_i\ x_iy_i\ 1\right]\left[a\ b\ c\ d\ q\right]^{\mathrm{T}} = -\sum_{\forall i} x_i x_i^2 \tag{1.48}$$

$$\sum_{\forall i} y_i \left[y_i^2\ x_i\ y_i\ x_iy_i\ 1\right]\left[a\ b\ c\ d\ q\right]^{\mathrm{T}} = -\sum_{\forall i} y_i x_i^2 \tag{1.49}$$

$$\sum_{\forall i} x_iy_i \left[y_i^2\ x_i\ y_i\ x_iy_i\ 1\right]\left[a\ b\ c\ d\ q\right]^{\mathrm{T}} = -\sum_{\forall i} x_iy_ix_i^2 \tag{1.50}$$

$$\sum_{\forall i} 1 \times \left[y_i^2\ x_i\ y_i\ x_iy_i\ 1\right]\left[a\ b\ c\ d\ q\right]^{\mathrm{T}} = -\sum_{\forall i} 1 \times x_i^2 \tag{1.51}$$

Adding Eq. (1.47) to Eq. (1.51), we get

$$A_e\left[a\ b\ c\ d\ q\right]^{\mathrm{T}} = B_e \tag{1.52}$$

where

$$A_e = \begin{bmatrix} S_1 y_1^2 & S_1 x_1 & S_1 y_1 & S_1 x_1 y_1 & S_1 \\ S_2 y_2^2 & S_2 x_2 & S_2 y_2 & S_2 x_2 y_2 & S_2 \\ \vdots & \vdots & \vdots & \vdots & \vdots \\ S_n y_n^2 & S_n x_n & S_n y_n & S_n x_n y_n & S_n \end{bmatrix}, \quad B_e = \begin{bmatrix} -S_1 x_1^2 \\ -S_2 x_2^2 \\ \vdots \\ -S_n x_n^2 \end{bmatrix} \tag{1.53}$$

$$S_i = y_i^2 + x_i + y_i + x_i y_i + 1, \; i = 1, 2, \ldots, n \tag{1.54}$$

The obtained elliptic coefficient vector is

$$\begin{bmatrix} a & b & c & d & q \end{bmatrix}^{\mathrm{T}} = \left(A_e^{\mathrm{T}} A_e\right)^{-1} A_e^{\mathrm{T}} B_e \tag{1.55}$$

Figure 1.13 shows the ellipse fitting results. The entire ellipse fits well. However, the fitting results of half ellipse showed obvious deviation. If you can predict the range of the scatter distribution and perform a symmetrical mirror operation on the data, you may get better results because of the added constraints.

Consider a method of determining a circle based on three coordinate points on a plane, as Fig. 1.14. Let the coordinates of three points be (x_1, y_1), (x_2, y_2), (x_3, y_3), Consider two midpoint coordinates as

$$x_{1,2} = \frac{x_1 + x_2}{2}, \; y_{1,2} = \frac{y_1 + y_2}{2}, \; x_{1,3} = \frac{x_1 + x_3}{2}, \; y_{1,3} = \frac{y_1 + y_3}{2} \tag{1.56}$$

The corresponding two equations for median perpendicular are

$$(y_2 - y_1)(y - y_{1,2}) + (x_2 - x_1)(x - x_{1,2}) = 0 \tag{1.57}$$

$$(y_3 - y_1)(y - y_{1,3}) + (x_3 - x_1)(x - x_{1,3}) = 0 \tag{1.58}$$

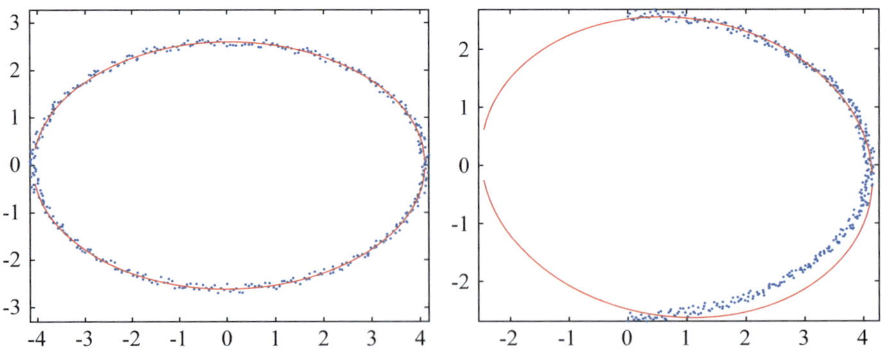

Fig. 1.13 Elliptic fitting results

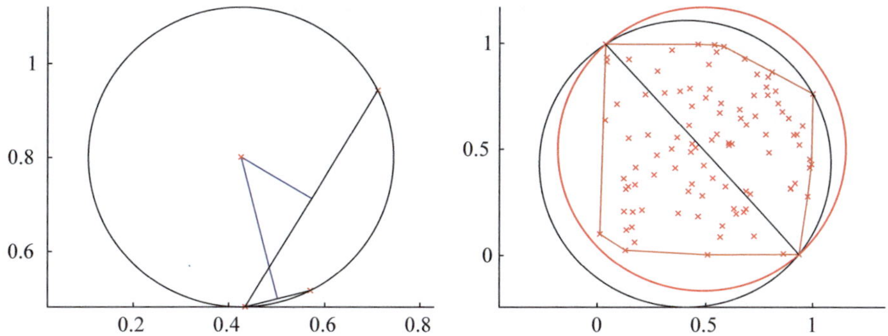

Fig. 1.14 Circle determination and circumscribed circle

Solve the system of equations corresponding to Eqs. (1.57) and (1.58), and apply matrix inversion to get the corresponding solution, that is, the center coordinates. The radius can be determined based on the distance from the center of the circle to these three points.

For calculating the determination of circumscribed circles of many points. As Fig. 1.14, determine the circumscribed polygon according to the method in Fig. 1.10. Select three points based on the vertices of the circumscribed polygon to calculate the circumscribed circle, and finally retain the circle with the smallest radius. This circle is like the black circle in Fig. 1.14, with a radius equal to 0.6764.

Another idea is to select the farthest pair of points from the vertices of the polygon, such as the endpoints of the black line segment in Fig. 1.14. Determine the circle whose diameter corresponds to this line segment (radius of $r_c = 0.6685$), whether it can contain all points, let the center of the circle be (x_c, y_c). If it cannot include all points, consider the points outside the circle and calculate the following function

$$F = (x_c - x_i)^2 + (y_c - y_i)^2 - r_c^2,\ i = 1, 2, ..., n \tag{1.59}$$

where, (x_i, y_i) represents the coordinates of all random points. Find the point corresponding to the maximum value of function F, and form three points with the endpoint of the previous line segment to determine a center information. At this time, the obtained circle can contain all points. With help of this idea, readers can design corresponding calculation programs. This latter idea requires less calculation.

1.5 Imaging and Recognition

Image acquisition can be accomplished by the imaging system. Computational optical imaging systems emerging in recent years can also accomplish this task [4]. The hardware of a computational imaging system is not a complete imaging

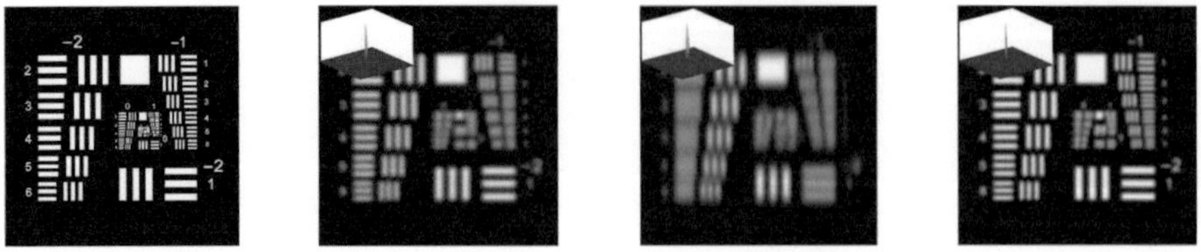

Fig. 1.15 Imaging results simulated by using point spread function

structure. Its essential components are light sources and cameras, such as coherent diffraction imaging systems.

1.5.1 Imaging System

The mathematical model of the imaging system can be expressed as

$$i(x, y) = o(x, y) \otimes h(x, y) + n(x, y) \tag{1.60}$$

where, i is imaging result. o is measured object. h is the point spread function of the system. For an ideal imaging system, the function h has a pulse shape. The smaller the effective response range of the function h, the higher the resolution of the imaging system. n is noise term. The symbol $\otimes$ is convolution operator.

Converting Eq. (1.60) to the Fourier domain allows convenient discrete calculations as

$$I(u, v) = O(u, v)H(u, v) + N(u, v) \tag{1.61}$$

The calculation basis for Eq. (1.61) is the Fourier transform and its properties. This allows the convolution to be calculated using the discrete Fourier transform algorithm. Figure 1.15 shows the imaging results of three different point spread functions. In the calculation, the point spread function is generated using a Gaussian function.

1.5.2 Computational Optical Imaging System

Computational optical imaging systems can take a very simple form. It is an imaging method that combines hardware with computing technology. Here is a description of the measurement and calculation methods of the coherent diffraction imaging system, as Fig. 1.16. In object plane, $f(x, y)$ is converted via diffraction process $\mathcal{D}$, the function of light filed $f'(x, y)$ in recording plane can be received. Without considering noise, the following relation can be found

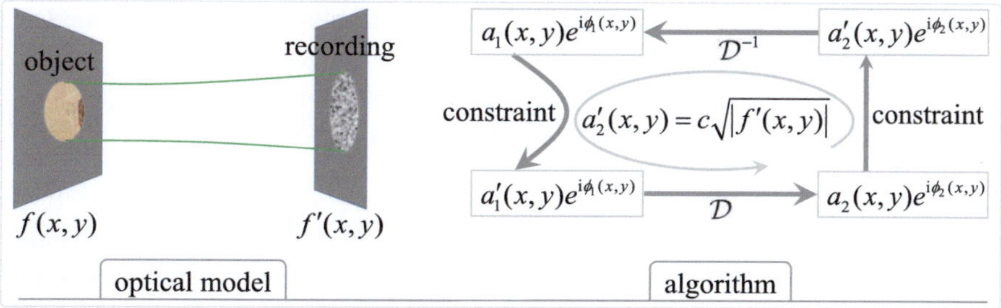

Fig. 1.16 Coherent diffraction imaging model and phase recovery algorithm

$$f'(x, y) = \mathcal{D}\{f(x, y)\} \tag{1.62}$$

The camera records the intensity of diffracted light field $f'(x, y)$, and it can be expressed as $a_2' = c_0\sqrt{|f'(x, y)|}$. Here function $f(x, y)$ is unknown. At this point, the iterative algorithm converges slowly. To speed up iteration, small holes can be used to limit the measurement range of the object surface [1]. Essentially reducing the number of unknowns. The underdetermined system of equations becomes positive definite or overdetermined.

1.5.3 Modeling Complex Optical Systems

The optical experimental system is complex, and its adjustment requires many considerations. For discrete computing, its modeling is very important. Matrix optics is the basis for modeling many optical systems. It is also known as Collins formula [5], ABCD matrix, linear canonical transform. The effect of optical elements on the beam can be represented by a matrix, as Fig. 1.17.

In this model, r_1 and r_2 are distances from intersection of light rays to optical axis. θ_1 and θ_2 are angles between light ray and optical axis. As Fig. 1.17, they are elements of a 2×1 vector. The matrix M is as a bridge, to connect optical elements

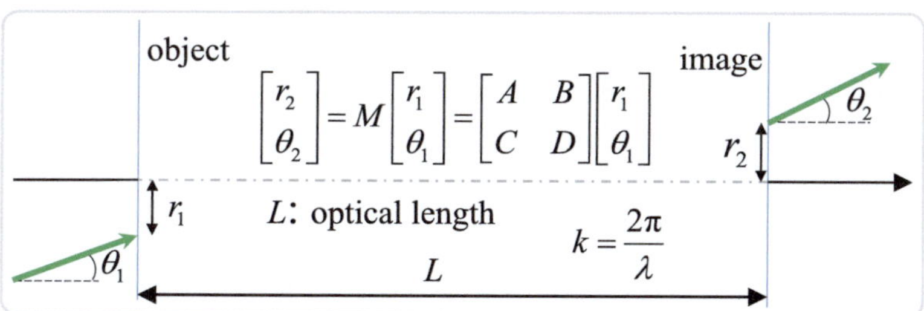

Fig. 1.17 Axially symmetric optical system with paraxial approximation

or processes connected in series. Some cascaded elements can be system modeled by the product of matrices as

$$M_{\text{eff}} = M_n M_{n-1} \cdots M_1 \tag{1.63}$$

where, M_1 is matrix corresponding to the optical element or optical process closest to the target. M_n is matrix of optical elements/processes close to the detection plane. Each matrix is multiplied in order to obtain the overall equivalent matrix. An important prerequisite for the establishment of this matrix is that the optical elements need to be coaxial and concentric. This requirement is very strict. It is also the reason why optical system adjustment is difficult.

The matrix M corresponds an optical process, which can be represented by the following integral

$$\begin{aligned} Q(u, v) = \frac{\exp(\mathrm{i}kL)}{\mathrm{i}\lambda B} \iint q(x, y) \\ \times \exp\left\{ \frac{\mathrm{i}k}{2B} \left[A\left(x^2 + y^2\right) + D\left(u^2 + v^2\right) - 2(xu + yv) \right] \right\} \mathrm{d}x\mathrm{d}y \end{aligned} \tag{1.64}$$

where, $q(x, y)$ is Complex value function of light field characterized by target to be measured. $Q(u, v)$ is light field function in the measuring plane. Matrix elements A, B and D serve as coefficients of phase term. The element C is not included in Eq. (1.64).

Equation (1.64) can describe many optical processes, such as Fourier transform, fractional Fourier transform, Fresnel diffraction and other optical processes. Using matrix M can easily obtain integration formulas for complex optical processes. The formulas for matrix M of some common optical processes are listed in Table 1.2.

Taking the Fourier transform implemented by a thin lens as an example, its equivalent matrix can be expressed as

Table 1.2 Matrix representation of optical processes

Models	Expressions	Parameter meaning
Thin lens	$\begin{bmatrix} 1 & 0 \\ -1/f & 1 \end{bmatrix}$	f: focal length
Diffraction in homogeneous media	$\begin{bmatrix} 1 & d/n \\ 0 & 1 \end{bmatrix}$	d: distance n: refractive index
Propagation in cavity	$\begin{bmatrix} g_1 & L \\ (g_1 g_2 - 1)/L & g_2 \end{bmatrix}, g_i = 1 - L/R_i,\ i = 1, 2$	R_1, R_2: spherical radius at left and right sides L: cavity length

$$M_{\text{eff}} = \begin{bmatrix} 1 & f \\ 0 & 1 \end{bmatrix} \begin{bmatrix} 1 & 0 \\ -1/f & 1 \end{bmatrix} \begin{bmatrix} 1 & f \\ 0 & 1 \end{bmatrix} \tag{1.65}$$

The inverse transform can be used for computational imaging. Here, the formula for the inverse process of Eq. (1.64) is given

$$\begin{aligned} q(x, y) = & \frac{\exp(-\mathrm{i}kL)}{-\mathrm{i}\lambda B} \iint Q(u, v) \\ & \times \exp\left\{-\frac{\mathrm{i}k}{2B}\left[A\left(x^2 + y^2\right) + D\left(u^2 + v^2\right) - 2(xu + yv)\right]\right\} \mathrm{d}u\mathrm{d}v \end{aligned} \tag{1.66}$$

Equation (1.66) is important in implementing image reconstruction algorithms in some computational imaging. Comparing Eqs. (1.64) and (1.66), we can find that Eq. (1.66) can be obtained, by using $-L$ and $-B$ to substitute L and B in Eq. (1.64).

Applying the convolution relation, Eq. (1.64) can be written as

$$\begin{aligned} Q(u, v) = & \frac{\exp(\mathrm{i}kL)}{A} \exp\left[\mathrm{i}k\frac{C}{2A}\left(u^2 + v^2\right)\right] \\ & \times \mathcal{F}^{-1}\left\{\mathcal{F}\left[q\left(\frac{x}{A}, \frac{y}{A}\right)\right] \exp\left[-\mathrm{i}\pi\lambda BA\left(f_x^2 + f_y^2\right)\right]\right\} \end{aligned} \tag{1.67}$$

According to Eq. (1.67), Eq. (1.64) can be calculated by applying the discrete Fourier transform algorithm.

1.5.4 Image Recognition

Target recognition is a commonly used image analysis tool. Because the target has changes such as translation, zoom, rotation, and occlusion, target recognition becomes difficult, or the recognition success rate is not high.

Correlation operations are introduced here. It can recognize moving targets. Its calculation formula is as follows

$$\begin{aligned} C_{1,2}(x, y) &= f_1(x, y) \odot f_2(x, y) \\ &= \mathcal{F}^{-1}\left\{\mathcal{F}\left[f_1\right] \mathrm{conj}[\mathcal{F}(f_2)]\right\} \end{aligned} \tag{1.68}$$

where, $C_{1,2}$ is to compute cross-correlation result of functions f_1 and f_2. The symbol $\odot$ is cross-correlation operator. The function conj is to calculate the conjugate of a complex function. Both cross-correlation and convolution can be calculated using a discrete Fourier transform. Convolution is order-independent for two input functions. The results of the related operations are related to the order of the input functions.

Figure 1.18 shows an example. The function $f_1(x, y)$ is standards pattern. The function $f_2(x, y)$ is patterns with displacement, rotation and scaling. It can be seen

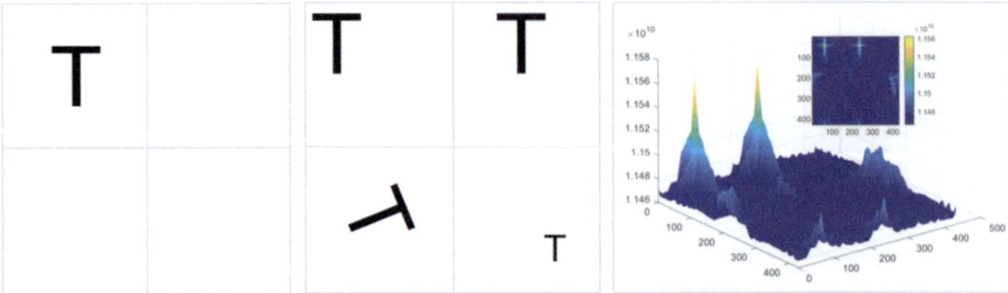

Fig. 1.18 Results of cross-correlation operation. The letter "T" is shifted, rotated, and resized. Two peaks in 3D plot are the detected positions

from the figure that correlation operations can identify the translation of the target. Insensitive to rotation and scaling. The peak of the cross-correlation results corresponds to the target position. You can determine the target position by finding the maximum value of the surface. In Eq. (1.68), replace the Fourier transform result with the phase corresponding to the amplitude function, that is, the amplitude function is set to a constant, so that the correlation result obtained is an ideal peak [6]. Sharp peaks avoid some secondary peak interference when determining peak locations. For rotation and scaling cases, some complex algorithms are needed, such as deep learning.

1.6 Optimization Methods

This section introduces some optimization methods that are important for finding optimal solutions. During application, it is necessary to abstract the problem and determine the input data and key parameters of each optimization algorithm, as well as update rules for key parameters.

1.6.1 Genetic Algorithm

The calculation process of genetic algorithm includes four stages: population initialization, fitness evaluation (screening), replication, and termination. Population initialization gives a preliminary estimate of the target solution. In fitness estimation, the chromosome fitness that characterizes the genetic quantity is estimated based on a preset fitness value. The replication stage includes three possible states: crossover, mutation, and acceptance of solution. In selection control, the solution with the best performance and the highest adaptability of the current population has the greatest probability of being used to obtain a new solution (inheritance). Obtaining methods include crossing and mutation. The new chromosomes are evaluated by the fitness function in the next round of iterative genetic algorithm calculations. When the set conditions are met, the iterative calculation of the genetic algorithm stops.

1.6.2 Particle Swarm Optimization

Particle swarm optimization originates from the analysis of birds foraging. When a bird is looking for objects, it will approach food or habitat. For habitat, it can determine location based on memory information. At the beginning of the foraging stage, the bird does not know where the food is. Through the information transmission of bird populations, birds approach the target area based on the information (Fig. 1.19).

Here a model can be abstracted. The initial state of the flock is dispersed. Based on the fitness function, birds can judge the quality of their current position. There is an exchange of information among the flock. Each bird can monitor food information from many locations within its reach, and can determine the best position or direction based on stored information. With information sharing, each bird knows the best food location the population has found. The mathematical process of the above process is called particle swarm optimization.

Similar to bionics, foraging for birds is similar to mathematical optimization. This model can be used to estimate the optimal solution. The entire optimization process is complex, but its rules are definite and simple. After multiple iterative calculations and position updates, the particles in the model approach the optimal solution.

In this algorithm [7], the position of the particle is an approximate solution to the optimization problem. Assuming there are M particles in the search space, the position of the particle is represented by p_m. m is the serial number of the particle. The velocity of motion of the particle is expressed by v_m. Evaluation of particle positions is determined using a fitness function. The current optimal solution is denoted as $p_{b,m}$. The optimal solution found by the population is denoted p_{gg}. In each iteration, a new position needs to be calculated based on the particle's velocity. The updated calculation formula for particle velocity and position is as

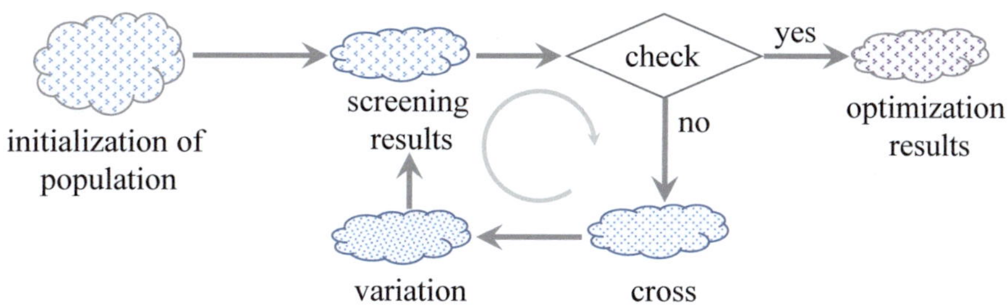

Fig. 1.19 The calculation process of genetic algorithm

$$v_m^{(k+1)} = v_m^{(k)} + l_1 r_1 \left(p_{b,m}^{(k)} - p_m^{(k)}\right) + l_2 r_2 \left(p_{gg}^{(k)} - p_m^{(k)}\right) \tag{1.69}$$

$$p_m^{(k+1)} = p_m^{(k)} + v_m^{(k+1)} \tag{1.70}$$

where, superscript k is number of iterations. r_1 and r_2 is random values between 0 and 1. l_1 and l_2 are learning factor, and it is related to the information transmission ability of the particle. The total number of particles is related to the optimization problem and corresponds to the number of optimal solutions. Maximum velocity $v_{\max}$ limits the maximum movement distance of a particle in iterative calculation, and its value is related to the convergence speed and accuracy. The stopping condition of the algorithm can be determined by the maximum number of iterations or the minimum error.

If the particles search at a fixed speed, you may miss the optimal solution. As the position optimization progresses, the particles will approach the optimal solution, which reduces the speed and prevents the particles from exceeding the range of the optimal solution. To do this, an inertia factor can be introduced. Speed updates become

$$v_m^{(k+1)} = w \cdot v_m^{(k)} + l_1 r_1 (p_{b,m}^{(k)} - p_m^{(k)}) + l_2 r_2 (p_{gg}^{(k)} - p_m^{(k)}) \tag{1.71}$$

Equation (1.71) is called as the standard particle swarm optimization algorithm. The inertia factor w is important, and related to the overall speed of finding the optimal solution. By correlating the value of w with the gradient of the fitness function, adaptive adjustment of parameters can be realized.

1.6.3 *Simulated Annealing Algorithm*

Simulated annealing algorithm can solve large-scale optimization problems [8]. It has the advantages of simple rules, convenient application, high calculation efficiency, and less influence by initial conditions. This algorithm is suitable for parallel computing and has the ability to jump out of local optimal solutions.

When a substance cools from a higher temperature to the crystallization temperature, the random motion of the particles helps to form a state with the least ability, making the crystal have good mechanical properties. During the optimization process of simulated annealing algorithm, Let a solution be x_i, its function value is $f(x_i)$. The energy of the state i is E_i, the corresponding control temperature is t. Metropolis criterion is

$$M = \begin{cases} 1, & \textit{if } f(x_j) \le f(x_i), \\ \exp\{[f(x_i) - f(x_j)]/t\}, & \text{otherwise.} \end{cases} \tag{1.72}$$

Equation (1.72) gives the probability from state i to j is M.

The main calculation steps of the simulated annealing algorithm are as follows.

(1) Set initial temperature. The initial temperature is taken as T_0, and it can be regarded as a rough evaluation solution.
(2) Cooling function. To simulate the temperature decrease process of material annealing, choose a monotonic decreasing function, as

$$T_{k+1} = \frac{T_k}{1 + T_k \ln(1 + \delta)/g(T_k)} \tag{1.73}$$

where, g is system function. δ is temperature control coefficient, and has positive and small amount.
(3) Status update. Construct an updated solution set and determine the next state based on the Metropolis criterion.

Loop the above two steps until the calculation reaches the stop condition of the program.

1.6.4 Gravitational Search Algorithm

The gravitational search algorithm originates from the gravitational model [9]. It has strong global search capabilities, all particles are subject to gravity, and has global attributes. Think of a particle as a small mass object. The fitness function of the particle position is used to express the feasible solution to the problem. By the attraction of each particle, the small-mass particle approaches the large-mass particle. Find the largest mass particle among them, that is, converge to the optimal solution.

In the search space, suppose there are N particles, and the position of the n-th particle is z_n. Because for an image model, z_n has two components.

At the time t, from the particle n_1, the gravitational force $f_{n_1,n_2}(i, t)$, exerted on particle n_2 can be expressed as

$$f_{n_1,n_2}(i, t) = G(t)\frac{m_{n_1}(t) + m_{n_2}(t)}{\varepsilon + r_{n_1,n_2}(t)}[z_{n_1}(i, t) - z_{n_2}(i, t)],\ i = x, y \tag{1.74}$$

where, $G(t)$ is the current gravitational constant. m_{n_1} and m_{n_2} are inertial mass of particles n_1 and n_2. ε is small constants with non-zero denominator. r_{n_1,n_2} is the Euler distance between two particles. $z_{n_1}(i, t)$ and $z_{n_2}(i, t)$ are position components of two particles.

In Eq. (1.74), the gravitational constant $G(t)$ is

$$G(t) = g_0 \exp(-\alpha t/T) \tag{1.75}$$

where, g_0 is initial value of gravitational constant. It needs to be determined based on the size of the search space range, and there is a monotonically increasing relationship between them. α is constant, and controls the convergence rate of the algorithm. For α, too small a value will cause slow convergence, and too much value will fall into a local optimal solution. The appropriate values of parameters g_0 and α should be determined from actual conditions. T is total number of iteration.

The expression of distance $r_{n_1,n_2}(t)$ is determined by L-2 norm as

$$r_{n_1,n_2}(t) = \left\| z_{n_1}(t) - z_{n_2}(t) \right\|_2 \tag{1.76}$$

In order to increase the random search ability of the algorithm, the formula for the resultant force is

$$f_{n_2}(i,t) = \sum\nolimits_{\forall n \& n \neq n_2} R(n) f_{n,n_2}(i,t) \tag{1.77}$$

where, $R(n)$ is random value in the interval [0, 1].

The acceleration component of a particle is

$$a_{n_2}(i,t) = \frac{f_{n_2}(i,t)}{m_{n_2}(t)} \tag{1.78}$$

The inertial mass of particle, $m_n(t)$, is related with fitness distribution function, as

$$m_n(t) = \frac{u_n(t)}{\sum_{\forall n} u_n(t)}, \ u_n(t) = \frac{F_{\text{fit}}(n,t) - w(t)}{b(t) - w(t)} \tag{1.79}$$

where, $F_{\text{fit}}(n,t)$ is fitness function value. $w(t)$ is minimum fitness values for all particles. $b(t)$ is maximum fitness values for all particles.

The update formula for the particle's velocity and position is

$$v_n(i,t+1) = R(n) v_n(i,t) + a_n(i,t) \tag{1.80}$$

$$z_n(i,t+1) = z_n(i,t) + v_n(i,t+1) \tag{1.81}$$

where, the time interval is considered to be a unit of time.

1.6.5 Ant Colony Algorithm

Ant colony algorithm originates from the observation and simulation of ants foraging [10]. Due to information transmission, more ants gather on the optimal channel for transporting food. There will be much fewer ants on other paths. You can use this algorithm to determine the optimal path for an optimization problem.

Ant colony algorithm achieves good results in determining the optimal path. For example, for the travel agent problem, a businessman starts from a city, requires each city to visit once, and returns to the original city after visiting all destination cities. The optimization calculation problem is selecting the shortest path.

Suppose there are N places where M ants need to reach. The m-th ant has been to a certain place, and updates the visit record of the location T_m. Every ant travels to the next location are affected by a probability $p_{n_1,n_2}(m, t)$. The probability value p_{n_1,n_2} is related to the information status of the current location n_1 and the next possible location n_2. Its calculation formula is

$$p_{n_1,n_2}(m,t) = \begin{cases} \frac{I_{n_1,n_2}(m,t)E_{n_1,n_2}(m,t)}{\sum_{\forall n_1} I_{n_1,n_2}(m,t)E_{n_1,n_2}(m,t)}, & n_2 \in P_{\text{all}} \\ 0, & \text{otherwise} \end{cases} \tag{1.82}$$

where, $I_{n_1,n_2}(m, t)$ is information function of the m-th ant at current location n_1 and location n_2. $E_{n_1,n_2}(m, t)$ is expectation function from the place n_1 to the place n_2. P_{all} is collection of all locations.

The expectation function $E_{n_1,n_2}(m, t)$ can be determined by

$$E_{n_1,n_2}(m,t) = d_{n_1,n_2}^{-1} \tag{1.83}$$

where, d_{n_1,n_2} is the distance from n_1 to n_2.

The information function $I_{n_1,n_2}(m, t)$, needs to be updated after the ant colony reaches a location, for changing visits to various locations. By iteratively applying the above calculation steps, the ant colony can find the optimal path. Here, we will add a description of the information update function, namely

$$\begin{aligned} I_{n_1,n_2}(t+1) &= (1-\rho)I_{n_1,n_2}(t) + \Delta I_{n_1,n_2}(t) \\ \Delta I_{n_1,n_2}(t) &= \sum\nolimits_{\forall m} \Delta I_{n_1,n_2}(m,t) \end{aligned} \tag{1.84}$$

where, ρ is information change factor, valued in the interval [0, 1]. $\Delta I_{n_1,n_2}$ is the change in information volume, with initial value 0. $\Delta I_{n_1,n_2}$ has an updated value as

$$\Delta I_{n_1,n_2} = \begin{cases} S/F, & n_1 \to n_2 \\ 0, & \text{otherwise} \end{cases} \tag{1.85}$$

where, S is total amount of information. The values of F have three kinds of case, such as 1, the distance between n_1 and n_2, path length from n_1 to n_2.

1.6.6 Sparrow Search Algorithm

The sparrow search algorithm originates from the sparrow's foraging process and the behavior of avoiding enemies [11]. This algorithm describes two behaviors of sparrows: (1) The discoverer is responsible for observing and guiding the group closer to the food. (2) Follow the sparrow and rely on the discoverer to survive. For foraging and food quantity constraints, the algorithm includes cooperation and competition models.

Relatively speaking, the sparrow search algorithm has slightly more rules to follow. The details are as follows:

(1) Discoverer and follower sparrow. The discoverer has high adaptability and has a guiding role, detecting food location information and protecting the safety of the group.
(2) Safety rules. When a member discovers a danger, other members are notified. When the information exceeds the security threshold, all sparrows quickly escape from their current location.
(3) Individual roles change. The identities of discoverers and followers of sparrows change. Those with the ability to discover the location of food will have the identity of discoverers. The ratio of discoverers to followers is kept fixed.
(4) Food competition and energy conservation. Sparrows compete for food, and sparrows that discover food first will gain more energy. Low-energy sparrows will explore new locations for food to increase their energy.
(5) Population renewal. Followers are willing to choose discoverers with strong ability to discover food. There is a competition for resources between different sparrows.
(6) Avoid enemy rules. When threatened, the sparrow will fly towards a safe area. Sparrows at the edge of the population have a greater risk factor.

Let the number of sparrows be N, sparrow location information be z, the position is initialized to

$$z_{n,i} = l_i + (u_i - l_i)R_n \tag{1.86}$$

where, n is the label of sparrow. $i = x, y$ is index of coordinates. l_i and u_i is lower and upper boundaries of corresponding dimensions. R_n is random value between 0 and 1. The fitness function of sparrow is f.

The location update formula of the discoverer is

$$z_{n,i}^{(t+1)} = \begin{cases} z_{n,i}^{(t)} \exp(-n/\alpha k_m), & \text{if } R < S_T \\ z_{n,i}^{(t)} + rV_1, & \text{if } R \geq S_T \end{cases} \tag{1.87}$$

where, t is index for iteration. The value of α is random in the interval [0, 1]. k_m is Total number of iterations. R is warning factor, valued from 0 to 1. S_T is a safety threshold, located in the range [0.5, 1]. r is random numbers that obey normal distribution. V_1 is a 1×2 vector.

In the sparrow population, the proportion of discoverers ranges from 10 to 20%. The follower's location update formula is

$$z_{n,i}^{(t+1)} = \begin{cases} r\exp[(z_w^{(t)} - z_{n,i}^{(t)})/n^2], & \text{if } n > N/2 \\ z_b^{(t+1)} + \left|z_{n,i}^{(t)} - z_b^{(t+1)}\right| AV_1, & \text{otherwise} \end{cases} \tag{1.88}$$

where, $z_w^{(t)}$ is the position located by the worst performing sparrow. $z_b^{(t+1)}$ is the position for the best performing sparrow. A is a 1×2 vector, valued 1 or -1, in order to control searching direction.

When faced with potential dangers, a small number of sparrows were randomly selected for alert tasks. The rules for updating the position of these sparrows are

$$z_{n,i}^{(t+1)} = \begin{cases} z_e^{(t)} + g\left|z_{n,i}^{(t)} - z_e^{(t)}\right|, & \text{if } f_n < f_g \\ z_{n,i}^{(t)} + u\frac{\left|z_{n,i}^{(t)} - z_w^{(t)}\right|}{f_n - f_w + \varepsilon}, & \text{otherwise} \end{cases} \tag{1.89}$$

where, $z_e^{(t)}$ is current global optimal position. g is random factor that obey normal distribution, to control step size. The value of u is in $[-1, 1]$, to determine the direction of motion. f_n is fitness function value for current sparrow. f_g is global optimal fitness. f_w is global worst fitness. ε is small-valued constant.

1.6.7 Beetle Search Algorithm

The Beetle Antennae Search algorithm is a biological heuristic algorithm [12]. It originates from an algorithm inspired by the foraging process of longicorn beetles. Beetles look for food based on the smell of food. For the two long antennae of the long-haired beetle, if the left antennae senses the smell strongly, the beetle will fly to the left. Otherwise fly right. If the two directions are close and it is difficult to determine the size, the flies forward. According to this rule, it can find food. Figure 1.20 shows the beetles search model.

It includes the following constraints: (1) The beetles are symmetrically distributed on the head. (2) The distance between the roots of two tentacles is d_0. The step length of the movement is $k_n d_0$. For fine search, k_n is changed according to iteration number. (3) Before each flight, the head orientation of the beetle is random.

The function of beetles sensing food odor is $f(x)$. Calculating the function values of the left and right whiskers $f(x_l)$ and $f(x_r)$ is used to determine the direction of

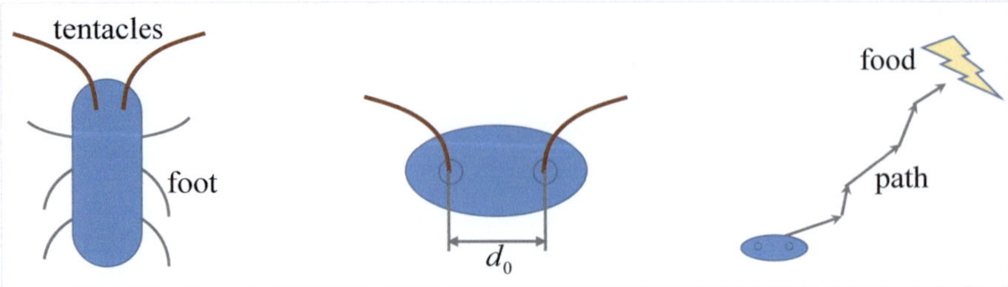

Fig. 1.20 Beetle search model

flight of the beetle. As Fig. 1.20, the beetle continues to approach the food in a broken path.

1.7 Image Analysis Method

Image analysis methods have less information than image processing methods. There is also a certain connection between them. Image analysis methods need to rely on some statistics to determine the attributes of an image.

1.7.1 Pixel Value Relation Between Adjacent Pixels

The pixel value relationship between adjacent pixels in an image can reflect the pixel distribution properties of the image. Figure 1.21 shows the visualization results for three image types.

The three types of images are road image, laser speckle map, and random image. Their total pixel count is 256×256. Pixel values are encoded by 8-bit data. The pixel value relationship of adjacent pixels is shown in the second row of subgraph in the figure. Here, 2000 points and their adjacent points to the right are randomly selected from the image and used as the coordinate axis data of the scatter plot. Points in non-random images are mainly concentrated near straight line $y = x$. The scatterplot data points of random images have uniform and scattered distribution characteristics.

The relationship between adjacent pixel values can also be expressed using a gradient, which is simpler. Table 1.3 shows the gradient values of three images. The gradient is calculated in four ways, and along the x and y directions, the minimum or maximum value of the absolute value of the gradient in the two directions is taken. The mean of the absolute values of the gradients in the four methods was counted as a reference for image evaluation. The gradient reflects the edge of the image. It is used as a reference for image quality evaluation in issues such as auto-focusing, and does not rely on the reference image.

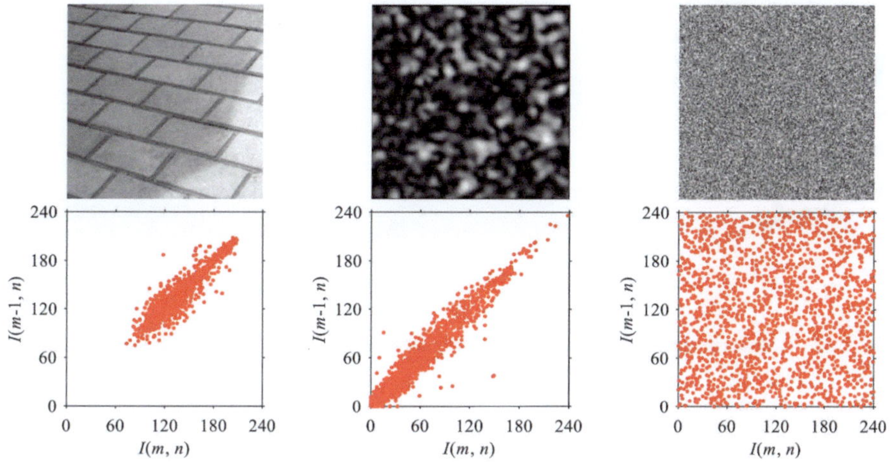

Fig. 1.21 Pixel value relation between adjacent pixels

Table 1.3 Mean of absolute values of gradient values of three images

Gradient	Road	Speckle	Random image
g_x	2.42	5.67	42.71
g_y	4.28	6.27	42.61
$g_{\min}$	1.82	3.31	25.44
$g_{\max}$	4.88	8.63	59.88

Histograms and information entropy are common tools for evaluating the pixel distribution of an image. They are dimensional-reduced expressions of image pixel information, and the data form is simple. Histograms can give a more comprehensive picture of pixel distribution than entropy. Faced with numerous image features, it is difficult for them to give more statistical information.

1.7.2 Line Detection

Line inspection is the simplest geometry analysis. Here, the road pattern in Fig. 1.21 is taken as the target and the line segments are detected. The binarization result of this graph is shown in Fig. 1.22a. The Fourier spectrum of this image is shown in Fig. 1.22b, where the amplitude of the Fourier transform result is shown on a logarithmic scale. It's a characteristic result where you can see two straight lines l_1 and l_2. The lines are related to the direction of the brick joints. Figure 1.22c shows the corresponding Radon transform results, from which it is difficult to see the direction information of the line.

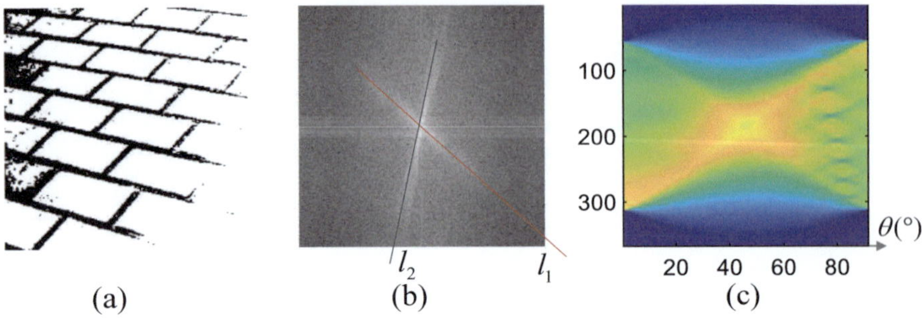

Fig. 1.22 Transform results of brick slit: **a** binarization, **b** Fourier spectrum, **c** Radon transform

Radon transform is very effective in detecting the direction of the line. Figure 1.23 shows an example of a Radon transform. Several line segments are unequally spaced, as Fig. 1.23a. For convenience of calculation, they are inverted as Fig. 1.23b, and subjected to Radon transformation, where the angle range is [0°, 180°]. The corresponding results are shown in Fig. 1.23c. It can be seen that the Radon transformation obtains very ideal results. Calculate the maximum values at different angles to obtain the red curve in Fig. 1.23c. The maximum value of the red curve corresponds to the direction of the line segment. And the intervals corresponding to the angle bright spots correspond to the intervals of the curves.

The actual image situation is very complex, and it is difficult to obtain ideal results by directly applying Radon transform, such as the situation in Fig. 1.22. Some preprocessing methods are needed to assist the Radon transform in determining the direction of the line segment, such as image binarization, erosion of morphological operations, edge detection, filtering, etc. The Hough transform can also be used for line detection.

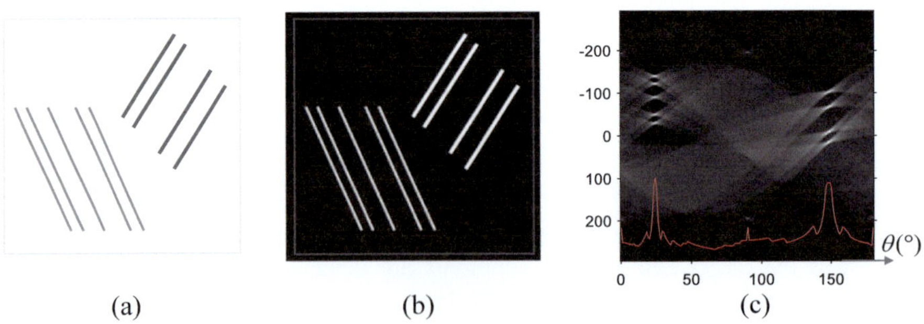

Fig. 1.23 Results of Radon transform: **a** line graph, **b** inverse color graph, **c** transform result

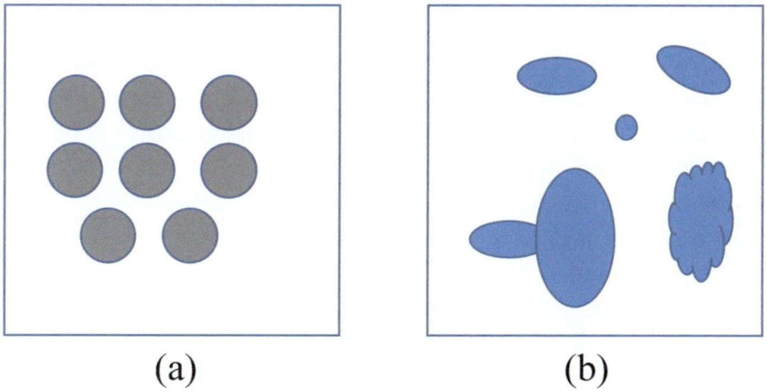

Fig. 1.24 Examples of using Euler numbers: **a** regular patterns, **b** special case

1.7.3 Euler Number

In binary images, Euler's number is a very important concept. Using Euler numbers, you can identify the number of some objects. The equation for Euler's number is

$$E = C - H \tag{1.90}$$

where, E is Euler number of binary image. C is the number of connected areas. H is the number of holes. In applications, the target can be regarded as a hole. Euler numbers can be used to count the number of objects in the image.

Figure 1.24 shows two sets of binary patterns. Figure 1.24a is a binary diagram of several regular patterns. The number of holes can be obtained by determining the Euler number and the number of connected areas. For the case in Fig. 1.24b, there is a case where targets are stuck, and the Euler number is a way to directly distribute the stuck targets. In this case, the target size can be reduced by etching, so that the two adhered targets will separate. After the targets are separated, the number of holes can be obtained by applying the Euler number. For targets of different sizes and shapes, the number of targets can also be counted using Euler numbers.

1.7.4 Extract Graphic Parameters

Extracting parameters of graphics is very important for image analysis. It is the process from data to graphics, and then from graphics to data. It involves some geometric methods and is also a channel to integrate with artificial intelligence methods.

Line extraction can use some edge extraction operators, which generally calculate the gradient of the image. For example, first-order operators, Sobel operator, Roberts operator and Prewitt operator; second-order operators, such as Laplacian operator, are sensitive to noise. Canny algorithm extracts image edges and has the characteristics of filtering and enhancement. It is a relatively comprehensive algorithm. Line segment detection can use the methods mentioned in Sect. 1.7.2.

The center and radius parameters of circles and ellipses can be obtained using the methods described in Sect. 1.4.4. For circular-like patterns, binarization can be used to calculate the area of the area. Use edge detection methods to get the perimeter. Use perimeter and area to determine the radius of a circle.

For other irregular shapes, after extracting point or line features, the shape parameters can be determined by using function fitting based on the vertices. Through graphic parameter extraction, images can be used in different image analysis fields, such as fingerprint, palm print, and iris images can be used for identification.

1.8 Image Preprocessing Method

Image preprocessing methods can be used to change image properties to make it more suitable for processing by the main algorithm. Preprocessing methods are very important to improve image analysis results and improve success rates. Algorithm design is related to the actual characteristics of the image. Its composition is uncertain, and sometimes algorithm parameters and content can be determined based on needs and attempts.

1.8.1 Image Enhancement

Image enhancement is very effective in improving low-light illumination images. In some image enhancement processes, the original pixel distribution is unchanged, and pixel values and colors are changed. It will cause visual distortion, and the visual effect of the image will be improved to a certain extent to meet specific application requirements. Image enhancement can be performed in the spatial domain or frequency domain. The frequency domain enhancement method mainly performs filtering and filters noise. The filtering method will be described in Sect. 1.8.2. In the case of underexposure or overexposure, the image gray distribution, or pixel histogram, is limited to a small range. Because the difference in pixel distribution is small, the visual effect will be blurred and there will be no gray level.

The simplest image enhancement method is called point arithmetic. It is adjusted for pixel values. We can use the following calculation relation as

$$I'(x, y) = F\{I(x, y)\} \tag{1.91}$$

where, I' is enhanced image. F is mapping function. For weakly exposed images, the mapping function needs to correspond to a function with a reciprocal greater than 1, such as $\exp(x)$. For the strong exposure function, F needs to choose a function with a reciprocal less than 1, as $\ln(x)$. Logarithmic functions are often used to adjust the Fourier spectrum because the Fourier spectrum of positive images has a large value at the origin, making it difficult to observe at other spectral positions when images at other positions directly display the amplitude. I denotes input image.

Here are two formulas for logarithmic and exponential functions for image enhancement, namely

$$I'(x, y) = \frac{1}{d}\ln[I(x, y) + 1] \tag{1.92}$$

$$I'(x, y) = \exp\{c \cdot I(x, y)\} - 1 \tag{1.93}$$

where, d and c are parameters, to control the range of pixel adjustments.

Figure 1.25 shows a curve showing the corresponding functional relationship between Eqs. (1.92) and (1.93). The corresponding slope of the red curve part in Fig. 1.25a and b is greater than 1, which has an enhancement effect on this part of pixels. wherein parameter values p_0 is for critical pixels. Figure 1.25c shows the relation between c, d and p_0. Their functional relations are as follows

$$d = \frac{1}{1 + p_0},\ c = \frac{1}{p_0} L_W(0, p_0) \tag{1.94}$$

where, L_W is Lambert W function. In the application, it should be noted that the mapped image cannot exceed the data range of the stored image, such as an 8-bit image, with the maximum value equal to 255.

The logarithmic function is equivalent to increasing the light source and virtual supplementary lighting. The exponential function is equivalent to sunglasses and reduces the intensity of light. For both low light and overexposure, the image can be enhanced using the following formula

$$I' = \frac{2^n - 1}{2}\left\{1 + \tan\left[\alpha\pi\left(\frac{I}{2^n - 1} - \frac{1}{2}\right)\right] \Big/ \tan\left(\alpha\frac{\pi}{2}\right)\right\} \tag{1.95}$$

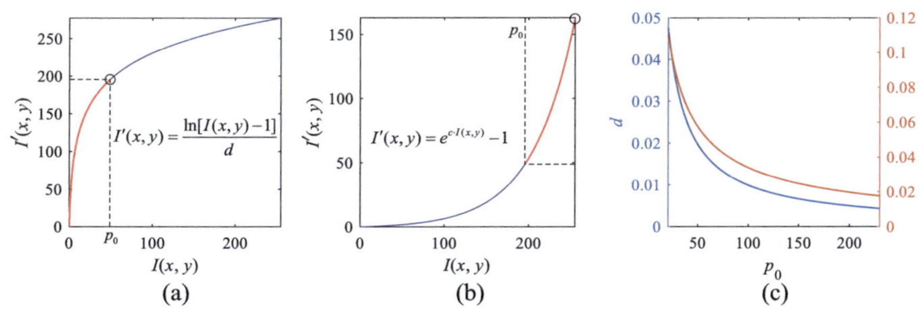

Fig. 1.25 Enhancement curves: **a** logarithmic function, **b** exponent function, **c** parameter relation

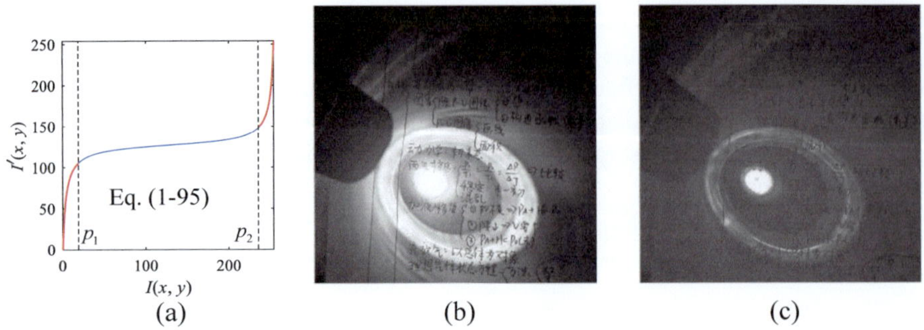

Fig. 1.26 Bilateral enhancement mapping relation: **a** mapping, **b** before, **c** after

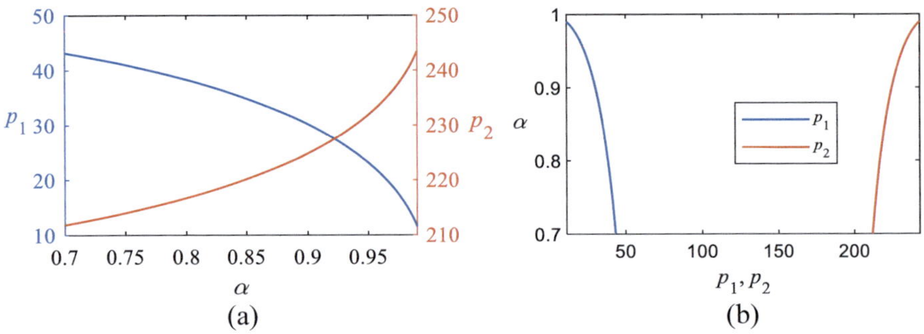

Fig. 1.27 The relation for p_1, p_2 and α: **a** α, **b** p_1 and p_2

where, n is number of bits for image. α is control factor, valued in the range [0, 1].

Figure 1.26 shows the mapping curve of Eq. (1.95). Figure 1.26a shows the mapping curve. Parameters p_1 and p_2 are critical pixel values, as boundary corresponding to dark and light areas. Figure 1.26b and c are the patterns before and after enhancement. Text information in the dark becomes visible. Figure 1.27 shows the critical pixel values p_1 and p_2, for parameter determination.

The histogram equalization method converts the gray histogram of the original image from a non-uniform distribution state to an approximately uniform distribution, expanding the dynamic range of pixel gray values. This method works well for low-contrast images, such as X-ray images, low-underexposed or overexposed images.

1.8.2 Image Filtering

Filtering removes or suppresses noise in an image. However, the noise in actual images is random, and it is theoretically difficult to completely remove the noise

[13]. It is an endless research topic. Improving performance by enhancing the anti-noise robustness of image analysis algorithms can allow a small amount of noise to exist in the image without affecting the final binary result.

Mean filtering uses an all-one matrix and the target image to perform spatial convolution to obtain the mean value in the neighborhood of pixels. Changing the matrix coefficients can adjust the weight coefficients of the neighboring pixels. For some highly noisy images, the median filtering method can obtain better visual enhancement effects [14]. The Laplacian operator can be used for high-pass filtering in spatial domain, which can enhance edge and detail information of the image.

The simplest frequency domain low-pass filter is defined as follows

$$L_{\text{low-pass}}(u, v) = \begin{cases} 1, & \rho < \rho_0 \\ 0, & \text{otherwise} \end{cases} \tag{1.96}$$

where, $\rho = \sqrt{u^2 + v^2}$. ρ_0 is cutoff frequency. This filter is also called an ideal low-pass filter. It has ringing because of sudden changes at the cutoff frequency, which creates a discontinuity in the frequency domain modulation function. It is easy to implement in an optical system. The pupil can be used to control the signal near the optical axis through the system at the Fourier surface.

A trapezium filter can be regarded as a variant of the ideal low-pass filter, and its definition is as

$$L_{\text{trapezium}}(u, v) = \begin{cases} 1, & \rho < \rho_0 \\ \frac{\rho-\rho_1}{\rho_0-\rho_1}, & \rho_0 \le \rho \le \rho_1 \\ 0, & \rho_1 < \rho \end{cases} \tag{1.97}$$

where, ρ_0 is cutoff frequency. ρ_1 is critical zero point. When $\rho_0 = \rho_1$, the trapezium filter becomes an ideal low-pass filter.

The Gaussian low-pass filter is

$$L_{\text{Gaussian}}(u, v) = \exp\left[-\frac{\rho^2}{2\rho_0^2}\right] \tag{1.98}$$

where, ρ_0 is cutoff frequency. At the cutoff frequency, the response coefficient of the filter is not equal to 0, and the filter function is continuous. This prevents the Gaussian filter from causing ringing. Outside the cut-off frequency, there are still signals that can pass through, but the modulation factor is small. As the frequency increases, the response coefficient decreases.

Butterworth low-pass filter is defined as

$$L_{\text{Butterworth}}(u, v) = \frac{1}{1 + [\rho/\rho_0]^{2n}} \tag{1.99}$$

where, ρ_0 is cutoff frequency. n is order.

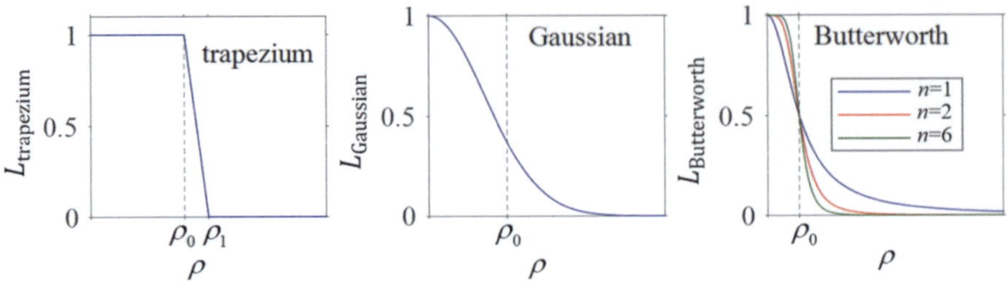

Fig. 1.28 Modulation function curves of three filters

Figure 1.28 shows the modulation function curves of the three filters. As the order n increases, the modulation function of the Butterworth filter approaches that of an ideal low-pass filter.

The determination of the three frequency-domain low-pass filters described above is independent of the image to be processed. Their adaptability is limited. The advantages are good robustness and simple program design. In applications, filters can be used in combination. For example, two different Gaussian filters can be subtracted to obtain a Gaussian difference function. In essence, it is a band-pass filter that restricts the passage of spectral components between two cut-off frequencies. The inverse function of a band-pass filter can form a band-stop filter. Using a band-stop filter would be very appropriate. If the noise is concentrated in a certain frequency band range. The high-pass filter can simply be generated from the low-pass filter such that their sum equals 1. High-pass filters improve high-frequency information, and image edges and details are highlighted.

Here are some filtering methods related to band-processed images. For example, guided filtering [15, 16]. Guided filtering applies the guidance map to filter the target image, so that the output image is similar to the target image, and the texture parts are similar to the guidance map. The output image is constructed by image derivatives and least squares ideas. In applications, the target image is often used as a guide map, so that the equivalent filtering process is related to the input image.

The 3D block matching algorithm (BM3D) matches the current pixel block with adjacent image blocks and combines similar blocks into a 3D matrix [17]. Filtering in three-dimensional space, and the result is inverted into two-dimensional space through an inverse process to obtain a noise-reduced image. This algorithm has significant noise reduction effect and has been verified in many imaging methods and optical image processing research. This algorithm uses block processing and has strong local processing properties.

Wiener filtering is an optimized linear filtering method [18] and has many applications in the field of image processing. It restores the original image from noisy images based on minimizing the mean square error. For additive noise, the filter function is

$$H_{\text{Wiener}}(\rho) = \frac{P_s(\rho)}{P_s(\rho) + P_n(\rho)} \tag{1.100}$$

where, P_s is power spectral density (PSD) function of signal. P_n is PSD for noise. Lucy-Richardson filtering algorithm is based on an iterative deconvolution method of maximum likelihood estimation, which can be used for image restoration and removes image blur and noise through iterative optimization [19].

In noise testing, there are many types of noise to choose from [13]. For example, Poisson noise, Gaussian noise, salt and pepper noise, uniformly distributed noise, etc. are commonly used. The addition form of noise can be additive or multiplicative, that is

$$I_n(x, y) = I_0(x, y) + a \cdot N(x, y) \tag{1.101}$$

$$I_n(x, y) = I_0(x, y)[1 + p \cdot N(x, y)] \tag{1.102}$$

where, I_n represents a noisy image and is used as input to the noise reduction algorithm. I_0 is original image, and serves as the true value to calculate the error. a and p are coefficients for additive and multiplicative noise, used to control the intensity of the noise. N is noise function. For some noise, it is often limited to random matrix data with a mean of 0 and a variance of 1.

1.8.3 Nonuniformity Correction

Due to the influence of uneven response factors such as imaging systems, detectors or light sources, fringes, granular noise, and irregular shape noise may occur in the obtained image. Some of these noises are irrelevant to the imaging target and belong to the inherent noise of the system. Some non-uniformity correction algorithms can mitigate the impact of noise on the image. The core goal of this algorithm is to improve image quality and provide more reliable image data for subsequent processing. Correction algorithms are mostly used in infrared imaging and medical and industrial testing.

Stripe noise can be removed or suppressed using a notch filter [20]. Noise reduction methods can be used to correct non-uniformity. For the inherent noise of the system, we can consider combining experiments and adopting no-load measurement to obtain the system's response to the illumination beam, which can be used as a reference to modify the response coefficient function of the system during imaging. In addition, combining some standard images can be used to better calibrate the non-uniform response distribution function of the system.

1.8.4 Image Segmentation

Image segmentation can obtain interesting processing areas and reduce the computational complexity of subsequent image analysis algorithms. Its challenge is how to adaptively find the required areas. Some image recognition algorithms can be considered upgraded versions of image segmentation methods.

Image segmentation is a marking process in which pixels belonging to the same area are assigned the same numerical value for numbering. Finally, the image is divided into multiple blocks based on the encoding matrix, called the mask matrix. Simple images can be segmented using pixel thresholds or color.

The area growing method counts pixels with similar properties together to form an area. Find a seed pixel in an area as a starting point for growth. In the neighborhood around the seed pixel, find pixels with the same or similar attributes as the seed pixel and merge them into the area where the seed pixel is located. By continuously expanding the search range around the algorithm, this algorithm can be calculated using a recursive program structure because the search length in each direction or search path is uncertain.

1.9 Image Conversion Methods

As an important tool for image analysis, image transform provides many ideas for algorithm design. In addition to spatial domain (or direct calculation of image pixels), many algorithms are performed in the frequency domain or transform domain to obtain some special effects.

1.9.1 Geometric Transforms

For geometric operations such as translation, rotation, and shearing, the following equations can be used to adjust coordinates

$$\begin{bmatrix} x_0' \\ y_0' \end{bmatrix} = \begin{bmatrix} a & b \\ c & d \end{bmatrix} \begin{bmatrix} x \\ y \end{bmatrix} + \begin{bmatrix} x_s \\ y_s \end{bmatrix} \tag{1.103}$$

where, x_0' and y_0' are coordinates after geometric transform. x_0 and y_0 are original coordinates. The 2×2 matrix elements, a, b, c, d are related to rotation and shearing. x_s and y_s are displacement along x and y. If it is purely translation, the 2×2 matrix is required to be identity.

Mirrors in the horizontal and vertical directions can be achieved by flipping the matrix. For the general mirror relationship, Fig. 1.29 shows the calculation model of the mirror relationship.

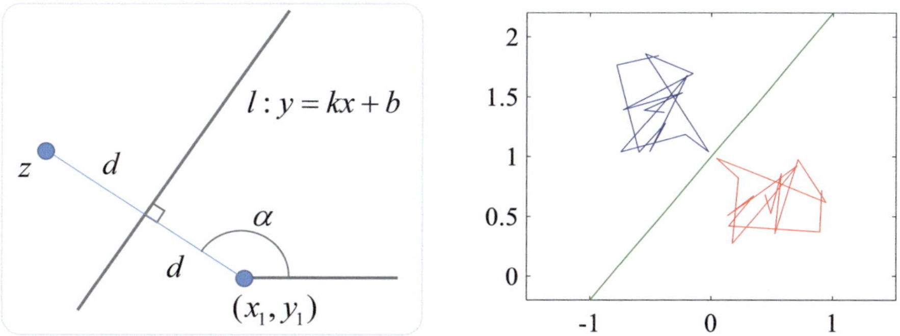

Fig. 1.29 Calculation of mirror relation

In order to avoid calculating complex expressions, the coordinates of the mirror points are calculated step by step here. Let the coordinates of a plane point be (x_1, y_1), the equation for the line l, where the mirror is located, is $y = kx + b$. The coordinates of the mirror point can be obtained through the following 4 steps.

(1) According to slope k, calculate angle $\alpha = \text{atan}(-1/k)$.
(2) If $kx_1 + b > y_1$, update angle α as $\alpha \leftarrow \alpha + \pi$.
(3) Compute distance from the given point (x_1, y_1) to line l as the following equation, $d = |kx_1 + b - y_1|/\sqrt{k^2 + 1}$.
(4) Move the point (x_1, y_1) along the direction defined by the angle α, and with the distance $2d$ to the mirrored point z, as $z = x_1 + \text{i}y_1 + 2d\,\exp(\text{i}\alpha)$. It is more convenient to use complex numbers here.

The graph on the right side of Fig. 1.29 shows a calculation result. Using the green straight line as the mirror, all turning points of the red fold line are mirrored to the turning points of the blue fold line. Apply the above method, mirror all coordinate points of the image, and finally the image is mirrored and flipped.

1.9.2 Fourier Transform

Fourier transforms are applied to fields such as image processing, image analysis and computational imaging. Its successful and widespread application stems from people's in-depth research on its theory, properties and fast algorithms. It is a mature and basic transformation tool.

Discrete cosine transform is used for block processing mainly because the main energy of the discrete Fourier transform result of digital images is concentrated at the origin, which is sometimes unfavorable for analysis and processing. Another advantage of the discrete cosine transform is that it is a real-domain transform. The discrete cosine transform can be regarded as the real part of the Fourier transform matrix.

Another transformation related to the Fourier transform is the Hartley transform, which is the sum of the real and imaginary parts of the Fourier transform [21]. The Hartley transform is also a real-domain transform, and the interesting property of this transform is that its inverse transform is itself. In addition, the eigenvalues of the Hartley transform are two different values, ± 1. This transformation can be achieved by an optical system.

The fractional Fourier transform is a generalization of the Fourier transform in the order dimension [22]. The fractional Fourier transform can be optically realized by a system composed of lenses. It has found many applications in the field of optical information processing. Its properties and discrete algorithm have also been studied in depth. Similar to wavelet transform, fractional Fourier transform also has the ability to jointly analyze time–frequency.

1.9.3 Wavelet Transform

Wavelet transform is a transform analysis method that combines time–frequency representation. It is related to the short-term (window) Fourier transform, and the size of the transform window changes with frequency. Through transformation, the local characteristics of the signal can be highlighted, the temporal (spatial) frequency can be locally analyzed, the signal (image) can be characterized on a multi-scale basis through expansion and translation, and the details of the image can be extracted. It is known as a mathematical microscope. It solves the difficult problem of Fourier transform processing time-varying signals.

Wavelet transform can provide frequency-domain characterization in image analysis and image processing. It has multiple forms of wavelet mother functions and provides multiple information processing options. In order to facilitate the application of wavelet transform, people have developed a special wavelet transform toolbox.

1.10 General Conclusions

This chapter describes basic programming methods that can be used for image analysis. This chapter includes image acquisition, image processing methods, program structure, combination of numbers and shapes, imaging and recognition, optimization methods, image analysis methods, image preprocessing methods, and image transformation methods. Using these contents, readers write some calculation programs for intelligent image analysis.

Acknowledgements This work was supported by the National Natural Science Foundation of China (12374271, 62375065, 62405079, 62575091); Fundamental Research Funds for the Central Universities (HIT.OCEF.2024055, AUGN5750300325); the Fund of National Laboratory on Adaptive Optics, China (FNLAO-24-MS-O07); Natural Science Foundation of Heilongjiang Province of China (LH2024A009); Key Laboratory of Image Processing and Intelligent Control (Huazhong University of Science and Technology), Ministry of Education, Wuhan, China (IPIC2025-05); Key Laboratory of Advanced Medical Imaging and Intelligent Computing of Guizhou Province (AMIIC250204); the China Postdoctoral Science Foundation-CCTEG Joint Support Program (2025T011ZGMK).

References

1. Faulkner HML, Rodenburg JM (2004) Movable aperture lensless transmission microscopy: a novel phase retrieval algorithm. Phys Rev Lett 93(2):023903
2. Zheng G, Horstmeyer R, Yang C (2013) Wide-field, high-resolution Fourier ptychographic microscopy. Nat Photon 7(9):739–745
3. Liu Z, Yang M, Liu W, Li S, Gong M, Liu W, Liu S (2012) Image encryption algorithm based on the random local phase encoding in gyrator transform domains. Opt Commun 285:3921–3925
4. Liu Z, Zhou X, Liu S (2024) Computational optical imaging: principle and technology. Springer Publisher
5. Collins SA (1970) Lens-system diffraction integral written in terms of matrix optics. J Opt Soc Am 60(9):1168–1177
6. Zhu B, Liu S (2001) Multifractional correlation. Opt Lett 26(9):578–580
7. Kennedy J, Eberhart R (1995) Particle swarm optimization. In: IEEE international of first conference on neural networks, pp 167–171
8. Kirkpatrick S, Gelatt CC, Vechi MP (1983) Optimization by simulated annealing. Science 220(4598):671–680
9. Naserbegi A, Aghaie M, Minuchehr A, Alahyarizadeh G (2018) A novel exergy optimization of Bushehr nuclear power plant by gravitational search algorithm (GSA). Energy 148:373–385
10. Dorigo M, Stützle T (2019) Ant colony optimization: overview and recent advances. Springer
11. Xue J, Shen B (2020) A novel swarm intelligence optimization approach: sparrow search algorithm. Syst Sci Control Eng 8(1):22–34
12. Khan AH, Cao X, Li S, Katsikis VN, Liao L (2020) BAS-ADAM: an ADAM based approach to improve the performance of beetle antennae search optimizer. IEEE-CAA J Autom Sin 7(2):461–471
13. Shen C, Bao X, Tan J, Liu S, Liu Z (2017) Two noise-robust axial scanning multi-image phase retrieval algorithms based on Pauta criterion and smoothness constraint. Opt Express 25(14):16235–16249
14. Liu Z, Guo C, Tan J, Wu Q, Pan L, Liu S (2014) A diffraction model of direction multiplexing method for hiding multiple images. J Mod Opt 61(14):1127–1132
15. He K, Sun J, Tang X (2013) Guided image filtering. IEEE Trans Pattern Anal Mach Intell 35(6):1397–1409
16. Qiao Z, Wen X, Zhou X, Qin F, Liu S, Gao B, Liu W, Chi D, Liu Z (2023) Adaptive iterative guided filtering for suppressing background noise in ptychographical imaging. Opt Lasers Eng 160:107233
17. Dabov K, Foi A, Katkovnik V, Egiazarian K (2007) Image denoising by sparse 3-D transform-domain collaborative filtering. IEEE Trans Image Process 16(8):2080–2095
18. Haykin S, Totterdill P (2001) Adaptive filter theory. Prentice Hall
19. Biggs DSC, Andrews M (1997) Acceleration of iterative image restoration algorithms. Appl Opt 36(8):1766–1775
20. Gonzalez RC, Woods RE, Eddins SL (2009) Digital image processing using MATLAB. Gatesmark Publishing
21. Liu Z, Zhang Y, Liu W, Meng F, Wu Q, Liu S (2013) Optical color image hiding scheme based on chaotic mapping and Hartley transform. Opt Lasers Eng 51(8):967–972

22. Ozaktas HM, Zalevsky Z, Kutay MA (2001) The fractional Fourier transform with applications in optics and signal processing. Wiley

Chapter 2
Target Recognition

Zhe Wang, Zhengjun Liu, and Yutong Li

Abstract This chapter introduces the basic concepts, common methods, and application examples of object recognition. The goal of object recognition is to automatically identify the objects of interest within an image through computational methods, and analyze and determine information such as their location, shape, and category. The content covered includes image feature extraction, template matching, region segmentation, and the foundational application of machine learning methods in recognition.

2.1 Basic Concepts of Target Recognition

Object recognition can essentially be considered a specific application of image recognition, with its core task being to detect and extract specific target objects from a given digital image. Image recognition and image processing are closely coupled at the technical level, together forming a complete computer vision processing workflow. Before implementing image recognition, it is essential to preprocess the original image. Image processing refers to the application of a series of mathematical transformations and algorithmic operations that convert the raw image data into an enhanced image more proper for analysis.

In the system architectures, both the input and output of the image processing module are digital image data. Image recognition, on the other hand, is an advanced visual analysis performed on the basis of preprocessing. Its key technologies include feature extraction based on edge detection and region growing, feature representation using geometric invariants and texture descriptors, and target recognition through

Z. Wang · Z. Liu (✉) · Y. Li
School of Physics, Harbin Institute of Technology, Harbin, China
e-mail: zjliu@hit.edu.cn

Z. Wang
e-mail: wanzhe17879349918@163.com

Y. Li
e-mail: yutong_li@hit.edu.cn

Z. Liu and Y. Li (eds.), *Intelligent Analysis of Optical Images*, Scientific Computation,
https://doi.org/10.1007/978-981-95-3720-4_2

classification algorithms, such as pattern matching, discriminant functions, decision trees, and graph matching. From a research perspective, image recognition mainly focuses on image classification modeling and structural description, with the core task being the construction and optimization of feature space, i.e. feature extraction. Thus, the input to an image recognition system is preprocessed image data, and the output is the target category label and image structure analysis results. It is important to note that, during feature extraction, image segmentation techniques are often required for more accurate local feature analysis [1].

2.1.1 The Difference Between Image Recognition and Object Detection

Image recognition is one of the most fundamental forms of image analysis tasks. It involves classifying the entire image, that is, determining whether the image contains a certain type of object or identifying the overall category of the image, as Fig. 2.1a. This type of task does not require determining the specific location of objects within the image. For example, when the system receives an image, it might conclude that the image contains a cat or belongs to the landscape category. Such tasks do not need to focus on the precise distribution of the objects in the image.

In contrast, object detection is a more advanced image analysis task. Object detection not only involves determining which categories of objects are present in the image but also requires specifying their spatial locations within the image, as Fig. 2.1b. A common output format is to draw a bounding box around each detected object, marking its location while also providing the object's category information. Object detection is more practical in real-world applications, as it can be used for automatic recognition and localization in multi-object scenarios, adapting to complex background variations and object occlusions.

The main difference between the two lies in the richness of the output information. Image recognition focuses solely on whether an object is present, while

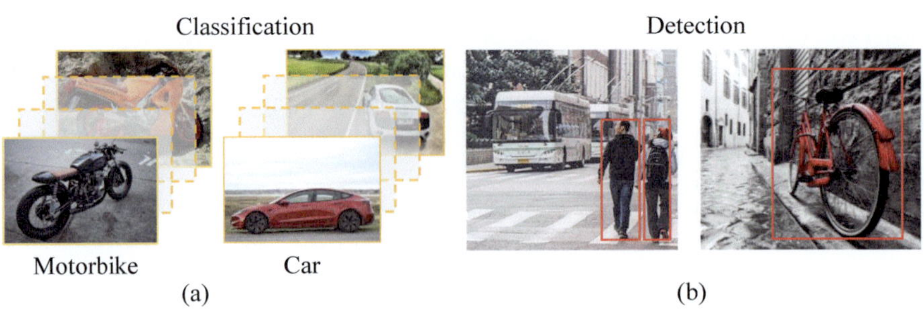

Fig. 2.1 The tasks of image recognition and object detection: **a** certain type, **b** positioning

object detection also considers where the object is located. In engineering applications, the process of object detection is often more complex, involving both global understanding of the image as well as the ability to assess local regions.

2.1.2 *The Basic Process of Target Recognition*

The implementation process of an object recognition system typically consists of several consecutive processing steps, with each step playing a crucial role in information transformation. The overall process begins with image acquisition, followed by a series of processing, analysis, and decision-making steps, ultimately outputting the target's location and category information, as Fig. 2.2. Understanding this process helps to grasp the basic logical structure of object recognition and provides insights for writing recognition programs in subsequent steps.

The first step in object recognition is image acquisition. In practical applications, image data can be collected using a variety of devices, such as standard visible light cameras, infrared imagers, CT or MRI imaging systems, remote sensing satellites, and more. The acquired images are typically digital images stored in the form of matrices on a computer, with each pixel value representing the brightness or color information at a specific location in the image. Depending on the imaging conditions, the images may be affected by issues such as noise interference, brightness variation, or blurring.

To enhance the analyzability of the image, preprocessing operations are required after image acquisition. The goal of image preprocessing is to amplify the difference between the target and background, reduce the interference from noise, and optimize the contrast and clarity. Common preprocessing methods include grayscaling,

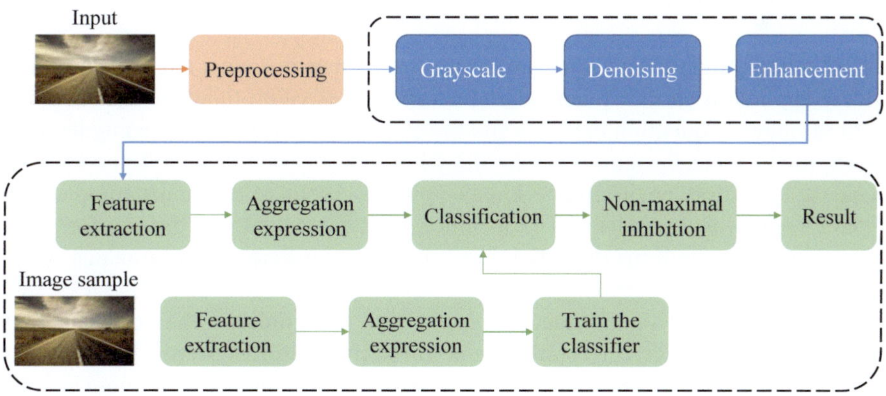

Fig. 2.2 Target recognition flowchart

histogram equalization, filtering and denoising, and image enhancement. These operations improve the overall quality of the image, allowing the subsequent recognition module to more accurately extract features [2].

Feature extraction is the core step in object recognition. Features can be understood as information in the image that helps to distinguish between target and non-target regions. Common features include edges, corners, shape contours, color distribution, and texture patterns. During feature extraction, it is necessary to select the appropriate feature type based on the specific requirements of the recognition task and extract them from the original image using algorithms. For example, in infrared images, intensity contrast might be the primary basis for recognition, whereas in visible light images, color and texture information are more crucial.

Once feature extraction is complete, target discrimination and classification are required. For simple tasks, rule-based methods such as thresholding and template matching can be used; for more complex scenarios, trained classifier models such as K-nearest neighbors, support vector machines, and decision trees can be introduced. Regardless of the method used, the goal of discrimination is to determine whether a target exists in the image based on the extracted features and to provide the category label and confidence level.

In the final output stage, the system displays the recognition results in a structured manner. A common output format is to enclose the recognized targets in bounding boxes within the image, including their location information, and label the category and recognition confidence. In engineering implementations, these output results can be used for visualization, subsequent task decisions, or data storage. For multi-object scenarios, the system outputs a set of recognition results, usually containing the location and category information for multiple targets.

2.1.3 The Application Background of Target Recognition

Object recognition, as an important research direction in image analysis, has been widely applied in various practical engineering fields. With the continuous enrichment of image acquisition methods and the ongoing improvement of computational capabilities, object recognition technology has not only demonstrated good performance in laboratory environments but also plays a crucial role in many complex and dynamic real-world scenarios. The accuracy, robustness, and real-time performance of recognition algorithms directly impact the overall efficiency and practical value of the system.

In the field of intelligent transportation, object recognition technology is widely used in autonomous driving systems and traffic monitoring systems. By capturing road images with onboard cameras, the system can automatically recognize vehicles, pedestrians, traffic signs, traffic lights, and other information, thereby enabling environmental perception and assisting in driving decisions. For example, when the system detects a pedestrian crossing the road ahead, it can issue a warning or autonomously reduce speed, enhancing driving safety.

Medical image processing is another important application of object recognition. In medical images such as X-rays, CT scans, MRI scans, and ultrasound images, recognizing lesion areas or organ boundaries can assist doctors in disease diagnosis, surgical planning, and treatment evaluation. For example, in lung CT images, automatically recognizing the position and shape of nodules is of practical significance for early lung cancer screening.

In remote sensing image analysis, object recognition is commonly used for land cover classification, disaster monitoring, and military reconnaissance. By recognizing targets in satellite images, the system can identify the distribution of buildings, roads, vegetation, water bodies, and other areas. After sudden disasters or special events occur, object recognition results can be used to quickly assess the affected areas and assist in emergency response deployment.

Security surveillance systems also heavily rely on object recognition technology to achieve intelligent features. Typical tasks include face recognition, behavior monitoring, and abnormal object detection. In video surveillance scenarios, the system can automatically recognize individuals entering designated areas or detect suspicious activities, triggering alarm systems to enhance security.

With the development of deep learning methods, the performance of object recognition technology has improved significantly, enabling it to handle complex conditions such as varying lighting, occlusions, multiple targets, and dynamic changes, far beyond the simple pattern recognition of earlier systems. Current object recognition technology is continuously evolving, expanding to more image modalities and sensor types, and is also integrating with tasks like semantic understanding, behavior recognition, and 3D reconstruction to form a more complete intelligent visual analysis system [3, 4].

The application background of object recognition is vast, driving continuous technological innovation, while also imposing higher demands on the interpretability, computational speed, and versatility of algorithms. For learners, mastering the basic principles and typical methods of object recognition helps in understanding the application logic of these technologies in practice, and provides the theoretical foundation and technical preparation for future research and development.

2.2 Target Features

The core of the object recognition task lies in distinguishing targets from non-targets, and the foundation for making this distinction is features. Features refer to information that reflects certain important attributes of the image, which can be used to describe the appearance, structure, or local variations of the target. In the image recognition process, feature extraction serves as the bridge between the raw image and the classification decision. A good feature not only accurately expresses the key shapes of the target but also should possess a certain level of stability, maintaining invariance under conditions such as changes in lighting, scale, or image rotation.

Image features can be classified into various types, including shape features, texture features, color features, and statistical features. In practical applications, multiple features are often selected or combined based on the task requirements. For example, in infrared images, gray intensity and local brightness contrast are fundamental criteria for judgment; whereas in natural images, contour structure, color distribution, and texture details often provide more discriminative information.

This section will introduce several representative types of image features, including edge features, corner features, contour features, and region descriptors. These features are widely used in specific recognition tasks such as template matching, feature point tracking, target segmentation, and shape discrimination.

2.2.1 Edge Features and Gradient Information

Edges are among the most important local structural features in grayscale images, typically corresponding to discontinuities in image properties such as grayscale, color, or texture across spatial domains. Essentially, edges mark the end of one region and the beginning of another, reflecting abrupt transitions in the image structure.

Edges in images can be classified based on the form of grayscale variation. Ideally, edges appear as instantaneous jumps in grayscale values at specific positions, known as step edges, as Fig. 2.3a. Their profiles exhibit an approximately vertical step structure. However, in practical image acquisition, due to limitations in sensor response speed, sampling rate, spatial resolution of imaging systems, and lighting conditions, edges often deviate from this ideal state and exhibit varying degrees of blurring.

In the ramp edge in Fig. 2.3b, the grayscale transition is no longer instantaneous but occurs gradually over a certain spatial range, forming an inclined slope. This

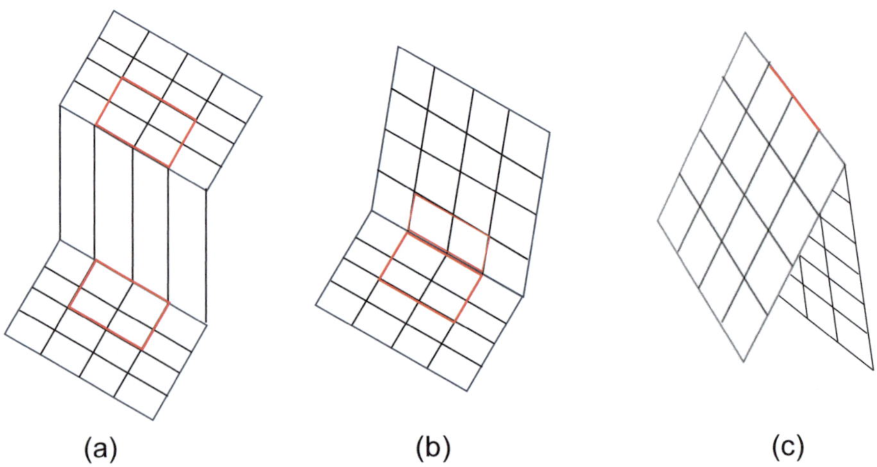

Fig. 2.3 Common edge types: **a** step edge, **b** slope edge, **c** roof edge

type of edge is more common, with its width varying depending on the degree of blurring, the more severe the blur, the wider the edge. In high-quality, sharp images, the transition region is narrower, closer to the ideal step form.

Another typical edge type is the roof edge, in Fig. 2.3c. Its grayscale profile shows a local maximum or minimum, with the edge located at the extremum point after a slow rise or fall in grayscale. Roof edges often appear in regions where both sides have similar brightness but the middle is slightly brighter or darker, such as fine stripes, elongated structures, or linear objects in images.

Different edge types exhibit distinct response characteristics in edge detection algorithms, and accurate modeling of them can improve detection precision and robustness. Edges possess two fundamental properties: direction and magnitude. Grayscale changes along the edge direction are usually gradual, while changes perpendicular to it are more abrupt. This behavior provides a clear mathematical basis for edge detection and extraction [5].

Edge detection often employs differential operators to compute image gradients, extracting regions with the most significant local grayscale variations. Gradients reflect the rate of grayscale change in spatial directions, typically calculated separately in horizontal and vertical directions. Common first-order differential operators include Sobel, Prewitt, and Roberts, which perform convolutional calculations within each pixel neighborhood by sliding across the image to derive gradient magnitude and direction. The magnitude describes the intensity of grayscale change, while the direction indicates its orientation. If the gradient magnitude at a point is significantly higher than the neighborhood average, that point can be considered a potential edge pixel.

In certain algorithms, second-order derivatives may also be introduced to determine edge locations by identifying zero-crossings in the second-order differences. Theoretically, such methods can more accurately localize edge centers, but they are more sensitive to image noise. Since differential operations inherently amplify high-frequency noise, preprocessing with low-pass filters (e.g., Gaussian smoothing) is typically required before edge detection to suppress noise interference and enhance the robustness of edge extraction.

Edge features play a crucial role in object recognition, particularly in scenarios where structural contours are indistinct, backgrounds are cluttered, or lighting conditions vary significantly. Stable edge information can effectively improve the localization accuracy and robustness of recognition systems. Edge extraction methods offer advantages such as computational simplicity and clear physical interpretability, making them widely applicable in tasks like contour detection, image registration, and shape analysis.

To further improve the continuity and accuracy of edge detection results, post-processing operations such as non-maximum suppression, double-thresholding, and edge linking are often employed to refine the initially extracted edges. These steps help produce more complete and structurally coherent object contours.

There are many algorithms for implementing edge detection. The following is a brief introduction to some commonly used edge detection operators.

(1) Gradient operator

Gradients correspond to first-order derivatives, and gradient operators are first-order derivative operators. For a continuous function $f(x, y)$, Its gradient at position x can be expressed as a vector

$$\nabla f(x, y) = [G_x, G_y] = \left[\frac{\partial f}{\partial x}, \frac{\partial f}{\partial y}\right]^{\mathrm{T}} \tag{2.1}$$

Its amplitude and direction angle can be expressed as

$$\mathrm{mag}(\nabla f) = \sqrt{[G_x^2 + G_y^2]} \tag{2.2}$$

$$\phi(x, y) = \arctan(G_y / G_x) \tag{2.3}$$

In practice, the convolution of small region templates is often used to approximately calculate the above partial derivatives. Since and each use a template, the two templates need to be combined to form a gradient operator. The common feature of these operators is that they perform certain weighting operations on the neighborhood of the object under investigation in the spatial domain, and the calculation result is taken as the edge strength. Its general mathematical model can be defined as

$$E_{i,j} = W * P_{i,j} \tag{2.4}$$

where, $E_{i,j}$ represents the edge intensity of point (i, j), $P_{i,j}$ is the neighborhood of scale $l \times l$ around pixel point (i, j), and W is the weight coefficient matrix, which is independent of the position of the pixel point.

The template form of the Sobel operator is shown in Fig. 2.4a, and the corresponding difference form of this template can be expressed as

$$\begin{aligned} G_x = &\{f(x+1, y-1) + 2f(x+1, y) + f(x+1, y+1)\} \\ &- \{f(x-1, y-1) + 2f(x-1, y) + f(x-1, y+1)\} \end{aligned} \tag{2.5}$$

$$\begin{aligned} G_y = &\{f(x-1, y+1) + 2f(x, y+1) + f(x+1, y+1)\} \\ &- \{f(x-1, y-1) + 2f(x, y-1) + f(x+1, y-1)\} \end{aligned} \tag{2.6}$$

Other common templates are shown in Fig. 2.4b, c. After completing the image gradient calculation, it is often necessary to compare the gradient amplitude with the preset threshold to determine whether a certain pixel position belongs to the edge area. If the gradient amplitude is greater than this threshold, it is generally considered that there is a significant gray level change at this pixel, and it can then be regarded as

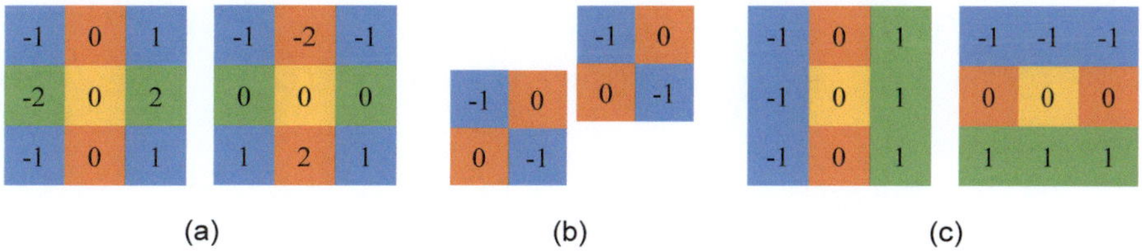

Fig. 2.4 Common gradient operator templates: **a** Sobel, **b** Roberts, **c** Prewitt

an edge point. The setting of the threshold has a direct impact on the edge detection results and is one of the key parameters for controlling the quality of edge extraction.

In images with noise, the threshold selection needs to strike a balance between two targets: on the one hand, it should avoid the real edges being missed due to an excessively high threshold; On the other hand, it is necessary to prevent false noise detections caused by overly low thresholds. Therefore, the setting of the threshold is essentially a process of compromise between losing valid edge information and introducing false edge responses. To enhance robustness, in practical applications, smooth filtering preprocessing and multi-threshold strategies are often combined, such as the double-threshold connection mechanism in the Canny operator, to obtain more complete and stable edge results.

(2) Laplacian operator

Laplacian operator is a second-order derivative operator. For a continuous function $f(x, y)$, The Laplacian value at position (x, y) is defined as follows

$$\nabla^2 f = \frac{\partial^2 f}{\partial x^2} + \frac{\partial^2 f}{\partial y^2} \tag{2.7}$$

In digital image processing, the Laplace values of functions can usually be approximately calculated through discrete templates. When constructing the template of the Laplacian operator, certain coefficient distribution requirements should be met: the coefficient corresponding to the central pixel in the template should be positive, while the coefficients corresponding to its adjacent pixels should be negative, and the sum of all coefficients in the entire template should be zero. This design can highlight the areas with drastic gray level changes in the image, effectively enhance the edge response, and at the same time suppress the influence of overall brightness shift on the result. Two common templates are as follows

$$\nabla^2 = \begin{bmatrix} 0 & 1 & 0 \\ 1 & -4 & 1 \\ 0 & 1 & 0 \end{bmatrix} \text{ or } \begin{bmatrix} 1 & 1 & 1 \\ 1 & -8 & 1 \\ 1 & 1 & 1 \end{bmatrix} \tag{2.8}$$

The difference forms corresponding to these two templates are respectively, as

$$\nabla^2 f(x, y) = \{f(x, y) - f(x+1, y) + f(x, y) - f(x-1, y)\} \\ + \{f(x, y) - f(x, y+1) + f(x, y) - f(x, y-1)\} \tag{2.9}$$

$$\nabla^2 f(x, y) = \{f(x, y) - f(x-1, y-1) + f(x, y) - f(x, y-1)\} \\ + \{f(x, y) - f(x+1, y-1) + f(x, y) - f(x-1, y)\} \\ + \{f(x, y) - f(x+1, y) + f(x, y) - f(x-1, y+1)\} \\ + \{f(x, y) - f(x, y+1) + f(x, y) - f(x+1, y+1)\} \tag{2.10}$$

In practical applications, if the output result of the Laplace operator contains negative values, their absolute values can be taken as needed for processing. The Laplacian operator has a strong response ability to isolated points, line ends and boundaries in the image, and can effectively enhance these structural features. However, for the boundary region where the gray level changes slowly, the Laplacian response often presents as a pair of adjacent positive and negative peaks, while the output is zero at the boundary itself. At this point, the zero-crossing point between the positive and negative peaks can be detected as the basis for determining the position of the boundary.

As a typical second-order differential operator, the Laplacian operator has high theoretical accuracy and can sensitively capture the second-order information of gray level changes in images. However, it is also highly sensitive to noise because noise often contains high-frequency components that are easily amplified in second-order differentials, leading to false detections. Therefore, directly using the Laplacian operator in a noisy background often leads to unstable edge positioning, seriously affecting the recognition effect. Due to this drawback, the Laplacian operator is used relatively infrequently in actual image processing and usually needs to be combined with smoothing preprocessing or other detection methods to enhance its robustness.

To improve the anti-noise ability of the Laplacian operator, the method of combining it with Gaussian smoothing operation is often adopted, which forms the Laplacian of Gaussian (LOG). The LOG operator first performs low-pass smoothing on the image through Gaussian filtering to suppress high-frequency noise interference, and then applies the Laplacian operator for edge enhancement, thereby significantly improving robustness while maintaining the sensitivity of edge detection.

The essence of the LOG operator is to perform a second-order differential operation on the image after Gaussian convolution processing. The mathematical description is as follows

$$\mathrm{LOG}(x, y) = -\frac{1}{\pi}\left(1 - \frac{x^2 + y^2}{2\sigma^2}\right)\exp\left(-\frac{x^2 + y^2}{2\sigma^2}\right) \tag{2.11}$$

Compared with directly using the Laplacian operator, the edge response output by the LOG operator is smoother and more coherent, and can effectively reduce the false edge response caused by noise. The edge detection results still exhibit typical zero-crossing characteristics, that is, the boundary positions correspond to the zero-crossing points of the Laplace response, and the edge positions can be determined by detecting the zero-crossing regions.

The spatial response of the LOG operator has a ring-shaped structure, which is suitable for detecting target contours with clear edges and gradual gray-level changes. It performs particularly well in scenarios with high requirements for detailed structure, such as medical images and remote sensing images. However, due to the fact that its calculation involves Gaussian filtering and the second derivative, the computational complexity is relatively high. In applications with strong real-time requirements, trade-offs still need to be made.

(3) Canny operator

Among various edge detection methods, the Canny operator stands out for its high detection accuracy and stability, making it widely applicable in diverse image processing tasks. By effectively preserving edge details while suppressing noise, it has become one of the most extensively used edge detection operators. The Canny algorithm integrates Gaussian filtering, gradient computation, non-maximum suppression, and thresholding into a systematic workflow, combining theoretical rigor with practical utility.

To mitigate high-frequency noise interference, the Canny algorithm first applies Gaussian smoothing to stabilize subsequent gradient calculations. The smoothed image is then processed using first-order differential operators (e.g., Sobel) to compute the rate of grayscale variation along horizontal and vertical directions, yielding gradient magnitude and orientation for each pixel. These data form the foundation for edge extraction.

Not all gradient transitions represent true edges. To ensure crisp and refined edge contours, the Canny algorithm employs non-maximum suppression (NMS). This step evaluates whether a pixel represents a local maximum along its gradient direction; if not, the pixel is suppressed, eliminating redundant responses along non-edge regions. This significantly enhances edge localization accuracy and thinness.

For final edge determination, Canny adopts a dual-threshold hysteresis mechanism. By setting high and low thresholds, strong edges are retained outright, while weaker edges are preserved only if connected to strong edges. This strategy balances edge continuity with noise resistance, effectively reducing fragmented edges and isolated noise points.

The Canny operator exhibits strong robustness and adaptability, making it suitable for edge extraction in most natural and grayscale images. Its results demonstrate high continuity, low false detection rates, and tunable precision via parameter adjustments. Consequently, it holds broad utility in practical applications such as image segmentation, object detection, and medical imaging.

In summary, edge features play a foundational role in object recognition, and different edge detection operators exhibit distinct characteristics in terms of precision, noise immunity, edge continuity, and computational complexity. To clarify the performance differences among these operators and highlight their respective

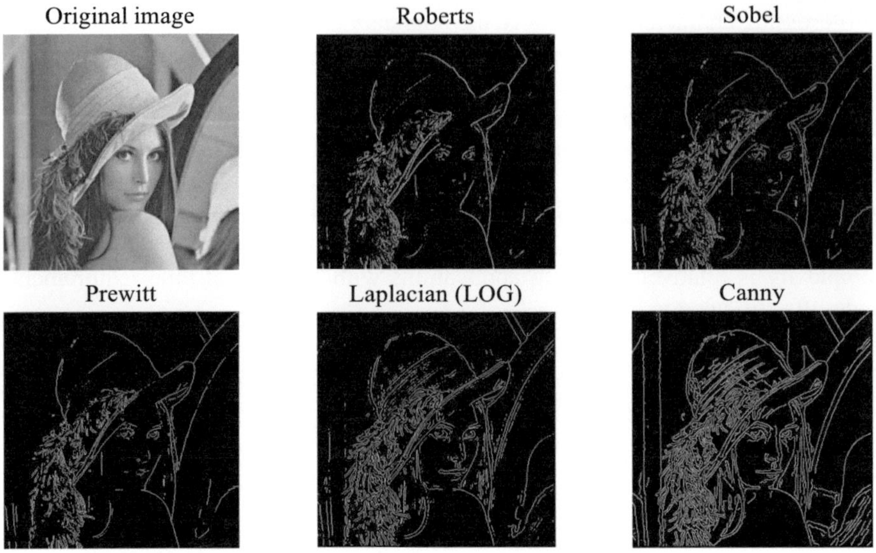

Fig. 2.5 Comparison of common edge detection operator results

strengths and limitations, we conducted a comparative analysis using real-world images. The results are illustrated in Fig. 2.5, and key properties are summarized in Table 2.1.

2.2.2 Line Detection

After completing edge extraction from an image, a common subsequent task is to identify whether the image contains targets with specific geometric structures, such as lines, circles, or ellipses. Among these, straight lines, as one of the most fundamental geometric shapes, hold significant importance in applications like industrial inspection, traffic image analysis, document recognition, and architectural contour extraction. The core objective of line detection is to accurately extract pixel clusters exhibiting linear distribution characteristics from known edge information and estimate their spatial parameters.

Traditional edge maps only provide localized boundary variation information and lack the ability to model overall geometric structures. To address this, parameter space transformation methods have been introduced into image analysis. These methods employ a "from image space to parameter space" conversion mechanism, transforming the problem of geometric model detection into an accumulation and voting process for specific parameters. Among such approaches, the most representative are the Radon transform and Hough transform [6].

Table 2.1 Summary of the performance of common edge detection operators

Operator	Advantages	Defects	Scenarios
Roberts	Simple calculation and fast edge positioning	It is sensitive to noise and has limited detection directions	Edge detection tasks for high-contrast images with distinct vertical edges or steep low noise
Sobel	Good noise resistance and relatively continuous edges	It is not sensitive to changes in gray scale and details may be lost	Edge detection tasks for images with a lot of noise and gradual gray-scale changes
Prewitt	A certain inhibitory effect on noise	Noise-sensitive and with poor anti-interference ability	Low-complexity grayscale gradient images without obvious noise
Laplacian	Accurately locate the step edge points in the image	Noise-sensitive, only applicable to noise-free images	It is used to normalize the detection results and calculate the variance, and can be applied to the fuzzy detection task
Canny	Not sensitive to noise and can detect weak edges	High-frequency edges are prone to smoothing, resulting in edge loss	It is used in practical projects such as lane line detection and is the most effective edge detection method

The Radon transform is an image transformation method derived from projection geometry, originally developed for tomographic reconstruction in medical imaging. In line detection tasks, it performs integral projections of edge points along various angles, generating a transform domain defined by angle and distance variables. In the transformed result, if a group of pixels is collinear in the original space, their response in the Radon domain will exhibit a distinct peak. This allows for the determination of whether straight-line structures exist in the image along specific orientations. A schematic diagram of this principle is shown in Fig. 2.6a.

The core principle of the Radon transform lies in the mapping between lines and points, converting from image space to parameter space. As Fig. 2.6b, a straight line in Cartesian coordinates can be expressed as

$$\rho = x\cos\theta + y\sin\theta \tag{2.12}$$

Then the Radon transform along the direction of angle θ in the two-dimensional space is defined as

$$R(\rho, \theta) = \iint\limits_S f(x, y)\delta(\rho - x\cos\theta - y\sin\theta)\mathrm{d}x\mathrm{d}y \tag{2.13}$$

where S represents the region where $f(x, y)$ is located, θ is the inclination angle of the straight line, and ρ is the distance from the origin to the straight line. The

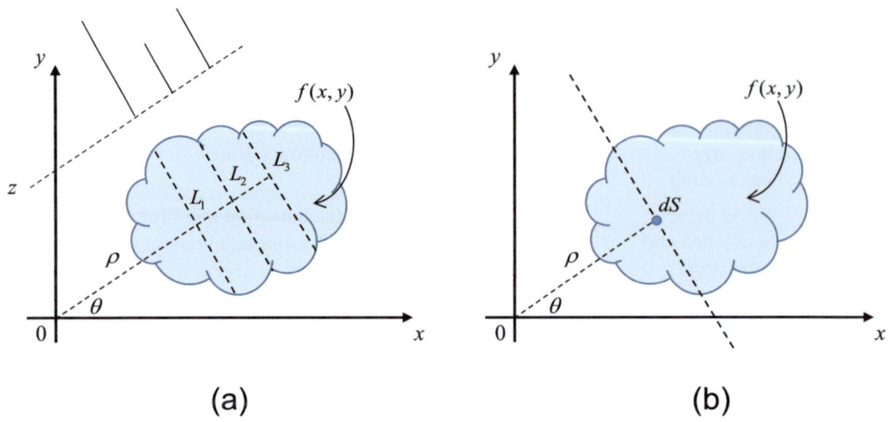

Fig. 2.6 Schematic diagram of Radon transform: **a** directional superposition schematic diagram, **b** Radon transform mapping

above equation represents the projection of $f(x, y)$ on a straight line in the specified direction, with the slope of the line being $\tan(\theta + \pi/2)$ and the intercept being $\rho/\sin\theta$.

For the Radon transform, we can understand it as the projection of the image on plane (ρ, θ), where each point on plane (ρ, θ) corresponds to a straight line on the plane. Lines with high gray values will form bright spots in the parameter space, while straight lines with low gray values will form dark spots.

To further understand the practical application effect of Radon transform in line detection, experimental analysis can be conducted through the line features in the image spectrum. Figure 2.7 shows a schematic of a set of line detection processes based on spectral images. The original data is an image with motion blur, and its two-dimensional Fourier transform result is shown in the left figure. In the spectrum, high-frequency textures with a linear distribution can be observed, which is due to the linear features in the frequency domain caused by directional blur in the image.

The middle figure shows the parameter space projection result obtained by applying the Radon transform to this spectrogram. During the transformation process, the spectrogram is projected and integrated from different angles, forming a Radon domain with angles and distances as coordinates. The brighter regions indicate a significant linear structure in a certain direction.

To further determine the direction of the straight line from the Radon domain, one-dimensional statistical analysis can be conducted on the pixel values in the transformation result graph to extract the aggregated features representing different Angle directions. The right side of the figure shows the characteristic curves obtained by three different statistical methods. The first method is to generate an Angle response curve by taking the median value of each column of the Radon image. The second method calculates the standard deviation of each column to measure the stability of the direction distribution. The third method adopts the difference and summation of each column to enhance the trend of change and suppress background noise.

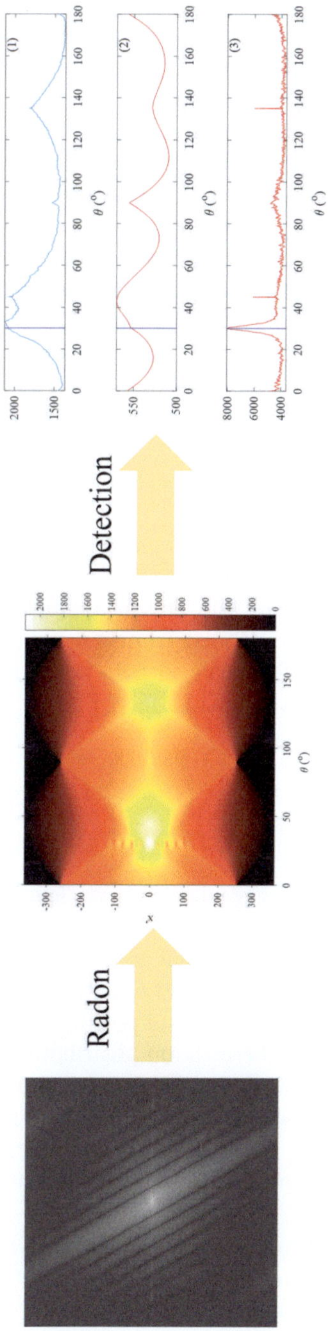

Fig. 2.7 Radon transform detection of straight line results

It can be seen from the results that the curve obtained by the difference summation method forms a distinct peak in the straight line direction, and the corresponding Angle is the main direction of the straight line structure in the image. This experiment demonstrates the effectiveness and visualization advantages of the Radon transform in detecting the direction of straight lines in images.

Compared with the Radon transform, the Hough transform is a more general parametric space method, proper for detecting arbitrary parameterizable geometries in images. For line detection, the Hough transform maps each edge point in the image to a curve in the parameter space. If different points belong to the same line, their corresponding curves will intersect in the parameter space. By conducting cumulative statistics on the intersection points in the parameter space, the set of straight lines with consistent parameters in the image can be located.

Both methods are essentially based on the voting mechanism of parameter space, but they differ in implementation methods and numerical stability. The Radon transform is more applicable for global structural analysis, with high operational accuracy but is sensitive to noise. The Hough transform has strong robustness and is appropriate for the detection of discrete edge points, but its computational complexity is relatively high.

In practical applications, the straight-line detection method is often closely integrated with the edge extraction process and can serve as a key link in the identification of high-rise structures. The following will present the detection effects of the two methods in different scenarios through experimental images and conduct a brief comparative analysis of their performance to help better understand the role and value of the parameter space method in target recognition.

2.2.3 Corner Feature

Corner points are local areas in an image where the gray level changes simultaneously in multiple directions, usually occurring at the intersection of two edges. Compared with edge points that only have unidirectional gradient changes, corner points have stronger local stability and identifiability, and thus play an important role in tasks such as image matching, target recognition and positioning, and motion estimation. Through the statistical analysis of the position, distribution and response intensity of diagonal points, it can also assist in inferring the geometric shape of the target area, thereby providing support for target recognition and classification. In images with complex structures or strong background interference, corner points are often more reliable than edge information and are an important basis for constructing local descriptors of images.

The basic idea of corner detection is that if an image window shows significant grayscale changes after being translated in multiple directions such as horizontal, vertical, and diagonal, then this area can be regarded as a corner area. With the description of the basic idea of diagonal points, the mathematical description of corner points is further defined as

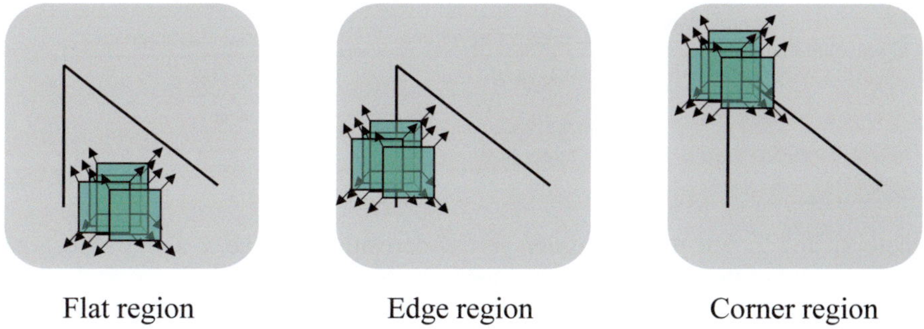

Fig. 2.8 The basic idea of corner detection

$$E(u, v) = \sum_{x,y} w(x, y)[I(x + u, y + v) - I(x, y)]^2 \tag{2.14}$$

where $E(u, v)$ is energy function, u, v represents the movement of the window, x, y is the coordinate of the pixel point, summed within the window range, and $w(x, y)$ is the weight coefficient corresponding to the difference of each pixel before and after movement, which can be median filtering or Gaussian filtering.

According to the mathematical description, the characteristics of Harris corner points can be summarized as Fig. 2.8. In flat areas, there is no gray-scale change when the window is moved in any direction; in edge areas, there is no gray-scale change when it is moved along the edge direction; at corner points, there is a significant gray-scale change when it is moved in any direction.

In actual calculations, the corner extraction process also relies on the estimation of image gradients. Once the gradient information of each pixel in the image is obtained, it is usually necessary to compare it with the set threshold to determine whether there are corner points or edge structures. If the gradient magnitude of a certain pixel is greater than a given threshold, it is considered that there may be corner points or edges at that position. The selection of the threshold directly affects the validity of the detection results. If the threshold is set too high, it may cause weak edges or detailed corners to be overlooked. However, if the threshold is too low, it is easy for noise or minor disturbances to be misjudged as structural features.

The following introduces the Harris corner detection algorithm. They conduct mathematical modeling of the gray-level variation distribution within a local neighborhood by analyzing the structural tensor of the image, thereby evaluating whether the region has corner characteristics.

Using the expansion of the Taylor series $I(x + u, y + v)$, Eq. (2.14) can be transformed into

$$E(u, v) = (u, v)\left[\sum_{x,y} w(x, y)\begin{pmatrix} I_x \\ I_y \end{pmatrix}\begin{pmatrix} I_x & I_y \end{pmatrix}\right]\begin{pmatrix} u \\ v \end{pmatrix} \tag{2.15}$$

Table 2.2 The characteristics of eigenvalue matrix

The eigenvalue values	Regional characteristics
$\lambda_1 \approx \lambda_2 \approx 0$	Smooth area
$\lambda_1 \gg \lambda_2 \approx 0$	Edge area
$\lambda_1 \gg \lambda_2 \gg 0$	Corner area

where I_x and I_y are the first-order partial derivatives in the x and y directions respectively. Let the matrix M be denoted as

$$M = \sum_{x,y} w(x, y) \begin{bmatrix} I_x^2 & I_x I_y \\ I_x I_y & I_y^2 \end{bmatrix} \tag{2.16}$$

Equation (2.15) can be simplified as

$$E(x, y) = (u, v) M \begin{pmatrix} u \\ v \end{pmatrix} \tag{2.17}$$

Matrix M is known as the structure tensor, which contains information about the changes in the gray-scale gradient of a local area of the image. By analyzing the characteristic values of M, it can be determined whether the gray-scale changes in this area have a large intensity in multiple directions simultaneously, thereby determining whether this area has corner characteristics. Let the two eigenvalues of matrix M be λ_1 and λ_2, which respectively represent the intensity of gradient changes of the image in the two principal directions. According to the magnitude relationship of feature values, image regions can be divided into three categories, as Table 2.2.

Directly calculating the eigenvalues is rather cumbersome in implementation. Therefore, the Harris algorithm introduces a simplified form of the corner response function to approximately determine the corners as

$$R = \det(M) - k[\mathrm{trace}(M)]^2 \tag{2.18}$$

where $\det(M) = \lambda_1 \lambda_2$, $\mathrm{trace}(M) = \lambda_1 + \lambda_2$, the constant k is an empirical parameter, usually ranging from 0.04 to 0.06. This function comprehensively considers the magnitude relationship of the two eigenvalues. Under the condition of meeting a certain threshold, the corner position is determined as the local maximum point of the response function R.

In noisy images, threshold configuration is inherently a trade-off strategy between losing genuine features and falsely detecting noise-induced pseudo-features. To address this, many corner detection algorithms incorporate adaptive thresholds or response-intensity-based ranking mechanisms to enhance detection robustness and precision.

As a critical component of image feature extraction, corner detection aims to accurately identify structurally significant responses within local regions. Methods based on structure tensor analysis evaluate intensity variations across different directions to construct discriminative response functions for effective corner localization. Among these, the Harris algorithm stands out for its computational efficiency and stable responses, making it widely adopted in practical image processing systems. However, in complex images, detection performance often suffers from noise, blur, or repetitive textures, leading to false positives or missed detections.

To balance true features against spurious responses, modern corner detectors typically employ adaptive thresholds or response-ranking filters, improving feature extraction robustness and generalization. This approach proves particularly vital for structural modeling in image recognition, especially in scenarios with occlusions, local deformations, or cluttered backgrounds, as it consistently delivers spatially well-distributed feature points with strong response magnitudes.

In object recognition tasks, corners serve not only as local structural markers but also as precise spatial references for subsequent geometric transformation estimation, object alignment, or feature matching. When combined with other features (e.g., edges or regional data), corners contribute to multi-scale, multi-angle feature representation models, forming a foundational element for robust localization and discriminative analysis in advanced recognition systems.

2.2.4 Hand Crafted Feature

Handcrafted Features refer to descriptive information extracted from images using expert-designed rules and algorithms based on physical and statistical properties. These features capture salient attributes such as texture, shape, color, and gradients to provide rich feature representations for object recognition. Compared to automatically learned features from modern deep learning, handcrafted features typically offer simpler computation, clearer physical interpretability, and easier debugging, making them particularly proper for scenarios with limited samples or constrained resources [7, 8]. Below are several commonly used handcrafted features and their applications.

(1) Local binary patterns

Local Binary Patterns (LBP) is a simple yet efficient texture descriptor that encodes local texture structures by comparing the gray values of neighboring pixels with the central pixel. Specifically, neighboring pixels within a certain range are assigned a value of 1 if their intensity is greater than or equal to the central pixel, otherwise 0. The resulting binary string is converted to a decimal number, representing the LBP value for that pixel. LBP exhibits rotation invariance and robustness to grayscale variations, effectively capturing fine-grained texture details. It performs well in texture classification, face recognition, and infrared small target detection, particularly in distinguishing targets with significant texture differences from the background.

(2) Scale-invariant feature transform

Scale-invariant feature transform (SIFT), is a classic method for detecting and describing local features, extracting key points and their surrounding gradient information with scale and rotation invariance. By constructing a Gaussian difference pyramid, SIFT detects extremal points as key points and computes gradient orientation histograms in their neighborhoods to form descriptors. These descriptors robustly represent local structural features, making SIFT suitable for tasks like object matching and image stitching. In object recognition, SIFT helps locate critical regions of targets, maintaining effectiveness under varying illumination or partial occlusion.

Speeded-up robust features (SURF) is an improved variant of SIFT, designed to enhance computational efficiency while preserving feature stability. SURF employs the determinant of the Hessian matrix for key point detection and uses integral images to accelerate filtering operations, rapidly constructing scale space. The SURF descriptor forms feature vectors by computing Haar wavelet responses in the key point neighborhood. Compared to SIFT, SURF offers faster computation, making it ideal for real-time object detection and recognition systems, particularly for low-contrast targets in infrared imagery.

(3) Color histogram

Color histograms statistically represent the distribution of pixel values across color channels in an image or region, reflecting global color characteristics. By dividing the color space into bins and counting pixel frequencies, they generate statistical features of color distribution. Color histograms are invariant to rotation and translation but sensitive to illumination changes. Although infrared images are typically grayscale, color features remain valuable in multispectral or fused images for distinguishing targets from backgrounds.

(4) Texture features

The Gray-level co-occurrence matrix (GLCM) describes the joint distribution of pixel pairs at specific distances and orientations, capturing spatial correlations in texture. Statistical measures such as contrast, energy, entropy, and correlation derived from GLCM provide quantitative texture descriptions. These features are crucial for recognizing fine structural details, especially in target segmentation and recognition within complex backgrounds.

The design of manual features focuses on the discriminative ability and robustness of the features, requiring them to remain becomingly under scale transformation, rotation, lighting and noise interference. They play a crucial role in traditional object recognition systems. They can be used as independent feature representations or often fused with other features to enhance the accuracy and robustness of recognition. Especially in infrared small target detection, combining multiple manual features can effectively enhance the discrimination between the target and the background and improve the detection effect.

However, the design and extraction process of manual features is highly dependent on the prior knowledge and experience of experts, which imposes certain limitations on them when dealing with the diversity and complexity of image scenes. Specifically, manual features are often optimized for specific image attributes or application scenarios, making it difficult to adapt to various lighting changes, perspective shifts, the diversity of target forms, and the interference of dynamic backgrounds. In large-scale data processing and complex environments, a single or small number of manual features are difficult to comprehensively capture the diverse features of the target, resulting in limited recognition performance. In addition, the design and parameter tuning of manual features require a significant amount of human and material resources. Moreover, when facing new targets or unknown environments, it is often necessary to redesign or adjust the feature extraction methods.

With the rapid development of deep learning technology, automatic feature learning methods based on neural networks have gradually become the mainstream in the field of object recognition. Deep learning, through a multi-layer network structure, can automatically extract multi-level and multi-scale abstract features from massive data, greatly enhancing the generalization ability and recognition accuracy of the model, as Fig. 2.9. However, manual feature calculation is simple, logically clear, and convenient for analysis and debugging. It still has obvious advantages in resource-constrained embedded devices or systems with high real-time requirements.

Overall, manual features, as an important cornerstone of object recognition technology, still hold a significant position in many practical applications due to their stability, computational efficiency and ease of interpretation. It not only provides theoretical support for understanding and designing complex target recognition systems, but also offers valuable experience and reference for the development of new technologies. In the future, manual features are expected to be deeply integrated with modern technologies such as deep learning, leveraging their respective strengths to achieve more efficient and robust object recognition solutions.

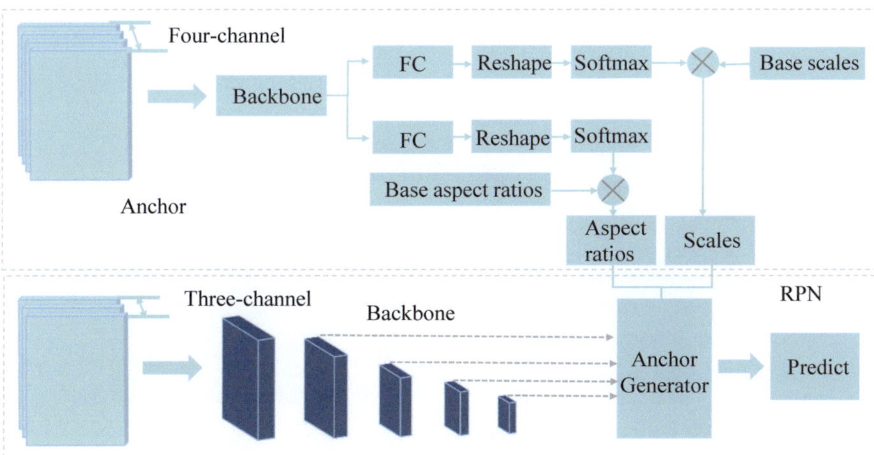

Fig. 2.9 Characteristics of deep networks

2.3 Template Matching and Structure Recognition

The key to target recognition lies in establishing an effective matching mechanism to achieve the correspondence between the input image and the known target. Among them, the recognition methods based on template matching and structure are two traditional and important recognition strategies, which are widely applied in tasks such as character recognition, face comparison, industrial inspection, and remote sensing image analysis. They describe the appearance and structural features of the target in different ways, and achieve target positioning and recognition through comparison and matching [9].

2.3.1 Template Matching Method

Template matching is a target recognition method based on similarity measurement. Its basic idea is to take the image sample of the target as a template, slide the search window in the image to be detected at a certain step size, calculate the similarity value between the window area and the template, and determine whether the target exists and its position based on this. This method does not rely on explicit feature modeling and is suitable for scenarios where the target shape and scale are appropriate and the background changes are small.

Template matching can represent finding a sub-region $R(x, y)$ in the input image $I(x, y)$ that is most similar to the template image $T(x', y')$, as Fig. 2.10, which can be expressed as

$$(x^*, y^*) = \underset{(x,y)}{\arg\max}\ S\left(I_{R(x,y)}, T\right) \tag{2.19}$$

Fig. 2.10 The template of matching process

where $S(\cdot)$ is the similarity measurement function. Some Common methods include the following categories.

(1) Square variance matching

The square variance matching method is a measurement approach based on the idea of error minimization. By calculating the square of the gray level difference between the template and the corresponding pixel of each candidate region in the image, and summing up all pixel differences, the position corresponding to the minimum difference is finally taken as the target matching point. Its essence is to find the closest luminance pattern between the template and the sub-regions of the image. The mathematical expression can be defined as

$$\mathrm{SSD}(x, y) = \sum_{i=1}^{w} \sum_{j=1}^{h} \left[I(x+i, y+j) - T(i,j)\right]^2 \tag{2.20}$$

where $I(x+i, y+j)$ represents the pixel value of the search window in the input image, $T(i,j)$ represents the pixel value of the template, and w and h are the width and height of the template. The final matching target point can be expressed as

$$(x^*, y^*) = \underset{(x,y)}{\arg\min}\ \mathrm{SSD}(x, y) \tag{2.21}$$

This method has a simple calculation process and is easy to implement. It is particularly suitable for application environments with high real-time requirements and stable lighting. However, since it directly compares the gray values of pixels, it is relatively sensitive to lighting changes and noise, and often needs to be combined with image normalization or filtering preprocessing to improve robustness.

(2) Mutual correlation matching

Cross-correlation is a similarity calculation method based on multiplicative accumulation. It measures the degree of matching by multiplying the template point by point with the image window and then summing up the product results. It is defined as

$$\mathrm{CC}(x, y) = \sum_{i=1}^{w} \sum_{j=1}^{h} I(x+i, y+j) \cdot T(i,j) \tag{2.22}$$

The matching target is the maximum correlation point

$$(x^*, y^*) = \underset{(x,y)}{\arg\max}\ \mathrm{CC}(x, y) \tag{2.23}$$

Compared with SSD, cross-correlation pays more attention to the areas in the image that are consistent with the gray scale mode of the template, especially in the environment where the image brightness is uniform and the noise is low, it has a good matching effect.

This method has a fast computing speed and can be accelerated in the frequency domain through the fast Fourier transform. Therefore, it still has certain practical value in some tasks that require large-scale search. However, its drawbacks are equally obvious: due to the failure to take into account the brightness offset and contrast difference between the template and the image window, the matching result is easily affected by the overall gray level.

(3) Normalized cross-correlation

To overcome the influence of gray-scale range differences, the mean and standard deviation are introduced to normalize the template and image window, which are defined as

$$\mathrm{NCC}(x, y) = \frac{\sum_{i=1}^{w}\sum_{j=1}^{h}\left[I(x+i, y+j) - \overline{I}\right]\cdot\left[T(i,j) - \overline{T}\right]}{\sqrt{\sum\left[I(x+i, y+j) - \overline{I}\right]^2}\cdot\sqrt{\sum\left[T(i,j) - \overline{T}\right]^2}} \tag{2.24}$$

The matching points can be defined

$$(x^*, y^*) = \underset{(x,y)}{\arg\max}\ \mathrm{NCC}(x, y) \tag{2.25}$$

By removing the mean and scaling magnitudes, ensure that the matching results only depend on structural similarity. This method can effectively eliminate the interference caused by changes in lighting conditions and performs more stably and reliably in most practical applications.

The output values of NCC are usually within the range of $[-1, 1]$, where 1 indicates a perfect match and -1 indicates a perfect opposite. Due to its excellent normalization properties, compared with SSD and CC, NCC is more stable in natural images and scenarios with changing lighting, and is suitable for template matching in complex backgrounds. It is one of the most widely used measurement methods in template matching. However, correspondingly, its computational complexity is also relatively high, requiring the calculation of pixel mean and standard deviation. Usually, it is optimized in combination with acceleration strategies such as integral graphs or parallel computing.

SSD, CC and NCC represent three typical approaches in template matching, respectively emphasizing the minimum difference, the maximum multiplication and the maximum normalized correlation. According to the different actual application requirements, the corresponding measurement methods should be reasonably selected to balance the calculation cost and matching accuracy. In tasks with high real-time requirements but ideal scenarios, SSDS and CCS still have their places to play. In applications with complex environments and variable image conditions, NCC often can better leverage its advantages.

Although the template matching method has strong intuitiveness and ease of operation in implementation, its application scope is somewhat limited. Especially when dealing with significant changes in target scale, posture, lighting conditions or deformation, traditional fixed templates often struggle to adapt, and their recognition performance is prone to decline significantly. Meanwhile, since the search process of the sliding window is essentially an exhaustive traversal, when the target search range is wide or the image resolution is high, the computational efficiency of the algorithm will also become a bottleneck.

To address the above issues, researchers have proposed a variety of improvement strategies to enhance the robustness and efficiency of the template matching method. For instance, multi-resolution matching utilizes the image pyramid structure to gradually reduce the search space layer by layer, thereby enhancing the positioning speed. Template set matching enhances the algorithm's adaptability to target diversity by constructing a template library containing multiple pose or scale variations. While statistical methods such as principal component template matching achieve dimensionality reduction and compression, they retain the main structural information of the template, effectively reducing the amount of matching computation. In addition, in recent years, there have also been methods attempting to combine template matching with learning strategies, such as introducing discriminative filters and template update mechanisms based on sample learning, as Fig. 2.11, which enables traditional methods to have a certain degree of adaptability in dynamic scenarios.

Overall, template matching, as a recognition method that requires no training and is based on intuitive similarity measurement, still has practical application value in tasks with clear structure, stable appearance, and controllable background interference. It remains a desirable choice in the early stage of the task for candidate screening, rough positioning, or as an auxiliary mechanism for other complex recognition algorithms.

2.3.2 Structural Identification Method

In addition to directly using the overall image template for point-by-point matching, there exists a more universal and semantically explanatory method in object recognition, namely structure-based recognition. This type of method no longer focuses on the overall gray-scale distribution or texture features of the image. Instead, it extracts the geometric structural elements with distinctiveness such as the edges, corners, and contours of the target, and combines their organizational relationships in space to construct a target model with abstract representation capabilities. Compared with the simple pixel similarity measurement, the structure recognition method is closer to the human visual understanding of shape and layout, and is suitable for recognition scenarios under non-ideal conditions such as scale changes, local occlusions or structural disturbances.

Essentially, structure recognition is a symbolic modeling process, and the key lies in how to extract stable structural features from images and encode these features in the form of graph structures or relationship expressions. Suppose the target is composed of several feature points. In the structure diagram $G = (V, E)$ formed by

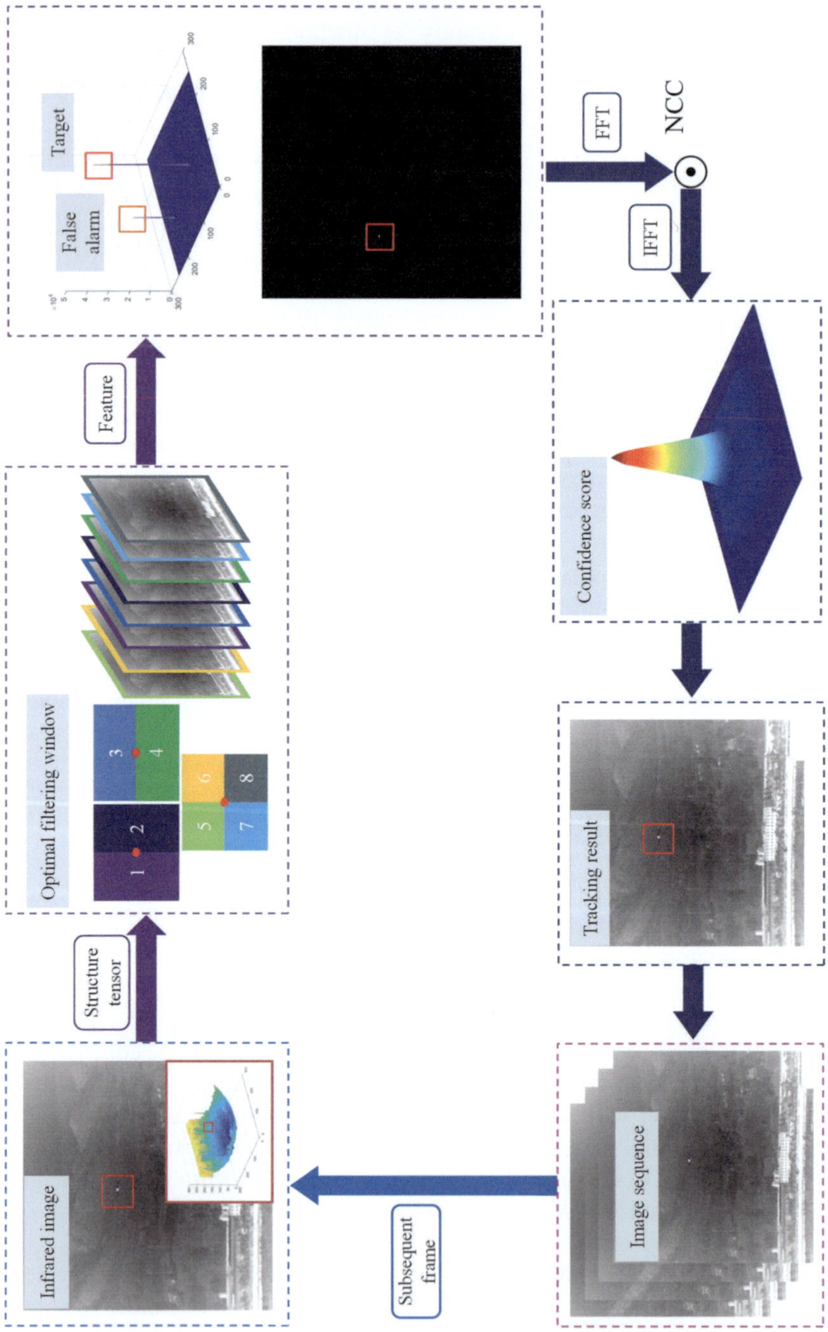

Fig. 2.11 Small target tracking process based on discriminative filters

these points, V represents the set of feature points and E represents the connection relationship among them. Each node and edge can carry additional attributes, such as coordinate position, direction Angle, relative distance, etc. The recognition task can be transformed into a graph matching problem, that is, to find the mapping relationship that maintains topological consistency between the target structure diagram and the template diagram, thereby determining whether they are of the same type of target.

Compared with traditional template matching, the advantage of structure recognition lies in its strong robustness to changes in image details. As long as the key structural layout in the image remains stable, the structural matching results are still reliable even if local deformation, occlusion or noise interference occurs. Take character recognition as an example. A handwritten letter A can be abstracted as a graph structure composed of three line segments, which have different shapes but fixed topological relationships. For instance, in the recognition of industrial drawings, a screw hole pattern can be modeled as a composite structure with a circle and edge line connection, and the recognition task is transformed into the matching and difference analysis between the drawings. In remote sensing image analysis, buildings may present as a combination of regular rectangular boundaries and corner points. These features are consistent in geometric structure. Even if the image itself has certain distortions or noise, its structural layout remains unchanged and it has strong transferability.

Structural recognition typically involves three core steps: The first is to extract structural features, such as obtaining the key points and connection relationships of the target through methods like edge detection, corner extraction, and contour analysis; Secondly, structural expression modeling is carried out, which can be done in the form of graph structures, relation matrices or symbol strings, etc. The last step is structural matching and recognition. By constructing a similarity function or a rule-based judgment mechanism, the recognition decision of the target is completed. In some systems, prior knowledge can also be integrated, such as the spatial combination constraints of target components and frequency characteristics, to enhance recognition accuracy.

This method is particularly suitable for target types in images where geometric relationships are significant and topological features are clear. For instance, in the recognition of industrial drawings, traffic signs, remote sensing buildings, medical organs, and handwritten symbols, structural recognition is often employed as a fundamental framework or combined with other methods to achieve more stable recognition results. In addition, structural recognition also has significant advantages in scenarios with limited data volume or high interpretability requirements.

The structure-based recognition method elevates image recognition from the pixel space to the structural space by constructing an abstract structural model of the target, reflecting the transition from data perception to structural understanding. Although deep learning methods have performed outstandingly in large-scale recognition in recent years, structural recognition still maintains irreplaceable value in specific tasks due to its high interpretability and good controllability. However, for many natural images and complex scenes, relying solely on structural modeling is still insufficient to cover all target variations, especially in cases where background interference is strong and target boundaries are unclear, where structural extraction itself may encounter difficulties. Therefore, in recognition tasks, it is often necessary to further introduce regional-level analysis and processing methods beyond structural description, such as image segmentation, connected region extraction, boundary refinement and morphological processing, etc., in order to obtain a more complete and stable target expression. These methods can not only provide clearer candidate regions for structural modeling, but also lay the foundation for the preprocessing of the recognition system.

2.4 Image Segmentation and Region Recognition

The prerequisite for target recognition is the ability to accurately locate and extract the target area in the image. Image segmentation is precisely the key step in this process. Its goal is to divide the image into several regions with consistent features, so that the target region can be effectively distinguished from the background or other irrelevant regions. In the task of object recognition, image segmentation not only determines the accuracy of subsequent feature extraction, but also affects the reliability and robustness of the final recognition [10, 11].

Unlike the local features such as edges, corners and structures introduced previously, image segmentation focuses on the overall contour and connected regions of the target. It provides the foundation for regional recognition methods, enabling recognition not only to rely on the description of points or lines, but also to be discriminated based on regional attributes such as the shape, area, and gray-scale distribution of the target. Area recognition is more in line with the way humans perceive the overall target and is also more suitable for some scenes with regular shapes, clear contours and obvious background contrasts.

This section will introduce common image segmentation methods and the expression approaches of regional features, and further introduce the application of morphological operations in segmentation to enhance the integrity and noise resistance of target extraction.

2.4.1 Basic Image Segmentation Methods

The goal of image segmentation is to divide an image into several regions with consistent internal attributes and relatively clear boundaries, so that each region corresponds to a semantic object or its component in the image. Based on the types of information utilized and the different segmentation strategies, image segmentation methods can be roughly classified into the following four categories: thresholding methods, region-based segmentation, boundary-based segmentation, and clustering or learning-based segmentation methods.

(1) Threshold segmentation

Threshold segmentation is the most direct and classic type of image segmentation method, and is widely used in image scenes with a strong contrast between the background and the foreground. This method assumes that the gray values of the image pixels follow a bimodal or multimodal distribution. By selecting one or more appropriate gray thresholds T, the image pixels are divided into two parts: the foreground and the background as

$$g(x, y) = \begin{cases} 1, f(x, y) > T \\ 0, f(x, y) \le T \end{cases} \tag{2.26}$$

where $f(x, y)$ is the pixel value of the original image, and $g(x, y)$ is the binary image after segmentation.

Thresholds can be set by human experience or obtained through automated algorithms. For instance, the Otsu method is an optimal global threshold selection method based on the maximum between-to-class variance. The core idea is to find a threshold that maximizes the inter-class variance between the foreground and background, thereby achieving the optimal distinction.

When the image is unevenly illuminated or the gray level of the target area varies greatly, local adaptive threshold segmentation is more effective. It dynamically adjusts the threshold according to the gray level statistical characteristics of different regions of the image, overcoming the problem of global failure of the fixed threshold.

(2) Regional segmentation

Region segmentation starts from one or more pixels in an image and forms regions through expansion. The core idea is that similar pixels should be classified into the same region. The region-growing method is the most typical representative. This method selects several seed pixels and recursively merges adjacent pixels into the current area based on similarity criteria such as pixel gray level, texture or spatial position. The commonly used similarity criteria are expressed as

$$\left|f(x_i, y_i) - f(x_j, y_j)\right| < \delta \tag{2.27}$$

where δ represents the allowable gray-level difference threshold. Region growing has good connectivity and region integrity, and is suitable for images with relatively uniform structures and blurred boundaries, such as the extraction of tissue regions in medical images.

In the region-growing method, the segmentation process is a bottom-up strategy that starts from a small region and expands outward. In contrast, there is another segmentation method that combines top-down and bottom-up approaches, namely the region splitting and merging method. It achieves dynamic segmentation of complex image regions through recursive division and conditional judgment, and is suitable for image scenes with diverse structural levels and significant differences in target scales.

In the specific implementation, the splitting process usually adopts a quadtree structure: the image is regarded as a two-dimensional spatial region. If a certain region does not meet the criteria, it is divided into four sub-regions. Repeat the judgment on the sub-regions until all sub-regions meet the criteria or reach the minimum division unit. The merging process starts from the bottom layer, judging the similarity between adjacent regions. If the merging conditions are met (such as the mean difference being less than a certain threshold), they are merged into a larger region. In this way, the entire process can ensure that neither details are missed nor over-segmented in the image. Formally, the criterion of regional consistency can be expressed as follows

$$\left|\mu_i - \mu_j\right| < T \tag{2.28}$$

where, μ_i and μ_j represents the area to be merged, and T is the allowable difference threshold.

The advantage of this method lies in its ability to adapt to the size and complexity of regions in the image, and it has good adaptability to images with uneven target distribution or large size differences. Especially when there are both large flat regions and small-scale detail structures in the image, region splitting and merging can dynamically adjust the segmentation granularity, avoiding misclassification or missed classification caused by a single strategy. At the same time, such methods also have some limitations: if the consistency criterion is not set reasonably, it may lead to excessive partitioning or merge failure; Moreover, the recursive structure of the quadtree has a relatively high computational complexity on large images and is not easy to meet the requirements of real-time processing. Therefore, in practical applications, this method is often combined with other techniques such as edge detection and morphological filtering to assist in the structured segmentation of complex targets.

(3) Boundary segmentation

Boundary-based image segmentation methods take the sudden changes in gray level or texture in the image as segmentation cues. The basic idea is that at the junction of the target and the background, the pixel values usually change drastically, so these edges can be used as the basis for regional division.

This method first extracts the edge information in the image through the existing edge detection means. These edge points usually present as a collection of pixels with significant local gradient amplitudes. After obtaining the edge map, the key to the segmentation process lies in how to organize the discrete edge points into coherent and closed boundary contours. This process often involves technical means such as edge connection, curve fitting, and contour closure to compensate for possible false detections of fractures or noise in edge detection. Once a closed or nearly closed boundary is obtained, the image can be divided into regions based on it. For example, the interior of each closed boundary can be marked to obtain the corresponding foreground area. Or perform a region filling operation based on the boundary to generate a binary mask image. This strategy of inferring the region from the boundary has achieved a transition from local changes to the overall structure.

The boundary-based segmentation method emphasizes the sensitive response to local gray-level changes in the image, has high positioning accuracy, and can better preserve the contour details of the target. However, at the same time, such methods are relatively sensitive to image quality. If there is strong noise or edge blurring, it may lead to discontinuous boundary detection, thereby affecting the accuracy of region division. Therefore, in practical applications, the boundary results are often optimized by combining morphological processing or subsequent regional analysis to enhance the overall segmentation robustness.

Compared with the region-based segmentation method, the boundary-driven strategy focuses more on edge representation and structural details. It is suitable for scenarios where the shape information of the target needs to be retained. At the same time, it can also serve as supplementary information for other segmentation methods to enhance boundary constraints or generate initial regions.

(4) Clustering segmentation

Image segmentation methods based on clustering fall within the category of unsupervised learning. Their core idea is to regard image pixels as data points, cluster them in a certain feature space (such as gray level, color, texture, spatial position, etc.), and group pixels belonging to the same category into the same region, thereby completing image segmentation.

Unlike methods based on thresholds or edges, clustering segmentation does not rely on the spatial structure or prior templates of images, but classifies them based on the feature similarity between pixels. Therefore, it has strong adaptability and flexibility. Common clustering segmentation methods include:

(1) K-means clustering segmentation

K-means is a commonly used unsupervised clustering method, which can be used to divide image pixels into several categories based on feature similarity. The basic idea is to regard each pixel in the image as a D-dimensional feature vector. Common features include gray values, RGB color components, spatial coordinates, etc. Pixel classification is achieved by measuring the distances between these vectors. During the segmentation process, several initial clustering centers are randomly selected first. Then, each pixel is assigned to the center class with the smallest distance. Based on the current classification result, the center position of each class is updated. This process is repeated and iterated until the clustering result stabilizes or reaches the upper limit of the number of iterations, as Fig. 2.12.

The K-means algorithm is simple to implement and has high computational efficiency. It is suitable for images with clear color distribution and uncomplex regional boundaries. However, this method requires the pre-setting of the number of clustering categories K, is relatively sensitive to the initial center, and when dealing with non-convex shapes or regions with complex textures, the segmentation results may have problems such as breakage or over-segmentation. For this reason, in practical applications, it is often necessary to combine other information for auxiliary judgment,

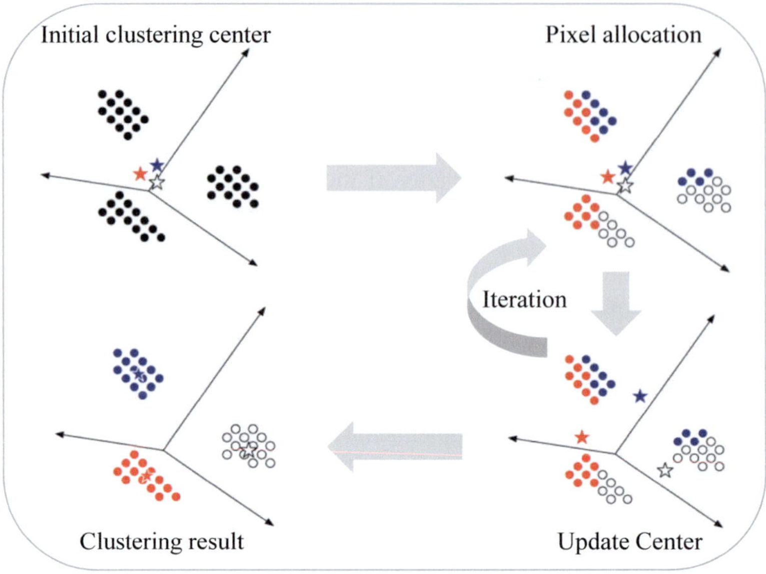

Fig. 2.12 K-means clustering process

or take it as the initial step of segmentation, and then carry out boundary correction and regional optimization.

(2) Mean shift clustering

Mean Shift is a clustering method for non-parametric density estimation. Its basic idea is to estimate the pixel density center through a sliding window in the feature space and cluster the pixels at the local maximum density position. Unlike K-means, Mean Shift does not require a preset number of clusters and is more suitable for processing natural images with color gradients and complex textures. It can automatically discover the natural regional distribution in the image and better preserve the regional boundaries. However, its computational complexity is relatively high and it is suitable for offline segmentation tasks.

(3) Graph theory method

Graph cuts and normalization cuts and other methods model images as graph structures: each pixel in the graph is a node, and edge weights are constructed between adjacent pixels based on gray level differences and color differences. The segmentation process is to find the optimal cut in the graph, dividing it into several subgraphs, each corresponding to a region in the image.

The advantage of graph cut methods lies in their ability to minimize the energy function on a global scale and achieve better segmentation results, especially suitable for images with overly complex textures or blurred target boundaries. However, its computational cost is relatively high, it is sensitive to image size and parameter

Settings, and usually requires the assistance of boundary constraints or interactive annotations for operation.

Image segmentation, as a key preprocessing step in object recognition, its core objective is to divide the image into semantically meaningful regions and extract candidate regions that may contain the target. Depending on the type of information and strategy on which segmentation is based, common methods include simple segmentation based on gray thresholds, region growing based on regional connectivity, boundary detection based on edge changes, and clustering segmentation and graph model methods based on statistical distribution and global optimization strategies, etc.

Different methods each have their own advantages and applicable scenarios. The threshold method is simple to implement and suitable for images with obvious contrast between the target and the background. The combination of region growing and splitting is applicable to the extraction of regions with regular structures and continuous gray levels. Boundary segmentation emphasizes contour retention and is suitable for tasks with high requirements for shape and structure. Clustering and graph cut methods, on the other hand, offer stronger expressive capabilities and can handle complex backgrounds and multimodal images.

In practical applications, image segmentation is often not used in isolation but as a front-end module in the recognition system, working in coordination with processes such as feature extraction and structural modeling. For instance, the segmented regions can serve as the input for shape feature analysis or be used to limit the search range of target detection, thereby reducing the computational complexity of subsequent recognition. Meanwhile, combining morphological operations for boundary optimization, noise suppression and connected region cleaning has become an important means to enhance the segmentation effect.

2.4.2 Morphological Operation

Image morphology is an image processing method based on set operation, which initially originated from the analysis of particle structure in microscopic images. The basic idea is to use a template called a Structuring element (SE) to perform sliding operations in the image. By comparing the pixels under the structuring element, the pixel values are changed at the origin position of the image. The implementation process is shown in Fig. 2.13a. If no pixel in the image matches the structural element, it is called a Miss. When at least one pixel of a structural element overlaps with an image pixel, it is called Hit. If all the pixels of a structural element match the image, it is called Fit. SE is usually a matrix or a graph similar to a core, as Fig. 2.13b, which analyzes and processes the shape, boundary and connectivity features in the image from the perspective of spatial structure. Morphological operations are often applied to binary images and can also be extended to grayscale images. They are used to complete various tasks such as regional denoising, contour smoothing, edge

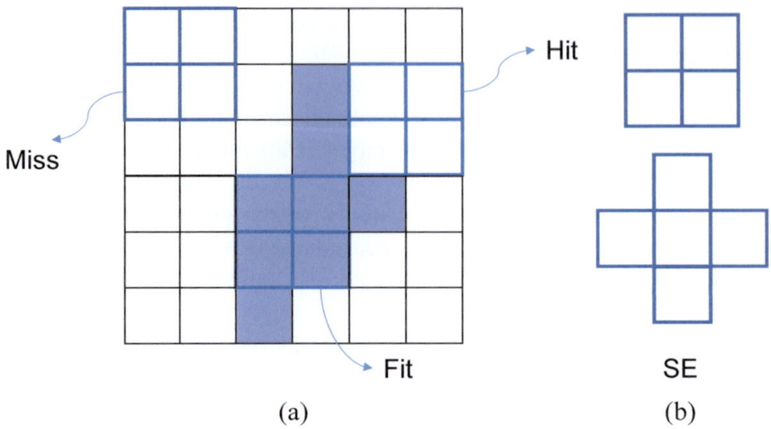

Fig. 2.13 Morphological operations: **a** sliding window matching, **b** structural element template

connection, and skeleton extraction, and have extensive application value in image segmentation and object recognition.

The two basic operations of morphology are erosion and dilation, which correspond respectively to the minimum inclusion and maximum coverage of a set. In A binary image, let the target area in the image be set A, the structural element be B, and its origin be at the center. Then, the corrosion operation can be defined as

$$A \ominus B = \{z | B_z \subseteq A\} \tag{2.29}$$

The schematic diagram of image corrosion is shown in Fig. 2.14, which involves sliding the structural element B in image A while retaining the positions z where all points of B are still completely contained within A. The result of corrosion is that the regional boundary contracts inward, which is suitable for removing small protrusions, isolated noise points or disconnecting narrow connections.

The dilation operation expands the target outward, which is defined as

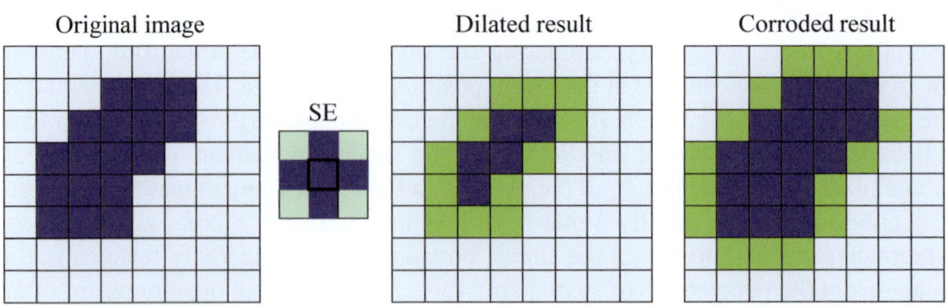

Fig. 2.14 Schematic diagram of image corrosion and dilation

$$A \oplus B = \{z | B_z \cap A \neq \emptyset\} \tag{2.30}$$

As long as there is a non-empty intersection between the structural element and region A, the current position is retained, thereby causing the region boundary to expand outward. Expansion is suitable for filling small holes, connecting adjacent areas or enhancing edge coherence. In actual processing, corrosion and expansion are often used in combination. The open operation formed by corrosion first and then expansion is used to remove noise in small areas while keeping the main structure unchanged, and is expressed as

$$A \circ B = (A \ominus B) \oplus B \tag{2.31}$$

The operation of first expanding and then corroding constitutes a closed operation, which is used to fill small gaps in the area or connect broken structures, and is expressed as

$$A \bullet B = (A \oplus B) \ominus B \tag{2.32}$$

These basic operations provide flexible control means in the process of image structure processing. Through different designs of structural elements (such as square, circular, cross-shaped, etc.), customized processing can be carried out according to the actual shape characteristics of the target.

After image segmentation, morphological operations are usually used as post-processing modules to optimize the structure of the segmentation results. For instance, open operations can effectively eliminate isolated small spots caused by lighting or noise, while closed operations can be used to connect boundary breaks and enhance the integrity of the target. In addition, the outer contour of the region can be extracted through the boundary extraction formula for subsequent shape analysis and feature extraction. More complex combined operations also include region filling, skeleton extraction, convex hull approximation, etc. These methods are widely used in pattern recognition, object detection, character analysis and remote sensing images.

Morphology plays an important role in structural optimization and interference suppression in the target recognition system. Its operation does not rely on the precise modeling of pixel values, but directly reflects the geometric characteristics of objects in the image through the analysis of regional shapes and spatial structures. Due to its simple principle, strong interpretability and high implementation efficiency, morphological operations remain an indispensable basic tool in image processing and object recognition to this day. Through the flexible design and multi-step combination of structural elements, morphology can not only adjust the image structure but also extract stable and reliable regional features in scenes lacking grayscale models, providing strong support for subsequent target recognition and classification (Fig. 2.15).

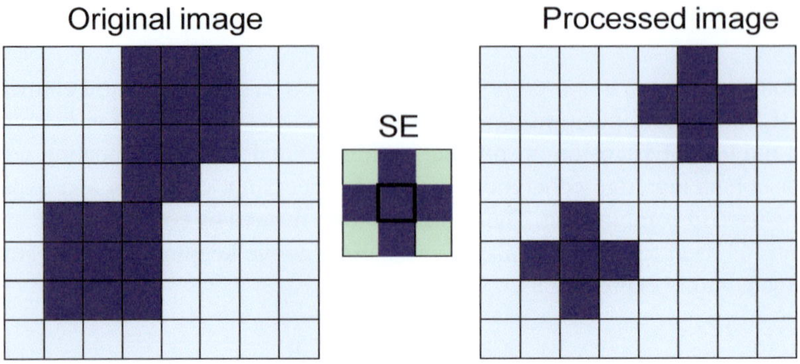

Fig. 2.15 Schematic diagram of opening operation

2.5 Overview of Target Recognition Methods

After completing image preprocessing, region segmentation and feature extraction, the core task of the object recognition system is to use these intermediate results to discriminate and classify the objects in the image. This section will introduce the basic process of object recognition and several common types of recognition methods, helping readers establish a preliminary understanding of the overall architecture of the recognition system.

2.5.1 Rule Feature Method

In the early target recognition systems, the recognition methods mainly relied on manually designed algorithms and feature extraction rules. This type of method first extracts artificially designed features such as the geometric shape, edge contour, color histogram, and regional area of the target, and then uses pre-set judgment criteria or template libraries to match and classify the input image (Fig. 2.16).

For instance, in traffic image analysis, the system can identify the geometric structure of traffic signs through preset template matching. In biological images, the edge contours of cell shapes or the size of nuclear regions are also often used as regular bases for classification. This type of rule and feature-based method has the advantages of clear logic and simple implementation, and is suitable for application scenarios where the target structure is relatively fixed and there is less interference. Once there are situations such as target rotation, scale change, occlusion or complex background in the image, the adaptability of such methods will significantly decline, making it difficult to ensure the accuracy and generalization ability of recognition.

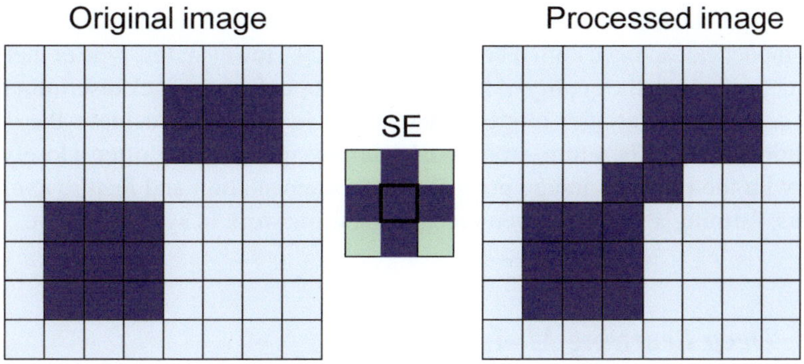

Fig. 2.16 Schematic diagram of closed operation

2.5.2 Traditional Learning Model

With the development of statistical learning theory, object recognition systems have gradually begun to introduce machine learning models to enhance classification performance. Unlike the early rule-based recognition, this type of method usually represents images in the form of feature vectors and trains a generalizable classifier model by learning the distribution patterns of these features in different categories. Common models include K-nearest neighbor (K-NN), support vector machine (SVM), Naive Bayes classifier, decision tree, random forest, etc. [12].

The core idea of these classifiers is to construct a function $f : \mathbb{R}^n \rightarrow \mathcal{C}$ in the feature space, mapped the n-dimensional feature vector $\mathbf{x}$ to a discrete category label $c \in \mathcal{C}$. For the image to be recognized, extract the feature vector X and input it into the trained classifier to obtain the predicted label as

$$\hat{c} = f(\mathbf{x}) \tag{2.33}$$

Take SVM as an example. Its goal is to find an optimal hyperplane to separate samples of different categories. In the case of linear separability, SVM enhances the robustness of classification by maximizing the inter-class interval. Its discriminant function can be written as

$$f(\mathbf{x}) = \operatorname{sign}(\mathbf{w}^{\mathrm{T}}\mathbf{x} + b) \tag{2.34}$$

where $\mathbf{w}$ is the weight vector and b is the bias term. The goal of model training is to learn this set of parameters to achieve the optimal classification.

Under the premise of good feature expression, traditional learning models have strong generalization ability, especially performing stably on medium and small-scale datasets. Their training processes are controllable and the implementation threshold is relatively low, so they are still widely used in engineering practice. For instance, in tasks such as texture classification, pattern recognition, and remote sensing image ground object classification, as long as the feature selection is reasonable, a relatively simple model can achieve high recognition accuracy. However, this type of method is highly dependent on the quality of features. If the input features lack discriminability, even if a complex classifier model is adopted, it is difficult to achieve the desired recognition effect. Therefore, traditional learning methods are often closely integrated with the previous image preprocessing, segmentation and feature extraction modules, forming a feature-driven recognition framework in system design.

2.5.3 Deep Learning Model

With the improvement of computing power and the accumulation of large-scale datasets, deep learning models have gradually become the mainstream methods in the field of object recognition. Unlike traditional machine learning that relies on manually designed features, deep learning automatically learns the hierarchical feature representations in images through multi-layer nonlinear transformations, directly extracting discriminative information from the original pixel data, significantly improving recognition accuracy and generalization ability [13, 14], as Fig. 2.17.

Deep learning models are usually constructed as network structures containing multiple layers of neurons, the most typical of which is Convolutional neural networks (CNN). CNN effectively captures the spatial structural features in images through local receptive fields, weight sharing and pooling operations, and gradually extracts the expressions from low-level edges to high-level semantics through multi-layer stacking. As shown in the multi-object recognition network in Fig. 2.18, for the input image I, after going through the layer-by-layer transformation of the network, a feature vector $\mathbf{h}$ is finally obtained. This vector is sent to the fully connected layer

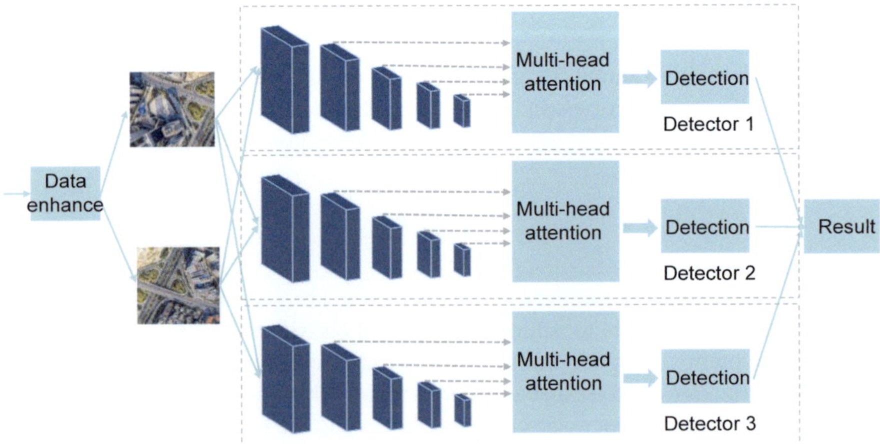

Fig. 2.17 Multi-level feature recognition through deep learning

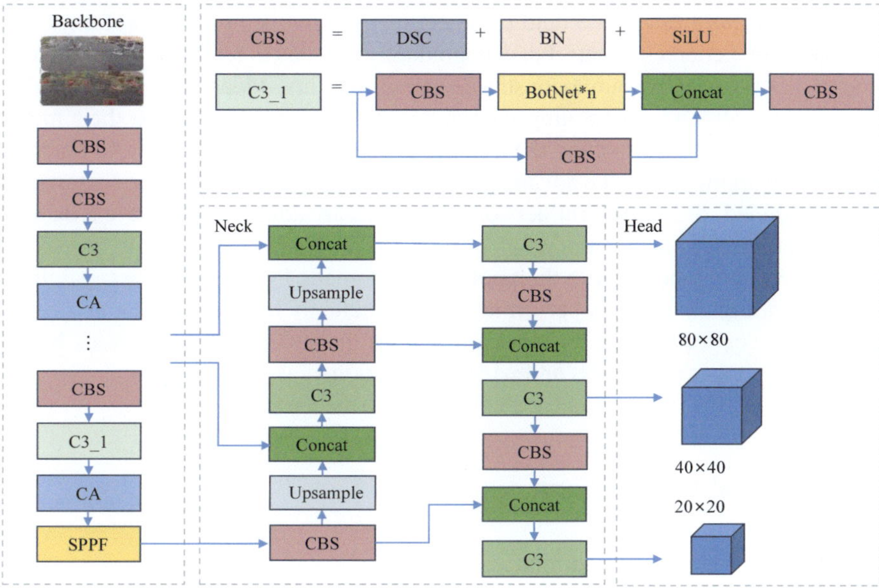

Fig. 2.18 Multi-object recognition network

or classification layer to predict the corresponding category label as

$$\hat{c} = f_{\theta}(I) \tag{2.35}$$

where f_{θ} represents the deep neural network mapping function with the parameter θ, which directly maps the input image to the output category space. Take the convolutional layer as an example, its calculation process can be formalized as

$$\mathbf{h}^{(l)} = \sigma\left(\mathbf{W}^{(l)} * \mathbf{h}^{(l-1)} + \mathbf{b}^{(l)}\right) \tag{2.36}$$

where $*$ is the convolution operation, $\mathbf{W}^{(l)}$ and $\mathbf{b}^{(l)}$ are the convolution kernel weights and biases of the 1 layer respectively, and $\sigma(\cdot)$ is the nonlinear activation function, which $\mathbf{h}^{(0)}$ is the input image.

The training process of deep learning models is based on a large amount of labeled data. The backpropagation algorithm is used to minimize the loss function between the predicted output and the real label, achieving end-to-end parameter optimization. Compared with traditional models, deep networks do not rely on manually designed feature extractors and can adaptively capture richer and more abstract image information, effectively overcoming challenges such as target deformation, occlusion, and lighting changes in complex environments.

In addition, with the continuous improvement of network structure and training strategies, such as the introduction of technologies like residual networks (ResNet), Attention mechanisms (Attention), and transfer learning, deep learning has made breakthrough progress in the field of object recognition. It not only performs well on large-scale general datasets, but also gradually extends to few-shot learning, real-time processing and resource-constrained environments. Meanwhile, deep models have a large demand for training data and computing resources, and their model structures are complex, making the training process difficult to explain. In practical engineering applications, the advantages of traditional features and deep features are often combined, or lightweight network structures are adopted to achieve a balance between efficiency and accuracy.

Deep learning models, with their powerful automatic feature learning capabilities and outstanding recognition performance, are gradually becoming the core driving force of object recognition systems, promoting the field to move towards higher precision, stronger robustness and wider application.

2.6 General Conclusions

This chapter systematically introduces the core basic knowledge and key processing procedures involved in the object recognition task, covering the extraction methods of local features such as structural features, edges and corners of the target in the image, the basic principles and methods of region segmentation, the expression of region geometry and grayscale features, as well as the specific application of morphological operations in structural processing. On this basis, an overview of common recognition methods was also provided, including traditional rule-based recognition approaches, classic classification models that rely on feature learning, and deep learning methods that have been widely applied in recent years. Through these contents, readers can not only understand the composition logic and information flow process of the target recognition system, but also master the basic ideas of image understanding and classification discrimination. Based on actual image processing tasks, readers can further attempt to construct an analysis module with target perception capabilities, achieving an integrated recognition process from image preprocessing, feature extraction to classification decision-making, laying a necessary foundation for subsequent algorithm design or system development.

Acknowledgements This work was supported by the National Natural Science Foundation of China (12374271, 62375065, 62405079, 62575091); Fundamental Research Funds for the Central Universities (HIT.OCEF.2024055, AUGN5750300325); the Fund of National Laboratory on Adaptive Optics, China (FNLAO-24-MS-O07); Natural Science Foundation of Heilongjiang Province of China (LH2024A009); Key Laboratory of Image Processing and Intelligent Control (Huazhong University of Science and Technology), Ministry of Education, Wuhan, China (IPIC2025-05); Key Laboratory of Advanced Medical Imaging and Intelligent Computing of Guizhou Province (AMIIC250204); the China Postdoctoral Science Foundation-CCTEG Joint Support Program (2025T011ZGMK).

References

1. Jagannathan A, Miller EL (2007) Three-dimensional surface mesh segmentation using curvedness-based region growing approach. IEEE Trans Pattern Anal Mach Intell 29(12):2195–2204
2. Ibrahim H, Kong NSP (2007) Brightness preserving dynamic histogram equalization for image contrast enhancement. IEEE Trans Consum Electron 53(4):1752–1758
3. Wang X, Li K, Chehri A (2023) Multi-sensor fusion technology for 3D object detection in autonomous driving: a review. IEEE Trans Intell Transp Syst 25(2):1148–1165
4. Zhou Q, Zuo J, Kang W, Ren M (2025) High-precision 3D reconstruction in complex scenes via implicit surface reconstruction enhanced by multi-sensor data fusion. Sensors 25(9):2820
5. Evans AN, Liu XU (2006) A morphological gradient approach to color edge detection. IEEE Trans Image Process 15(6):1454–1463
6. Hendriks CLL, Van Ginkel M, Verbeek PW (2005) The generalized Radon transform: sampling, accuracy and memory considerations. Pattern Recognit 38(12):2494–2505
7. Kumar N, Sharma M, Singh VP, Madan C (2022) An empirical study of handcrafted and dense feature extraction techniques for lung and colon cancer classification from histopathological images. J King Saud Univ Sci 34(8):5122–5131
8. Stanley RJ, Hagerty JR, Almubarak HA (2019) Deep learning and handcrafted method fusion: higher diagnostic accuracy for melanoma dermoscopy images. IEEE J Biomed Health Inform 23(4):1385–1391
9. Hashemi NS, Aghdam RB, Ghiasi ASB (2016) Template matching advances and applications in image analysis. Pattern Recognit Lett 84:70–79
10. Li Z, Liu G, Zhang D, Xu Y (2016) Robust single-object image segmentation based on salient transition region. Pattern Recognit 60:77–88
11. Han X, Zhong Y, Zhang L (2017) An efficient and robust integrated geospatial object detection framework for high spatial resolution remote sensing imagery. Remote Sens 9(7):666
12. Qian Y, Zhou W, Yan J, Li W, Han L (2015) Comparing machine learning classifiers for object-based land cover classification using very high resolution imagery. Remote Sens 7(1):153–168
13. Rawat W, Wang Z (2017) Deep convolutional neural networks for image classification: a comprehensive review. IEEE Trans Pattern Anal Mach Intell 39(12):2353–2373
14. Deng L, Yu D (2014) Deep learning: methods and applications. Pattern Recognit 47(9):3056–3070

Chapter 3
Autofocusing

Ziyang Li, Zhengjun Liu, and Yutong Li

Abstract In the early stages, focusing was mainly performed by manually adjusting the distance between the lens and the object, followed by evaluating the imaging quality through visual observation. Manual focusing was highly susceptible to external disturbances, which compromised accuracy and reduced operational efficiency. Autofocus technology is designed to determine the object-to-image plane distance or accurately position optical elements along the optical axis, making it an advanced, high-precision positioning technique. Driven by rapid technological progress, autofocus systems have evolved toward miniaturization and high-level integration. Today, autofocus systems are indispensable core components in portable imaging devices, including digital cameras, smartphones, and other mobile imaging platforms.

3.1 Overview of Autofocus Technology

Based on differences in focusing principles and the presence or absence of external signal transmission devices, autofocus technology can be divided into two categories: active and passive. Active autofocus operates by emitting a signal toward the target object, receiving the reflected signal, and analyzing it to determine the correct focus. This approach typically relies on external hardware components such as laser rangefinders, dedicated sensors, and dynamic focusing lenses for accurate positioning or focal length adjustment. Its key advantage lies in its rapid response, making it particularly well-suited for dynamic imaging applications.

Z. Li · Z. Liu (✉) · Y. Li
School of Physics, Harbin Institute of Technology, Harbin, China
e-mail: zjliu@hit.edu.cn

Z. Li
e-mail: hitliziyang@163.com

Y. Li
e-mail: yutong_li@hit.edu.cn

Z. Liu and Y. Li (eds.), *Intelligent Analysis of Optical Images*, Scientific Computation,
https://doi.org/10.1007/978-981-95-3720-4_3

3.2 Active Autofocus

Active laser ranging is a non-contact measurement technique that employs a laser source to estimate target distance by measuring time of flight, phase variation, or through triangulation principles. The time-of-flight method involves emitting a narrow laser pulse toward the target using a pulsed laser, with a photodetector detecting both the emitted and reflected pulses. By measuring the interval between the transmitted and returned pulses, the pulse flight time is obtained, and the distance is calculated using the speed of light. This technique provides the benefits of high-speed and non-contact measurement; however, it incurs higher costs and exhibits strong environmental dependency. Consequently, it is primarily employed in applications requiring distance accuracy from kilometers to centimeters, including LiDAR, rail transportation, security surveillance, and mobile photography.

3.2.1 Laser-Illumination Autofocus

Astigmatism-based autofocus has shown excellent detection accuracy and robustness across a wide range of microscopic imaging applications. The underlying principle involves analyzing astigmatic distortions resulting from the displacement of the object plane relative to the optical focal point to quantitatively assess defocus. Specifically, the method utilizes the unique astigmatic characteristics induced by axial displacement between the objective and the sample, extracting critical image parameters indicative of astigmatism and matching them to a pre-calibrated astigmatism–displacement response curve. This comparison allows precise estimation of the target's axial offset relative to the focal plane, thereby achieving highly accurate autofocus control [1]. This technique is well-suited for applications demanding exceptional focusing precision, offering significant advantages in thick-sample imaging, 3D reconstruction, and high-throughput scanning.

The laser-based split-image autofocus technique actively introduces distinctive image features into the detection scene, thereby enhancing both accuracy and robustness of active autofocus systems [2]. This technique utilizes a dedicated optical path for image separation, comprising a lens, a pair of prismatic light wedges, and a patterned mask. Each light wedge incorporates staggered wedge surfaces designed to split the dot array pattern into two segments. Upon passing through the lens, the beam enters the light wedges and undergoes refraction at both the upper and lower surfaces. Depending on the image position relative to the intersection point of the light wedges, the system operates in three states: far-focus, in-focus, and near-focus. As illustrated in Fig. 3.1a, under far-focus conditions, the light beam from point *A* initially enters the upper surface of the red-light wedge, where it refracts from a low- to a high-refractive-index medium inside the wedge. Subsequently, the beam exits from the lower surface of the wedge, undergoing another refraction from a high- to a low-refractive-index medium. Both refractions adhere to Fresnel's law. Ultimately,

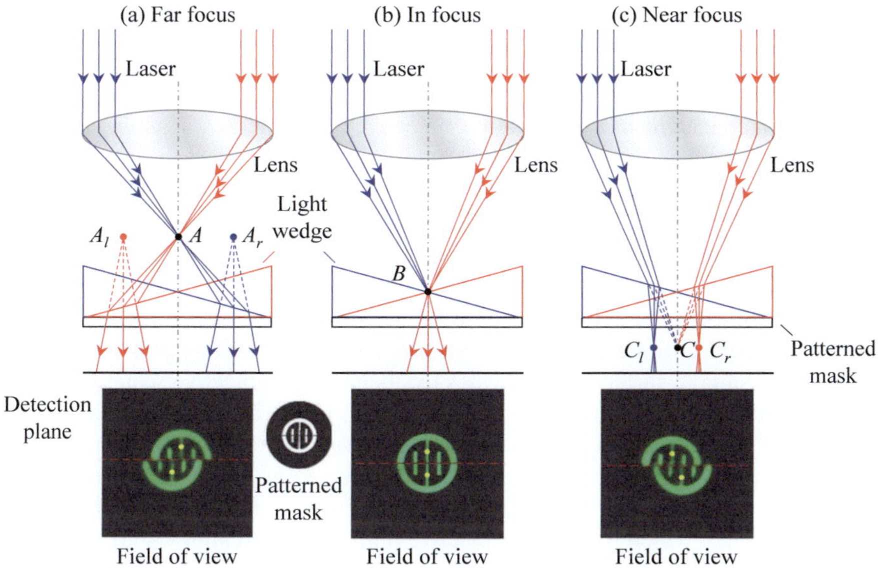

Fig. 3.1 Schematic diagram of the laser-illumination split-image system under **a** far-focus, **b** in-focus, and **c** near-focus states

the emerging beam produces a virtual image at position A_l. Similarly, the beam from point A passes through the blue light wedge and forms another virtual image at position A_r following a similar refraction process. In the near-focus state, virtual images are similarly generated by the laser through the light wedges, but their displacement is reversed compared to the far-focus condition. This observation demonstrates that a single object point yields two separate images after passing through the wedges, with opposite separation directions in far- and near-focus conditions. When accurately focused, the beams converge at the intersection of the two wedges, producing an image in which the split images exhibit a symmetric distribution pattern, as Fig. 3.1b.

An efficient autofocus strategy can be realized by constructing a mapping model that relates image features to their corresponding actual defocus distances. The fundamental concept of this approach is to leverage the predictable relationship between statistical or high-dimensional visual features observed in defocused images and their true defocus values. Once trained, the model can infer the current defocus distance from the features of a single defocused image, enabling real-time autofocus control. The mathematical formulation of this relationship is given as follows

$$\Delta z = \Theta(I_k), \ \forall k \in \{1, 2, \ldots, n\} \tag{3.1}$$

where Δz denotes the defocus distance, I_k refers to the defocused image, and $\Theta(I_k)$ represents the model describing the relationship between the split image and the defocus distance.

3.2.2 Point Autofocus

The point autofocus microscope functions essentially as a non-contact optical profilometer, consisting of a laser source, objective lens, objective scanning unit, and a spot-position sensor. Owing to its small measurement spot and exceptional vertical resolution, point autofocus microscopy provides distinct advantages for measuring intricate microstructures. The fundamental optical configuration and operating principle are illustrated in Fig. 3.2 [3]. A collimated laser beam emitted from the source is focused by the objective lens onto the sample surface. The reflected beam then traverses the objective and is focused onto the spot-position sensor through a focusing lens. The spot-position sensor monitors spot displacement resulting from height variations on the sample surface and provides feedback to the objective scanning mechanism, enabling real-time positional adjustment. As Fig. 3.2a, the system is in focus when the reflected beam is sharply concentrated at the center of the spot-position sensor. When the sample point height changes, the sensor detects a spot shift by a distance Δx, as Fig. 3.2b. The spot-position sensor relays this positional data to the objective scanner, which moves the objective until the spot realigns with the sensor center, as Fig. 3.2c. At this stage, the objective displacement Δz_2 equals the height variation of the sample surface Δz_1.

The spot-position sensor, as a critical component in point autofocus systems, plays a decisive role in determining measurement accuracy. Accordingly, the proper selection of a spot-position sensor is essential for developing high-performance point autofocus systems. In microscopic measurement systems, the most widely employed spot-position detection methods include quadrant detectors, position-sensitive detectors, and image sensors (e.g., CCD and CMOS) that incorporate image processing techniques. Quadrant detectors, widely used in optical microscopic measurements,

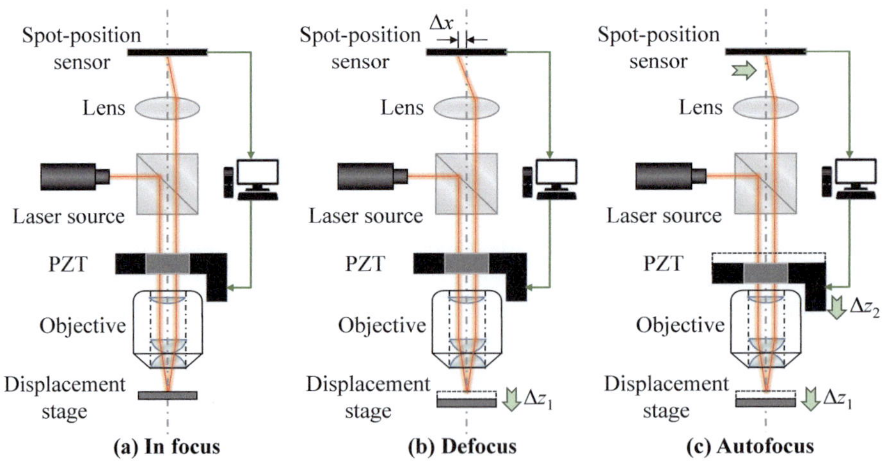

Fig. 3.2 Schematic diagram of the point autofocus microscopic system

comprise four structurally identical p-n junction photodiodes. The quadrants are denoted as Q_1, Q_2 Q_3, and Q_4, with corresponding photocurrents represented as I_1, I_2, I_3, and I_4. When a laser beam strikes the surface of a quadrant detector, each quadrant produces a corresponding photocurrent. The relative position of the laser spot center with respect to the detector's center is computed using the following equation

$$X = \frac{I_2 + I_3 - I_1 - I_4}{I_1 + I_2 + I_3 + I_4} \tag{3.2}$$

$$Y = \frac{I_1 + I_2 - I_3 - I_4}{I_1 + I_2 + I_3 + I_4} \tag{3.3}$$

The system is capable of performing dynamic scanning measurements of object surface profiles. The precision of point autofocus microscopy is primarily determined by the accuracy of the sensor in detecting the position of the focused spot. However, intricate micro-components often exhibit uneven surface morphology, leading to substantial variations in the reflected beams at different measurement points and, consequently, considerable fluctuations in the optical power detected by the sensor. The resolution of position sensors is highly sensitive to fluctuations in incident beam power; therefore, this imposes a limitation on the applicability of point autofocus microscopy for surfaces with large curvature or complex geometries.

3.2.3 Dynamic Tunable Lens Autofocus

As autofocus technology evolves toward non-contact, high-precision, and high-speed operation, the integration of tunable lenses into optical systems offers enhanced response speed and a more compact design. Consequently, dynamic tunable lens technology significantly improves the flexibility and integration capability of autofocus systems. The adoption of components such as liquid crystal lenses and electrically tunable lenses introduces a novel, non-contact focusing approach to replace conventional mechanically driven zoom and focusing systems. Liquid-crystal-based defocus ranging enables depth sensing and autofocus without mechanical displacement, representing an emerging trend toward miniaturization and high-level integration in autofocus systems. In this context, an all-optical autofocus system for microscopy is introduced, which combines an electrically tunable lens with total internal reflection infrared laser technology [4]. This system is implemented through the integration of an electrically tunable lens with a microscope objective, with the complete optical configuration illustrated in Fig. 3.3.

The autofocus system is fully integrated within a standard inverted fluorescence microscope. An electrically tunable lens is positioned adjacent to the rear focal plane of the objective lens. The infrared laser diode beam enters the microscope through a fiber tail, is redirected by mirrors and a dichroic mirror, traverses the electrically

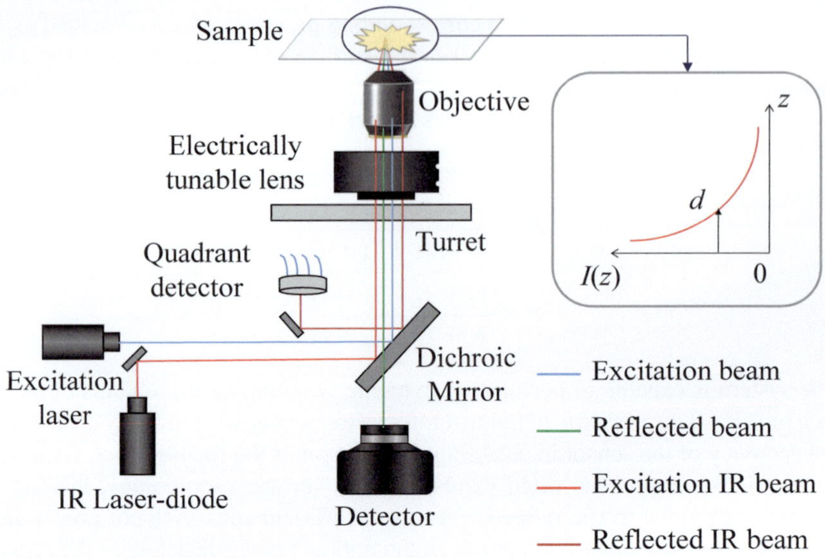

Fig. 3.3 Schematic diagram of the all-optical microscope autofocus system

tunable lens and the objective lens, and reflects off the sample surface. The reflected beam is redirected by a mirror onto a quadrant photodiode detector. The excitation laser follows its standard optical path, being reflected by the dichroic mirror, while emitted fluorescence is directed toward the detector. Variations in the reflection spot position on the quadrant detector provide a sensitive measure of axial displacement. A proportional–integral–derivative controller translates position errors from the quadrant detector into driving currents, dynamically adjusting the focal position of the objective-tunable lens assembly as voltage is applied. The approach employs the reflected infrared laser beam as a precise axial position sensor to dynamically modulate the focal plane, enabling real-time active autofocus performance.

The optical architecture of the zoom-enabled digital in-line holographic microscope incorporates an electrically tunable lens positioned between the condenser and focusing lenses, creating an optical pathway with active zooming capability, as Fig. 3.4 [5]. The key advantage of this configuration is that tuning the applied voltage dynamically modifies the refractive index profile of the tunable lens, which in turn controls the illumination beam wavefront curvature, achieving axial refocusing without mechanical movement.

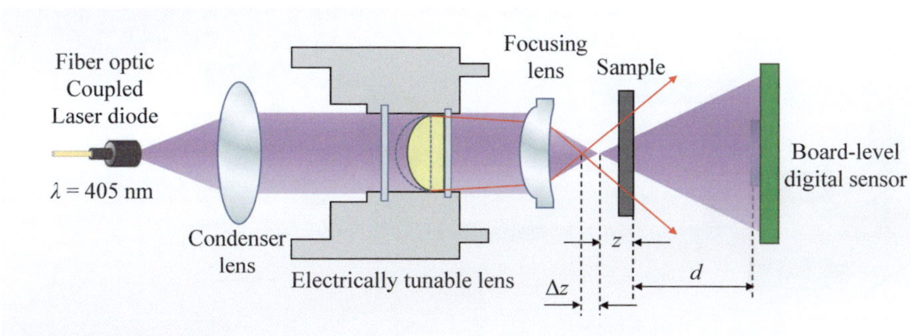

Fig. 3.4 Schematic diagram of the zoom-enabled digital in-line holographic microscope

This approach greatly enhances autofocus flexibility and precision, enabling fine control of the illumination focal point without mechanical displacement. It effectively eliminates challenges inherent in conventional lens-shifting methods, such as structural complexity, latency, alignment errors, and mechanical wear. Electronic, inertia-free, and contactless focus modulation delivers superior system stability and reliability while ensuring high repeatability and robustness during autofocus operation. Consequently, the tunable-lens-based zoom design offers a highly efficient, flexible, and precise focal adjustment strategy for digital in-line holographic microscopy, ideally suited for demanding applications such as biomedical imaging, microstructural analysis, and dynamic specimen monitoring.

3.2.4 Wavefront Modulation for Autofocus

This section presents a multimodal 3D super-resolution imaging approach that integrates adaptive optics with remote focusing, as illustrated by the experimental setup in Fig. 3.5 [6]. This method employs a deformable mirror for wavefront correction and non-mechanical axial focusing, mitigating challenges such as optical aberrations, axial drift, and uneven photobleaching in deep-tissue imaging, and achieving stable whole-volume imaging at sub-diffraction resolution.

In the adaptive optics system, the deformable mirror functions as the key element, enabling real-time correction of wavefront distortions induced by refractive index inhomogeneity while dynamically adjusting the incident wavefront for remote focusing, eliminating the need for mechanical movement of objectives or samples. Two continuous-wave lasers, independently modulated using acousto-optic tunable filters (AOTFs), are expanded and focused at the back focal plane of an oil-immersion objective for wide-field illumination, while an infrared laser employing total internal reflection ensures axial stabilization, minimizing Z-axis drift during imaging.

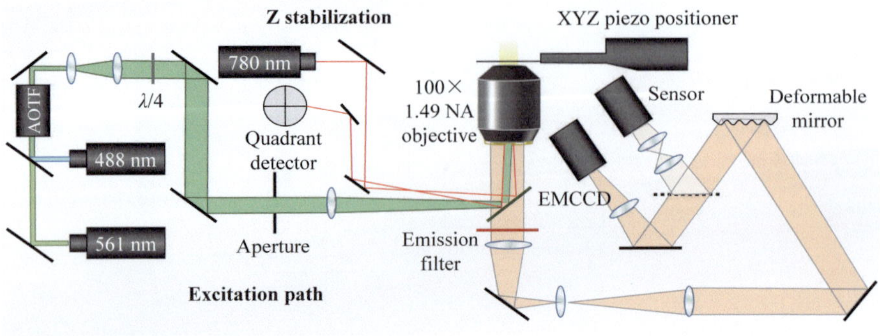

Fig. 3.5 Experimental schematic of multimodal 3D super-resolution microscopy enabled by remote focusing. AOTF: acousto-optic tunable filter; EMCCD: electron-multiplying CCD

Emitted fluorescence is collected through the same objective, filtered by a dichroic and emission filter, and passed into an adaptive optics module comprising a deformable mirror, wavefront sensor, and water-cooled camera for real-time correction and imaging. The deformable mirror dynamically shifts the focal plane along the Z-axis for sequential 3D stack acquisition, while periodic zooming minimizes localized photobleaching, ensuring uniform bleaching across layers and enhancing imaging stability and duration. The resulting 3D stacks are reconstructed via a hybrid approach integrating super-resolution optical fluctuation imaging and single-molecule localization microscopy, yielding sub-diffraction-scale resolution across the whole-cell volume. By enabling remote focusing through deformable mirror-based wavefront control, this approach obviates mechanical movement of samples or objectives, dramatically improving adaptability and precision for large-depth, high-resolution, deep-tissue imaging, and offering strong potential for cell biology, live imaging, and multimodal microscopy.

3.3 Passive Autofocus

In most microscopic imaging scenarios, active autofocus approaches that require auxiliary hardware are restricted by system compactness and often exhibit limited accuracy because they lack direct image-based feedback. Passive autofocus employs computational algorithms and image processing to analyze visual features, dynamically adjusting the lens or imaging system position to maintain a sharp and well-focused representation of the target [7].

3.3.1 Clarity Evaluation Function and Autofocus Procedure

The clarity evaluation function (CEF), also referred to as the image quality metric, constitutes a fundamental element of passive autofocus strategies. CEF serves as a quantitative descriptor of image sharpness, offering a rigorous numerical foundation for determining focus. In passive autofocus, the imaging system varies object or image distance, acquiring a sequence of images along different propagation planes and evaluating each using the CEF for quantitative analysis. The extremum of the CEF indicates the focal plane at which the image attains its highest clarity. When axial misalignment occurs, analysis of the CEF across captured images accurately identifies the optimal focus, facilitating precise compensation and real-time correction. Accordingly, the standard passive autofocus procedure typically proceeds as follows.

(1) Image acquisition phase

Capture a series of images across a specified axial scanning range encompassing the target focal plane. Each frame represents a distinct axial position of the sample, supplying critical data for the subsequent clarity assessment process. To enhance acquisition efficiency and focus precision, step size and image count should be optimized based on system resolution, depth of field, and sample-specific properties.

(2) Clarity evaluation phase

Apply a predefined CEF to quantify image sharpness across the captured image sequence. Specifically, image clarity is quantified using metrics such as gradient-based measures, spectral energy distribution, or texture contrast, providing robust indicators of imaging quality. The computed clarity values are then plotted against axial positions, generating a continuous clarity-position response profile.

(3) Focus determination phase

Analyze the clarity-position curve from the evaluation step to identify the position corresponding to the optimal metric value. The optimum is generally represented by a function maximum (or minimum), indicating the point of highest image clarity where the system is in perfect focus. Upon locating the optimal point, the corresponding frame is designated as the sharpest image, and its axial coordinate is recorded as precise focal plane data for feedback control and final image output.An ideal autofocus curve, illustrated in Fig. 3.6 [8], exhibits a unimodal profile characterized by a single global extremum.

It should demonstrate high discriminability around the focal region. For robust and adaptable autofocus performance, the curve should exhibit the following critical properties.

(1) Monotonic behavior around the focal point

The curve must maintain a clear monotonic trend on either side of the focal point, allowing accurate extremum detection and preventing convergence to local optima or ambiguous solutions.

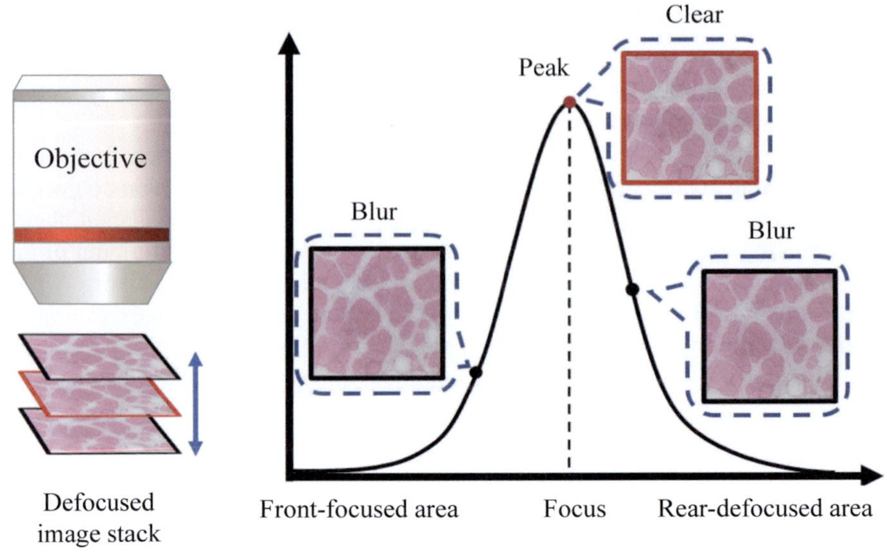

Fig. 3.6 Ideal autofocus curve

(2) Wide depth-of-field adaptability

The curve should retain strong responsiveness over a broad axial range, accommodating samples with large depth of field or complex morphology for versatile applicability.

(3) Noise robustness

The curve must maintain structural stability and clear detectability under adverse conditions, including background fluctuations, illumination changes, and mechanical vibrations.

(4) Global optimization compatibility

The curve should exhibit global continuity and avoid local extrema, enabling effective integration with global optimization strategies such as hill climbing, gradient descent, and particle swarm optimization for accurate focus determination.

3.3.2 Spatial/Frequency-Domain Autofocus

The effectiveness of a CEF strongly depends on the sample's structural attributes and the target application context. Tailoring the choice of CEF to the imaging task is essential for achieving high-precision autofocus.

Spatial-domain-based CEFs offer strong resilience to imaging noise, making them ideal for autofocus tasks under noisy or high-frequency disturbance conditions. In spatial-domain analysis, amplitude serves as a fundamental descriptor. The integrated amplitude metric, derived from the overall intensity distribution, reflects signal concentration and serves as a measure of image sharpness. Typically, sharper images generally exhibit a more concentrated amplitude distribution and stronger peak values. Furthermore, well-focused images display greater gray-level differentiation and enhanced local contrast relative to blurred counterparts. Consequently, histogram-based CEFs, including deviation metrics or variance-based measures, are widely adopted to characterize image contrast distribution. These approaches offer simplicity and fast response, making them particularly suited for real-time, stable passive autofocus implementations.

Edge sharpness is widely considered a key indicator of image clarity. Frequency-domain CEFs leverage gradient information, using first- or higher-order derivatives to quantify rapid gray-level changes and edge intensity. Common implementations include Roberts, Sobel, and Laplacian operators for edge-based sharpness analysis. Conventional CEFs include summed gradient, Laplacian gradient, squared gradient, Brenner gradient, and Tenengrad gradient, all of which characterize image focus using gradient magnitude or edge response metrics. Another major category includes frequency-spectrum-based measures, such as bandpass filtered power spectrum summation, which analyze energy distribution through the Fourier transform. Sharp images exhibit increased high-frequency energy in the spectral domain, making high-frequency content a key indicator of clarity.

Below is an overview of representative CEFs in both spatial and frequency domains. For an image $I_n(x, y; z_n)$ of size $L_x \times L_y$ from the defocused stack, the computation methods for its CEF values are summarized in Table 3.1 [7, 9–11].

For optimal reliability, CEFs should be selected according to sample characteristics, emphasizing noise robustness, strong responsiveness to focus changes, and computational efficiency. Frequency- and gradient-based CEFs provide strong axial resolution sensitivity, making them well-suited for low-light conditions or targets with uniform amplitude profiles. Nonetheless, their noise susceptibility and limited robustness often result in multiple local extrema under noisy conditions, raising the likelihood of incorrect focusing. Furthermore, in real-world imaging with complex textures, uneven illumination, or dynamic backgrounds, these approaches struggle to deliver accurate global focus estimation. Therefore, in challenging imaging conditions, integrating noise reduction, feature enhancement, or spatial-frequency fusion remains essential to enhance the robustness and practical utility of CEFs.

3.3.3 Gradient Sparsity-Based Autofocus

Traditional CEFs deliver robust performance in standard imaging systems but exhibit severe limitations under complex imaging conditions. In digital holographic

Table 3.1 Summary of CEFs and analytical expressions

CEF	Analytical expressions
Integrated amplitude, AMP	$\mathrm{AMP}(z_n) = \frac{1}{L_x L_y} \sum_{x=1}^{L_x} \sum_{y=1}^{L_y} I_n(x, y; z_n)$
Image variance, VAR	$\mathrm{VAR}(z_n) = \frac{1}{L_x L_y} \sum_{x=1}^{L_x} \sum_{y=1}^{L_y} \left[I_n(x, y; z_n) - \overline{I}_n(x, y; z_n)\right]^2$
Summed gradient, GRA	$\mathrm{GRA}(z_n) = \sum_{x=1}^{L_x} \sum_{y=1}^{L_y} \lvert\nabla I_n(x, y; z_n)\rvert$
Summed Laplacian, LAP	$\mathrm{LAP}(z_n) = \sum_{x=1}^{L_x} \sum_{y=1}^{L_y} \left\lvert\nabla^2 I_n(x, y; z_n)\right\rvert^2$
Squared gradient, SG	$\mathrm{SG}(z_n) = \sum_{x=1}^{L_x} \sum_{y=1}^{L_y} \lvert\nabla I_n(x, y; z_n)\rvert^2$
Brenner gradient, BRE	$\mathrm{BRE}(z_n) = \sum_{x=1}^{L_x} \sum_{y=1}^{L_y} \left[I_n(x+2, y; z_n) - I_n(x, y; z_n)\right]^2$
Tenengrad gradient, TEN	$\mathrm{TEN}(z_n) = \sum_{x=1}^{L_x} \sum_{y=1}^{L_y} \left[G_x^2(x, y; z_n) + G_y^2(x, y; z_n)\right]$
Bandpass filtered power spectrum summation, SPEC	$\mathrm{SPEC}(z_n) = \sum_{x=1}^{L_x} \sum_{y=1}^{L_y} \ln\left\{1 + \mathcal{F}\left[I_n(x, y; z_n) - \overline{I}_n(x, y; z_n)\right]\right\}$

Where ∇ denotes the gradient operator, while G_x and G_y correspond to Sobel operators for horizontal and vertical edge detection

microscopy, coherent light propagation introduces intricate diffraction and interference phenomena. Defocus image stacks retain rich high-frequency information, and some high-frequency components remain largely unaffected by defocus. Consequently, frequency-domain or edge-based CEFs fail to accurately reflect the focusing state in these scenarios. Furthermore, in in-line holography, twin-image artifacts frequently arise in reconstructed images, distorting the clarity curve, introducing numerous local extrema, and thereby complicating accurate focus detection. These challenges critically compromise the robustness and generalizability of current CEFs, particularly under low signal-to-noise ratio conditions or when imaging structurally complex samples.

To address these constraints, spatial-frequency hybrid CEF strategies have been introduced, jointly exploiting spatial and spectral information for robust focus assessment. These approaches combine spatial intensity features with frequency-domain energy metrics through multi-scale analysis, adaptive weighting, and feature fusion, thereby improving sensitivity to focus variation and robustness under coherent interference. These strategies substantially enhance autofocus robustness under challenging interference conditions, marking a critical advancement for complex imaging systems like holographic microscopy.

Gradient sparsity-based CEF, founded on the principle of edge sparsity in focused images, offers a highly accurate and broadly applicable focus metric [12]. This method supports accurate autofocus across diverse imaging modalities, including amplitude-only, phase-only, and complex amplitude reconstructions. It leverages natural image statistics, where smooth backgrounds dominate and edge features remain sparse and localized, providing an effective basis for sparsity-driven autofocus. The function is mathematically expressed as

$$\mathrm{SoG}(z_n) = \mathrm{S}(|\nabla I_n(x, y; z_n)|) \tag{3.4}$$

where S represents the metric used to quantify gradient sparsity. The core concept of the SoG function assumes that, at perfect focus, edge information in the reconstructed image becomes highly concentrated, resulting in the sparsest distribution of gradient magnitudes. Under such conditions, sharp edges are few and confined to structural boundaries, whereas background and smooth regions contribute negligible gradient responses. Consequently, gradient sparsity peaks at the true focus, causing the SoG metric to attain its maximum (or minimum) precisely at the optimal focal plane.

For real non-negative images, the Tamura coefficient (TC) and the Gini index (GI) are defined as

$$\mathrm{TC}(z_n) = \sqrt{\sigma\left[I_n(x, y; z_n)\right] / \langle I_n(x, y; z_n)\rangle} \tag{3.5}$$

$$\mathrm{GI}(z_n) = 1 - 2\sum_{i=1}^{L_x L_y} \frac{c_i}{\mathrm{sum}\left[I_n(x, y; z_n)\right]}\left(\frac{L_x L_y - i + 0.5}{L_x L_y}\right) \tag{3.6}$$

where σ denotes the standard deviation, $\langle\cdot\rangle$ represents the mean intensity, and c_i is the gray-level value of the i-th element in the image matrix $I_n(x, y; z_n)$ after ascending sorting. Under the gradient sparsity paradigm, alternative sparsity measures allow the design of two CEFs—Tamura of the gradient (ToG) and Gini of the gradient (GoG), derived from the Tamura coefficient and Gini index, respectively. Both functions capture gradient sparsity from complementary statistical perspectives, thereby offering a more reliable representation of image sharpness. TC and GI inherently exhibit scale invariance, ensuring consistency of sparsity evaluation under linear intensity variations caused by illumination changes or gain adjustments. Consequently, ToG and GoG deliver superior robustness in autofocus applications, especially under strong noise interference or unstable imaging conditions. In challenging systems like in-line holographic microscopy, ToG and GoG excel at discriminating genuine structural edges from artifact-induced gradients, enabling precise focus detection and robust determination of the optimal propagation distance for high-accuracy autofocus.

3.3.4 Fractional Fourier Autofocus

Fractional Fourier transform (FrFT) enables joint spatial-frequency analysis, offering an intermediate-domain framework that flexibly bridges the spatial and frequency domains for precise image sharpness assessment. The classical Fourier transform performs a π rad rotation of the signal representation from the spatial to the frequency domain, essentially acting as a fixed-angle coordinate transformation. FrFT extends this principle by introducing a fractional order, allowing arbitrary rotations between spatial and frequency domains, thus providing enhanced multi-scale feature representation. Specifically, by adjusting the rotation angle through the fractional order, the FrFT allows an image in a fractional domain to simultaneously contain local structural information from the spatial domain and global energy distribution features from the frequency domain. Consequently, FrFT-based methods retain spatial detail and capture frequency characteristics, making them ideal for challenging imaging scenarios with frequency interference or structural blur.

On this basis, a CEF based on amplitude differences in the FrFT domain is constructed [13]. The approach quantifies amplitude fluctuations of defocused stacks within the FrFT domain to deliver precise sharpness evaluation. By leveraging spatial-frequency synergy, this method ensures stable and robust focusing performance across diverse and challenging imaging environments, as Fig. 3.7.

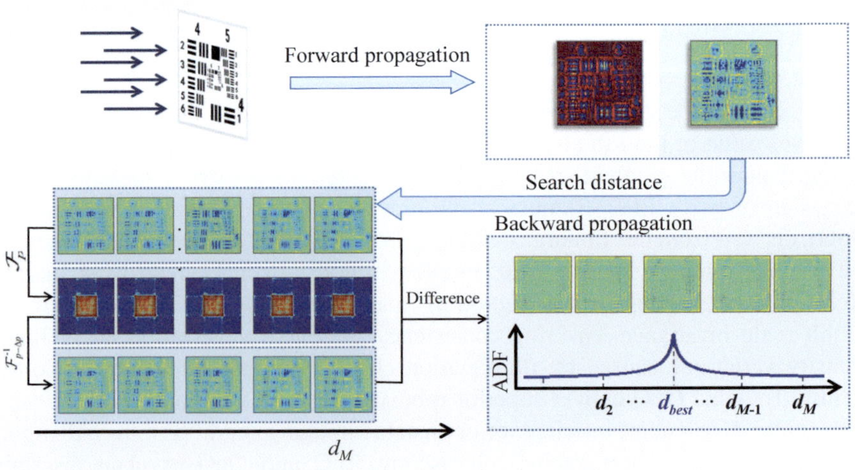

Fig. 3.7 Schematic diagram of the autofocus based on amplitude difference of FrFT

The basic procedure of the FrFT-based CEF is as follows: first, apply a fractional Fourier transform of order p to the complex amplitude of each image to obtain its fractional spectrum; then apply an inverse fractional Fourier transform of order $(p - \Delta p)$ to the spectrum to generate a new set of complex amplitude images. By calculating the difference between the original and inverse-transformed complex amplitudes, the degree of edge variation across fractional domains can be extracted and used as an effective measure of image clarity. This method has an explicit analytical expression in mathematics, which can be simplified as:

$$\mathrm{ADFrFT}(z_n, \Delta p) = \frac{1}{L_x L_y} \sum_{x=1}^{L_x} \sum_{y=1}^{L_y} \mathcal{F}_{\Delta p}\left[I_n(x, y; z_n)\right] + (\Delta p - 1) I_n(x, y; z_n) \tag{3.7}$$

Autofocus performance of the amplitude difference of FrFT (ADFrFT), ToG, GoG, and representative clarity metrics from Table 3.1 was systematically evaluated across different samples and imaging conditions. Figure 3.8a, b present the autofocus response for a USAF 1951 target and a guinea pig hair shaft section, whereas Fig. 3.8c, d depict results under Gaussian noise to evaluate robustness against noise perturbations.

Results confirm that the FrFT amplitude-difference metric consistently achieves precise focus detection across all samples, producing smooth, unimodal autofocus curves. Particularly under Gaussian noise, its focusing response remains virtually unaffected, maintaining high kurtosis and stability, significantly outperforming most traditional evaluation functions. More importantly, compared with other methods, ADFrFT introduces a fractional-order parameter Δp, providing the

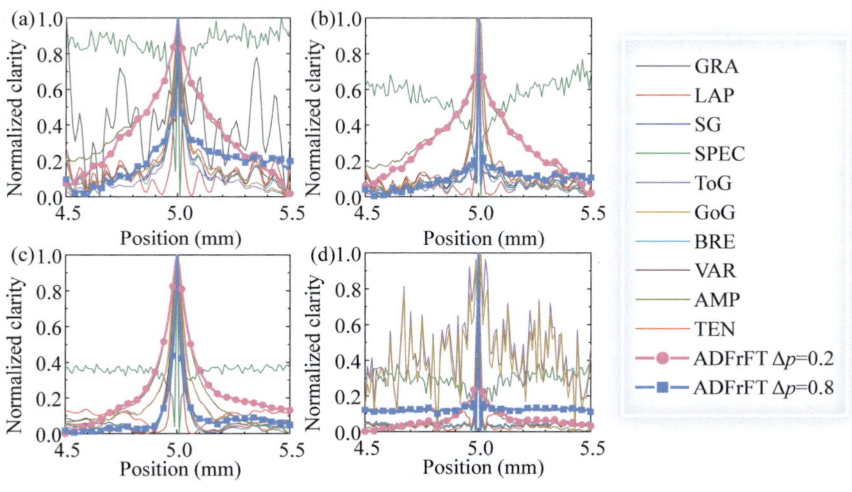

Fig. 3.8 Comparison of autofocusing curves: **a** USAF 1951 amplitude resolution target; **b** Guinea pig hair shaft section; **c, d** Corresponding to **a** and **b** with Gaussian noise of variance 0.05 added at the object plane

system with an additional adjustable degree of freedom. This flexibility allows seamless control between spatial and frequency domains, enabling adaptive tuning for diverse imaging scenarios: when Δp approaches zero, the transform leans toward the spatial domain, producing smoother autofocusing curves with fewer local extrema, thereby improving focusing stability for low-structure images. As Δp increases, the function gradually emphasizes frequency-domain features, significantly increasing the kurtosis of focus curves, improving focus sensitivity and axial resolution, making it suitable for images with clear textures or rich edges. In summary, this approach excels in accuracy, stability, noise resilience, and generalizability, positioning it as a highly promising solution for passive autofocus systems.

3.3.5 Color Images Autofocus

Color image autofocus represents a critical approach in microscopy, enabling automatic determination of the optimal focus for color samples to achieve high-resolution images with accurate color reproduction. The autofocus performance for color microscopic images is governed by intrinsic image characteristics as well as external factors including illumination conditions, dispersion properties, sample thickness variations, and system spectral response.

Existing color autofocus methods commonly project the RGB color space into a grayscale domain (e.g., HIS or YCbCr), subsequently applying conventional clarity metrics for focus evaluation. While computationally efficient, this transformation discards inter-channel semantics and contextual relationships, diminishing the sensitivity of clarity metrics to structural variations and compromising focusing accuracy. Particularly in cases of structurally complex samples or noisy imaging environments, grayscale conversion exacerbates uncertainty in blur assessment, weakening the discrimination ability of autofocus algorithms for color images and reducing image quality and focusing precision.

To address these limitations, a Tanimoto-coefficient-based pixel-difference strategy for color autofocus has been introduced, significantly enhancing clarity evaluation accuracy in color microscopy [14]. The approach leverages red and green channel information, performs normalization to mitigate brightness and contrast distortions induced by defocus, ensuring more reliable evaluation. Binary processing is then employed to emphasize edge and texture transitions, simplifying structural content and improving pixel-difference discriminability. A dynamic thresholding mechanism tailored to sample features is incorporated to refine image masking and improve focus response reliability. Proper threshold determination is pivotal, impacting defocus suppression, focus sensitivity, and the unimodality and stability of the autofocus curve. The method concludes by computing Tanimoto coefficients for normalized red and green channels to quantify inter-image pixel overlap. Tanimoto coefficient serves as a robust metric for pixel-distribution overlap across focus states, offering an indirect yet accurate measure of sharpness. Analysis of Tanimoto coefficient variation along the axial dimension facilitates construction of autofocus curves with high sharpness and resolution, ensuring robust and precise color autofocus across diverse imaging conditions. The system structure and processing workflow are illustrated in Fig. 3.9, and the complete computational procedure is expressed as

$$
\begin{aligned}
\mathrm{Tc}(C_1, C_2) &= \frac{C_1 C_2}{\|C_1\|_2^2 + \|C_2\|_2^2 - C_1 C_2} \\
C_1 &= \mathrm{NBTh}(I_{\lambda_1}) = \left\{x_i | P_{\lambda_1}(x_i), i = 1, \ldots, m\right\} \\
C_2 &= \mathrm{NBTh}(I_{\lambda_2}) = \left\{x_j | P_{\lambda_2}(x_j), j = 1, \ldots, n\right\}
\end{aligned}
\tag{3.7}
$$

where C_1 and C_2 correspond to the pixel matrices from two channels at the same spatial position in the defocus stack., the notation $\|\cdot\|_2$ represents the L-2 norm. In NBTh, N denotes normalization, B refers to binarization, and Th indicates the thresholding step. x_i and x_j are the residual elements of pixel sets $P_{\lambda_1}(x_i)$ and $P_{\lambda_2}(x_j)$ following these operations, whereas m and n specify the remaining pixel counts after processing.

Under monochromatic illumination, the Tanimoto coefficient-based pixel-difference autofocus approach transforms the intrinsic wavelength dependency of imaging into an axial distance dependency with high precision. Numerical and experimental analyses confirm that, even when conventional amplitude-difference-based autofocus strategies fail or lose precision, this method robustly discriminates clarity variations across defocus levels, highlighting its strong resilience. In color microscopy scenarios featuring intricate textures and strong defocus artifacts, the Tanimoto-based autofocus curve delivers superior unimodality and high peak sensitivity, effectively eliminating local extrema misidentification. Comparative analysis

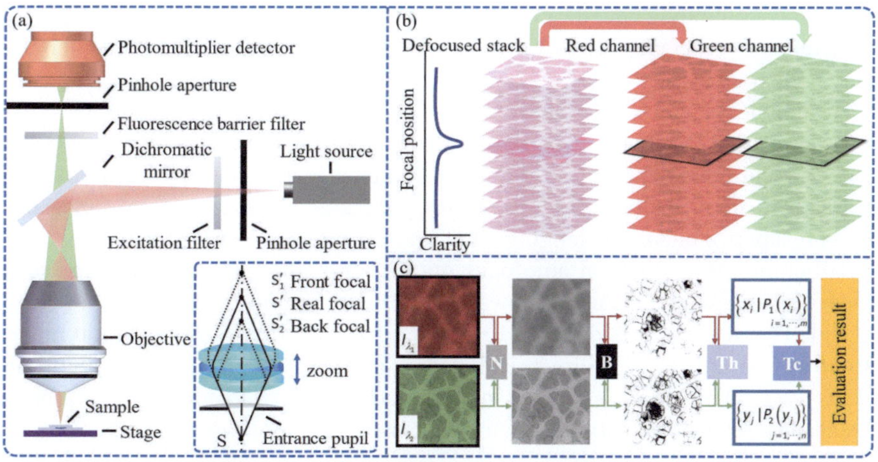

Fig. 3.9 Schematic diagram of the experimental setup and principle for autofocusing based on Tanimoto coefficient pixel difference: **a** experimental setup; **b** channels and focusing model; **c** binary operation. N: Normalization; B: Binary processing; Th: Threshold setting; Tc: Tanimoto coefficient computation

of curve sharpness, stability, and computational efficiency confirms the method's outstanding superiority for color microscopic autofocus applications.

3.3.6 Search-Optimized Autofocus

Passive autofocus fundamentally aims to locate the global optimum of an objective function within a one-dimensional search space, which translates into identifying the focus plane where the image sharpness evaluation function attains its maximum (or minimum) response during scanning. In practice, the exhaustive passive focusing algorithm is a fundamental approach, acquiring images at fixed small increments within the focus range and computing corresponding sharpness metrics. However, for wide-field, high-resolution imaging, the exhaustive search across numerous focal positions combined with dense feature extraction imposes a substantial computational burden, limiting both response speed and real-time performance, and creating a major bottleneck in focus detection efficiency.

To address the inherent inefficiency of exhaustive approaches, recent studies have introduced multiple search optimization strategies tailored for autofocus applications. These techniques leverage structural assumptions or prior knowledge of the objective function, allowing for faster and more accurate focus determination with significantly fewer evaluations. Representative approaches comprise the bisection method, Fibonacci search, hill-climbing algorithms, and approximation-based techniques.

The bisection method, a classic interval-halving strategy, iteratively partitions the search range, evaluates sharpness at two symmetric positions, and retains the better subinterval, continuing until the error tolerance is met. Its simplicity and rapid convergence render it especially effective for unimodal sharpness functions encountered in many focusing tasks.

The Fibonacci search, an enhancement over the bisection method, positions search points according to Fibonacci ratios at each iteration, as Fig. 3.10a [15]. It achieves more efficient interval reduction, resulting in higher focus accuracy and robustness. Although it often involves more iterations, its resilience to asymmetry and noise makes it superior under challenging conditions.

Hill-climbing, a common autofocus optimization approach, mimics the ascent process by iteratively advancing toward increasing function values until a peak is located. The schematic is presented in Fig. 3.10b [16]. Its implementation hinges on two key parameters: the direction of search and step size. Initially, the algorithm selects a search direction, acquiring images incrementally and computing sharpness at each step. When the sharpness metric increases, the algorithm proceeds along the same path; when it decreases, the system reverses direction and halves the step size, enabling finer resolution near the focus peak. This iterative refinement continues until the step size meets the precision threshold, yielding the optimal focus location. The hill-climbing approach is highly adaptive and well-suited for situations with unknown sharpness function gradients or limited computational resources. Unlike structured

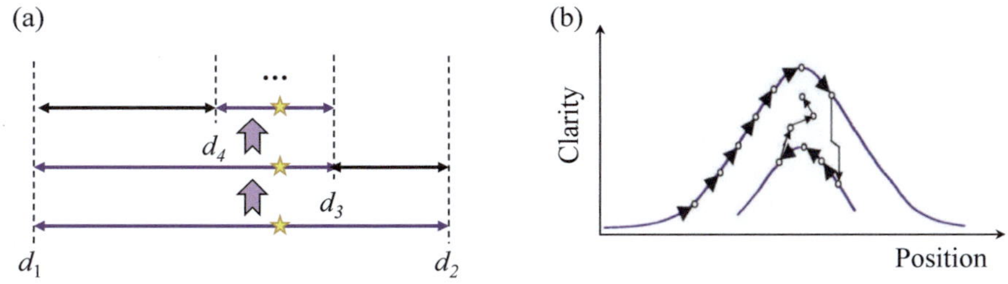

Fig. 3.10 Schematic diagram of auto-focusing method: **a** Fibonacci searching; **b** the hill-climbing

search schemes like bisection or Fibonacci, hill-climbing employs a flexible step-size adjustment mechanism that adapts dynamically to function curvature.

In contrast to conventional exhaustive autofocus approaches, the bisection and Fibonacci search strategies deliver notable gains in efficiency and applicability, particularly under large depth-of-field conditions or when rapid focusing response is critical. However, these methods fundamentally depend on the unimodal nature of the CEF and precise initialization of the search range, rendering them vulnerable in complex scenarios with multiple local optima or oscillatory focus-response curves, often degrading accuracy and robustness. Similarly, hill-climbing, a representative local search algorithm, provides flexible step-size adaptation but remains highly sensitive to initial direction and function morphology; in the presence of multiple local maxima or strong noise interference, it frequently converges to suboptimal peaks, compromising accuracy and reliability. Consequently, future improvements in search-optimized autofocus should emphasize the integration of globally convergent intelligent search paradigms to ensure adaptability to highly complex functional landscapes and significantly bolster robustness against local extrema and noise.

3.3.7 Intelligent Search-Based Autofocus

Figure 3.11 depicts the conceptual framework of a high-precision, fast autofocus strategy that integrates step-reduction search with particle swarm optimization [17]. Combining rapid coarse search with high-precision global optimization, this method aims to significantly enhance autofocus performance in digital holographic imaging. Initially, a defocused image stack is acquired, and an initial focal plane search range is defined. Two candidate axial distances within this range are reconstructed via the angular spectrum method, producing preliminary images. Two candidate axial distances within this range are reconstructed via the angular spectrum method, producing preliminary images.

Convergence to the precise focal region is achieved via iterative step-size reduction, progressively shrinking the search interval to a narrow zone around the true focal plane. At this stage, particle swarm optimization, guided by an amplitude-difference

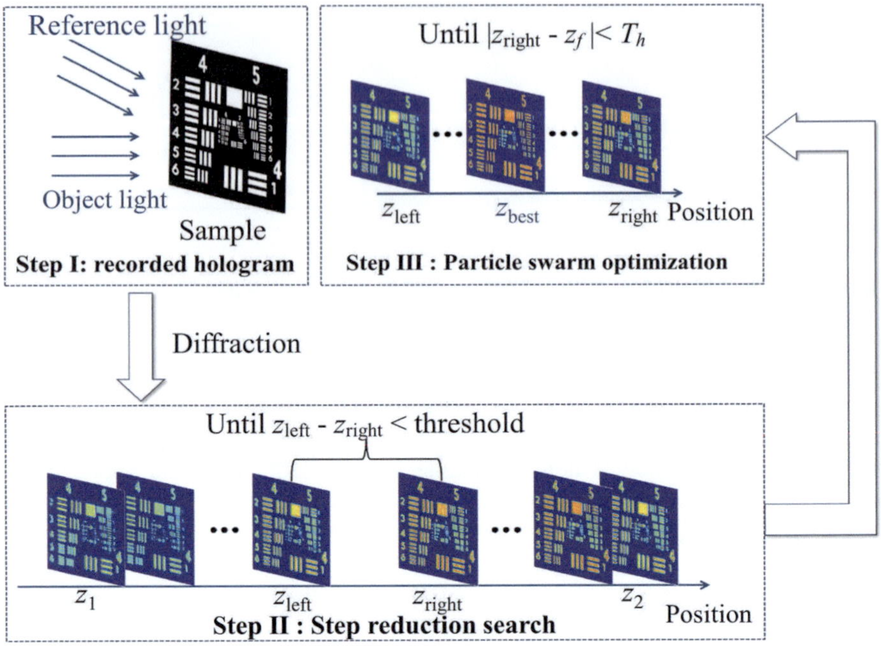

Fig. 3.11 Workflow of autofocus via stepwise reduction search and particle swarm optimization

clarity metric, performs fine-scale global search to pinpoint the true focus with high accuracy and robustness.

Compared to conventional autofocus techniques in digital holography, this approach strikes an optimal trade-off between speed and precision, offering clear performance advantages. First, a low-cost, noise-robust integral amplitude metric accelerates coarse focus localization, sharply reducing computational overhead while improving initial convergence efficiency. Second, a fine-stage optimization using an amplitude-difference metric and particle swarm algorithm ensures precise and robust detection of the focal extremum at sub-pixel resolution. This hybrid approach improves resilience to local maxima and prevents convergence to suboptimal solutions, guaranteeing sensitivity and positional accuracy even in challenging imaging conditions.

In coherent diffraction imaging, reconstruction quality is extremely sensitive to propagation distance accuracy, where minor deviations can significantly degrade image fidelity. Consequently, an ideal autofocus curve must be unimodal, steep, and highly sensitive, ensuring that its extremum precisely corresponds to the best focal position. To mitigate reconstruction degradation from propagation distance errors, a fast and precise autofocus algorithm based on multi-level subdivision search is introduced in Fig. 3.12, achieving an optimal balance between efficiency and reconstruction fidelity in coherent diffraction imaging [18].

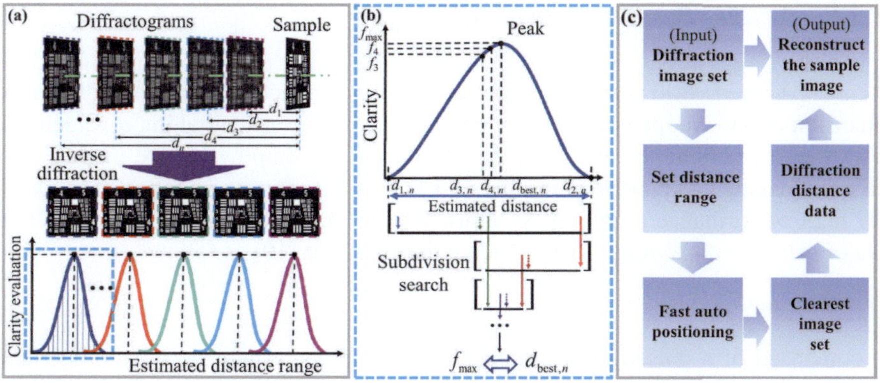

Fig. 3.12 Workflow of autofocus localization algorithm based on subdivision search

Conventional exhaustive search in Fig. 3.12a demands reconstruction and clarity evaluation at numerous candidate distances, imposing heavy computational costs and significant time delays, especially for high-resolution or multilayer imaging. To overcome this, a multi-position subdivision search mechanism is introduced in Fig. 3.12b. The method begins by defining an initial search interval $[d_1, d_2]$ and computing its midpoint d_3 as the current focal estimate. A supplementary offset $d_4{=}d_3{+}(d_2 - d_1)/S_n$, scaled by subdivision factor Sn, is introduced to refine local distance evaluation. Each iteration reconstructs images at d_3 and d_4, computing clarity metrics f_3 and f_4 for comparison. Based on these comparisons, the algorithm adaptively narrows the search interval for the next iteration. Iterations proceed until convergence to the optimal propagation distance d_{best}, finalizing the focus estimation.

Figure 3.12c illustrates the full workflow from diffraction-based localization to final image reconstruction. Initial search limits are determined via coarse estimation and iteratively refined through the hierarchical subdivision process. Ultimately, the accurately estimated distance enables high-fidelity sample reconstruction via iterative phase retrieval. Additionally, interval-based recording across stages accelerates phase retrieval convergence and enhances algorithmic stability. Experiments demonstrate that this method achieves approximately 30 times speedup over exhaustive search under identical conditions, drastically reducing computational overhead. The algorithm's simplicity and minimal hardware dependency make it highly practical without adding system complexity. Offering strong noise resilience and reconstruction robustness, this approach excels in multi-distance, multi-layer diffraction imaging scenarios demanding both high precision and efficiency, demonstrating exceptional flexibility and adaptability.

3.3.8 Active–Passive Hybrid Autofocus

Active–passive hybrid autofocus integrates the speed of active laser ranging with the precision of passive image-based evaluation, delivering substantial improvements in both focusing efficiency and accuracy. The approach employs active modules for coarse focusing and passive algorithms for fine-tuning, forming a robust framework for high-performance autofocus systems. As Fig. 3.13, the system comprises two major components: a ranging unit and an imaging unit [19] as follows.

(1) Ranging module

It composed of a pulsed laser, transmit/receive optics, avalanche photodiode, and signal processing unit. It rapidly acquires depth data and estimates the ideal focal plane, driving compensating optics for initial coarse focusing.

(2) Imaging module

It integrates zoom and compensating optics, an image sensor, control electronics, and precision actuators to enable image capture and fine-focus adjustment.

(3) Coarse focusing stage

A pulsed laser illuminates the target, with echoes collected and analyzed in real time for depth estimation. Using geometric optics, the system predicts the focal plane and drives the compensating optics close to the target position, achieving rapid coarse alignment.

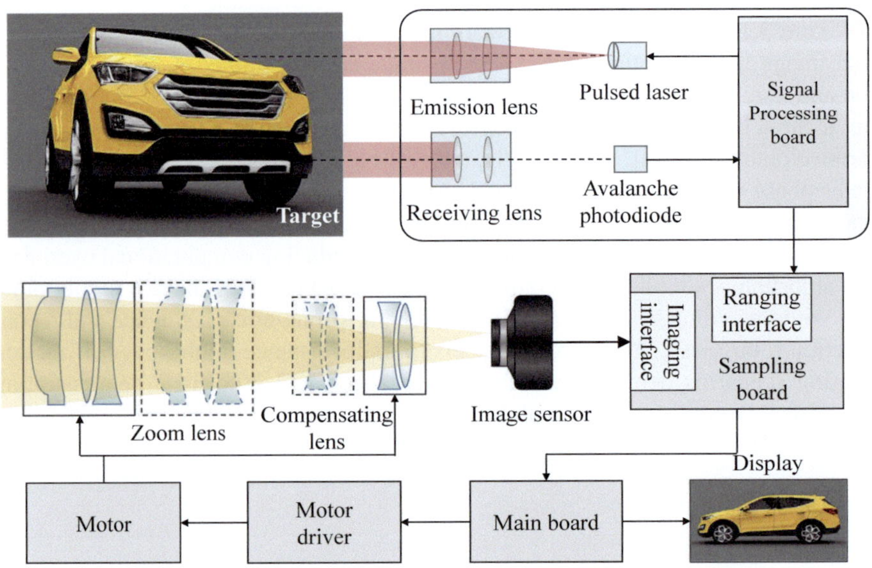

Fig. 3.13 Schematic diagram of a hybrid active–passive autofocusing imaging system

(4) Fine focusing stage

A retina-inspired sampling strategy prioritizes central high-resolution data while reducing peripheral sampling, optimizing computational load. The controller evaluates sharpness via gradient, contrast, or frequency metrics, iteratively adjusting optics for precise focus.

This hybrid architecture ensures both speed and accuracy: laser-based depth acquisition accelerates coarse positioning, while passive image-based algorithms refine precision, guaranteeing superior image fidelity. Additionally, the system maintains robustness under variable or low-light conditions, broadening its applicability to challenging imaging environments. Moreover, the retina-inspired nonuniform sampling approach enhances computational efficiency without compromising central image sharpness, mimicking human visual optimization. Through joint optimization of performance and adaptability, this approach shows strong potential in advanced applications, including biomedical imaging, defect inspection, high-throughput microscopy, and nanoscale structure analysis.

3.4 Deep Learning-Driven Autofocus

Active autofocus techniques typically require modifications to microscope hardware, resulting in high costs and reduced system compactness. Passive autofocus relies on evaluating multiple defocused frames over a defined range, making it computationally expensive and time-consuming. Incorporating deep learning into autofocus systems offers a promising route to enhance both speed and precision.

3.4.1 Single-Shot Autofocus

This section presents a learning-based single-frame autofocus strategy, developed to achieve robust autofocus performance across diverse microscopy scenarios [20]. The approach is broadly compatible with advanced imaging modalities such as super-resolution, quantitative phase, and multi-contrast microscopy, highlighting its versatility and strong application prospects. Its core concept lies in employing a neural network model to learn effective features from a single defocused image to accurately predict the current defocus distance of the microscope. Under spatially coherent illumination, defocused images still retain rich structural information critical for learning-based analysis. These defocused images encode ample frequency and edge characteristics, providing a robust basis for feature extraction and learning by neural networks. Through tailored network architecture and training, the method learns a robust mapping between image features and axial defocus, eliminating the need for prior structural knowledge or multi-frame image stacks, and enabling true single-frame autofocus.

Implementation employs off-axis LED illumination to flexibly modulate spatial coherence conditions. Operationally, the system acquires defocus stacks over a ±60 μm axial range in 1 μm increments around the focal plane. For each sample region, two complementary datasets are acquired: a coherent stack using single-LED illumination and an incoherent stack via multi-LED illumination. While coherent illumination preserves structural cues in defocused frames, incoherent illumination enhances clarity predominantly at the true focal plane. Consequently, the true focal reference is established from the incoherent image dataset.

Specifically, image sharpness is quantified by integrating the high-frequency components of the radially averaged logarithmic power spectrum for each frame in the incoherent image stack, enabling the generation of a sharpness-versus-axial-position curve. The maximum of this curve corresponds to the focal position, which serves as the ground-truth label for the training sample, as Fig. 3.14a. This calibration approach is readily scalable to large datasets.

For transparent samples, asymmetric illumination is applied to obtain incoherent stacks, enhancing phase contrast for improved visibility. In the training stage, each defocused image captured under coherent illumination serves as an independent training instance for predicting its associated true focal position. The adopted architecture is a fully connected Fourier neural network that leverages single-frame frequency-domain representations as input features. The training pipeline begins by applying a 2D Fourier transform to each coherent image, extracting the magnitude of the central spectral region, and reshaping it into a one-dimensional feature vector for network input.

The network performs defocus estimation from single frames, allowing every image in the stack to act as an independent training sample for efficient model learning. Once trained, the network predicts focal distances from any single defocused image under matching illumination conditions, achieving automatic focus correction, as Fig. 3.14b.

This autofocus method demonstrates strong versatility, accommodating diverse microscopic samples. It can be implemented in conventional bright-field microscopes without replacing existing optical components—requiring only the addition of an off-axis illumination module, thus significantly reducing system complexity and integration costs. From a network design perspective, the method incorporates physical priors of the imaging system, explicitly modeling single-scattering behavior under spatially coherent illumination. The network is designed under the assumption that critical defocus cues reside within the single-scattering region induced by off-axis illumination, discarding extraneous features beyond this region to enhance both training efficiency and predictive accuracy.

Relative to standard convolutional neural networks (CNNs), this fully connected Fourier neural network achieves comparable prediction accuracy while cutting the number of trainable parameters and memory usage by 2 and 3 orders of magnitude. This structural compactness accelerates both training and inference while improving deployment flexibility and scalability. This structural compactness accelerates both training and inference while improving deployment flexibility and scalability.

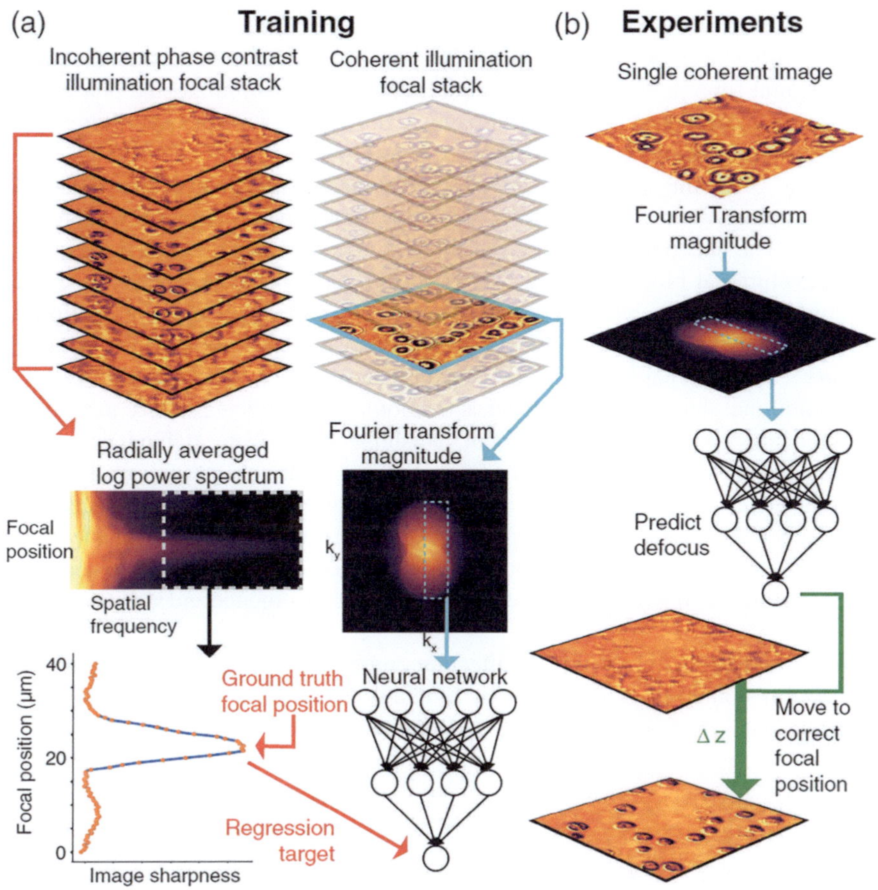

Fig. 3.14 Workflow of a single-frame autofocus algorithm based on a learning mechanism: **a** training; **b** experiments

3.4.2 Incoherent Illumination Microscopy Autofocus

Light-field microscopy under incoherent illumination faces inherent challenges in acquiring quantitative phase information. To address this issue, this section introduces a lightweight deep learning framework designed to simultaneously achieve autofocus and quantitative phase reconstruction, featuring high speed and low complexity [21]. Training data are acquired under both incoherent and hybrid coherent illumination, allowing the network to exploit complementary information from these modalities, thereby improving adaptability to diverse imaging conditions.

To mitigate pixel drift from system vibrations and the low similarity among defocused images, a perceptual loss function is incorporated to retain essential structural features and boost sensitivity to high-frequency details, thereby enhancing the accuracy and robustness of focus and phase retrieval. In the inference stage, the model achieves real-time performance, restoring both in-focus amplitude and phase maps from a single defocused bright-field image at 10 fps under incoherent illumination, providing rapid autofocus and quantitative phase imaging.

As Fig. 3.15, the network comprises three main components: an encoder for down-sampling, four residual blocks for feature refinement, and a decoder for up-sampling. The encoder is composed of three convolutional stages, each integrating two convolutional blocks and a 2 × 2 max-pooling layer to progressively extract and compress features, building a hierarchical multi-scale representation. Following down-sampling, four residual blocks refine deep representations, improving the network's ability to capture focus-dependent features. Each block includes two 3 × 3 convolutional layers with nonlinear activations and employs identity mapping to fuse input and output, reducing gradient vanishing and ensuring stable training. The decoder comprises three stages of transposed convolutions, progressively reconstructing spatial details from compressed feature maps. Skip connections link encoder and decoder stages during up-sampling, preserving fine details and enhancing the fidelity of reconstructed outputs. Batch normalization and nonlinear activations are integrated into all convolutional layers to accelerate convergence and stabilize feature learning.

In summary, this lightweight architecture combines multi-scale spatial encoding with deep residual refinement, offering a reliable framework for precise amplitude and phase recovery from single defocused inputs.

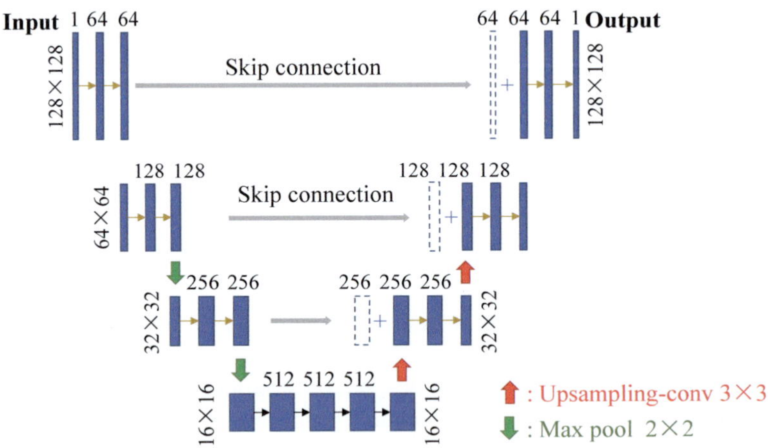

Fig. 3.15 Schematic of the network architecture for autofocus in the incoherent illumination microscopy system

3.4.3 Extended Depth-of-Field Autofocus

Conventional approaches typically train neural networks using predefined in-focus images as supervisory ground truth, where the axial distance between the sample and the sensor is determined either through prior system calibration or by applying an independent autofocusing algorithm. However, these methods exhibit significant depth-of-field dependency and suffer severe performance degradation beyond the focal range. However, these methods exhibit significant depth-of-field dependency and suffer severe performance degradation beyond the focal range. For instance, in high-resolution imaging of pathological tissue sections, a defocus as small as 4 μm can drastically reduce image resolution and introduce severe distortion in subcellular details, ultimately compromising downstream image analysis, cell identification, and diagnostic accuracy. This observation underscores the vulnerability of conventional single-focal-plane-trained neural networks to defocus, preventing them from achieving robust and reliable imaging outside the depth-of-field range. Therefore, improving robustness and generalization in defocused regions, while mitigating reliance on precise focusing, is essential to overcoming this limitation.

This section introduces a deep learning-based extended-focus holographic image reconstruction method. It enables both autofocusing and phase retrieval from a single hologram intensity image, eliminating the need for multiple image stacks. Compared with conventional approaches, it markedly accelerates reconstruction while substantially extending depth of field, improving tolerance to focus deviation and optical misalignment [22]. The key innovation is the use of a convolutional neural network trained on a dataset that combines standard in-focus patches with holograms exhibiting various random defocus levels, supervised using their corresponding in-focus amplitude and phase reconstructions. This training paradigm empowers the network to infer structural features and recover phase information from blurred holograms, delivering high-fidelity amplitude and phase reconstructions even under significant defocus. Furthermore, the approach reduces reconstruction complexity from the conventional $O(nm)$ to $O(1)$, where n is the number of object points in the 3D volume and m is the size of the discrete focusing search space, achieving a dramatic improvement in computational efficiency.

As Fig. 3.16, the network adopts a U-Net-inspired design, comprising a down-sampling path (encoder) and a symmetric up-sampling path (decoder), enabling efficient multi-scale feature extraction and accurate image reconstruction. By incorporating hierarchical down-sampling, the network learns to disentangle true image and twin-image features across multiple scales, significantly improving its capacity to capture and reconstruct multi-frequency spatial details.

Experimental evaluations demonstrate that the extended-focus holographic reconstruction approach achieves simultaneous phase retrieval and autofocusing, expanding the depth of field by more than an order of magnitude over conventional methods and substantially enhancing system robustness to defocus variations. These findings offer compelling evidence for advancing microscopic image reconstruction and highlight the transformative potential of deep learning in transcending

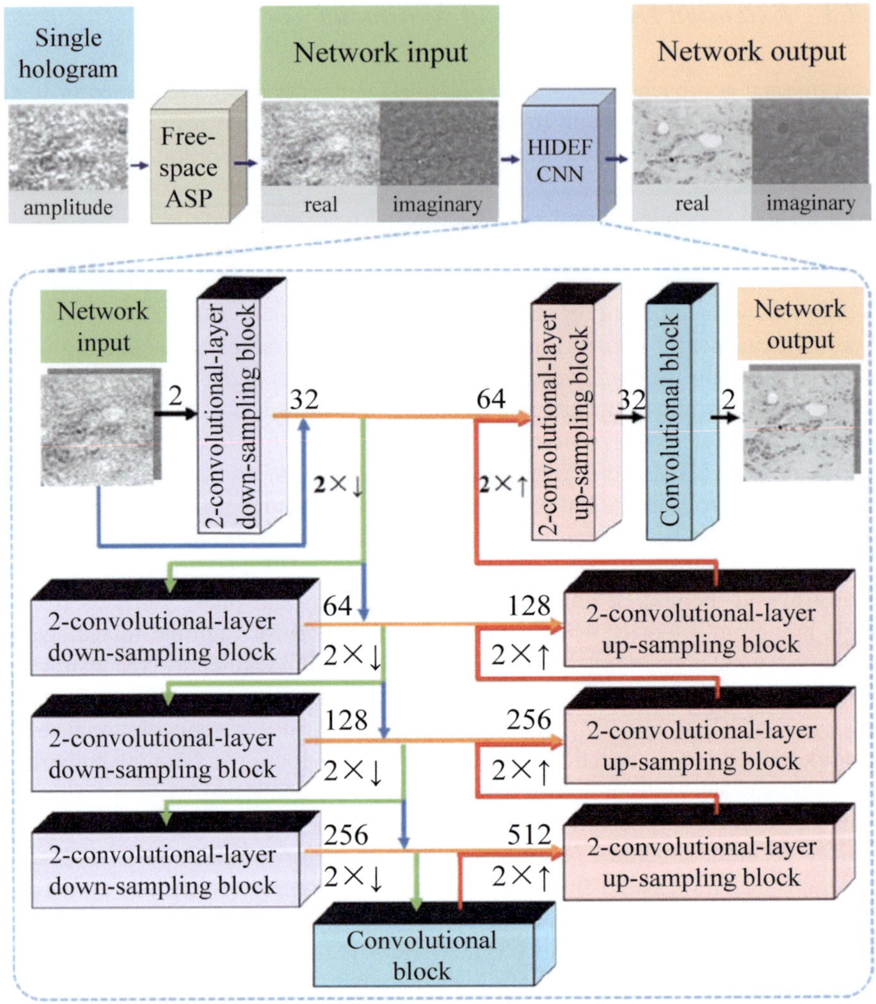

Fig. 3.16 Neural network architecture for extended-focus holographic image reconstruction. ASP: Angular spectrum propagation

the inherent constraints of conventional physics-based models. Furthermore, the extended-focus holographic reconstruction method exhibits high scalability, and its framework and principles can be broadly adapted to other imaging modalities such as fluorescence microscopy.

3.4.4 *Autofocus for Multicolor-Illumination Whole-Slide Imaging*

Whole-slide imaging aims to deliver high-resolution digital reconstructions of pathological sections, providing an alternative to conventional manual microscope observations, thereby enhancing diagnostic efficiency and standardization. Traditional whole-slide imaging systems require expensive hardware and intricate focusing mechanisms, limiting their accessibility in resource-constrained environments and primary care settings. This section introduces a low-cost, easily deployable design for whole-slide imaging systems, centered on a robust autofocusing method that leverages deep learning to directly predict defocus levels, effectively addressing the reliance on multi-frame acquisition or complex algorithms in conventional approaches [23].

In the acquisition phase, the system employs a programmable LED array for sample illumination. Red and green LEDs illuminate the specimen from opposite angles before bright-field capture, facilitating multicolor multiplexed image acquisition. The acquired images are processed by a lightweight MobileNetV3.based neural network, which performs feature extraction and accurately predicts defocus distance. Its convolutional modules effectively capture both local and global structural features, enhancing computational efficiency and real-time inference performance. In the training phase, the network is trained on an extensive dataset of annotated defocused images, allowing it to robustly predict spatial offsets between the sample and the focal plane across various tissue types and lighting conditions. This approach minimizes hardware cost and system complexity while delivering strong generalization and real-time capability, offering a practical autofocusing solution for digital pathology applications. The algorithmic workflow and network architecture are illustrated in Fig. 3.17.

In the image preparation phase, comprehensive focal range datasets comprising incoherent and partially coherent illumination stacks were acquired to facilitate deep learning model training and inference. For each lens ring position, bright-field and partially coherent illumination images were simultaneously captured to construct the incoherent illumination stack and the multicolor, multi-channel illumination stack. The Brenner gradient computed from the incoherent illumination stack was used as the CEF to determine each image's true focal position and corresponding defocus distance. The frame with the highest Brenner gradient was taken as the focal position for the field of view, and all defocus distances were calculated accordingly to generate training labels. In the training phase, multicolor multiplexed images were cropped into 512×512 patches, using only the red and green channels as inputs to reduce color crosstalk and redundant information from the blue channel, enhancing accuracy and robustness. During inference, each multicolor multiplexed image was split into 12 non-overlapping 512×512 tiles and processed in batch mode by the trained network.

Experimental results indicate that this method achieves high-precision defocus distance prediction across diverse sample types. Additionally, the method maintains

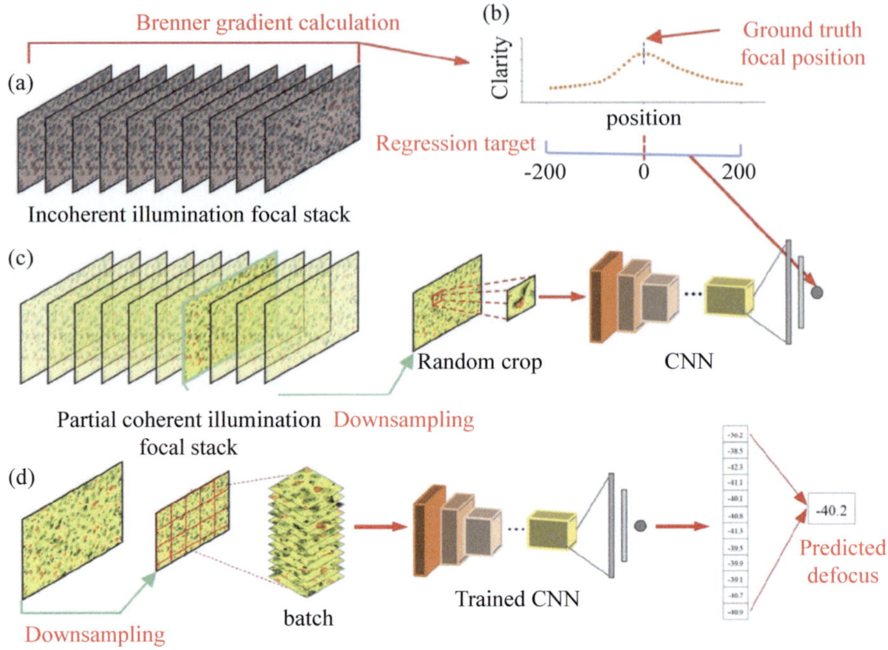

Fig. 3.17 Algorithmic workflow and network architecture for whole-slide imaging with multicolor multiplexed illumination

strong generalization to unseen sample types, reflecting the robustness and transferability of the learned features. These characteristics provide the model with strong robustness in real-world scenarios, ensuring adaptability to diverse microscopic autofocusing tasks and highlighting its practical value and scalability.

3.4.5 *Autofocus for Light-Sheet Fluorescence Microscopy*

Light-sheet fluorescence microscopy (LSFM) offers a minimally invasive, high-throughput solution ideal for capturing 3D structural details of large tissue volumes with subcellular resolution. LSFM operates by forming an ultrathin light sheet within the specimen to illuminate an optical section while fluorescence signals are collected through an orthogonal detection pathway, enabling rapid 3D imaging with minimal phototoxicity, as depicted in Fig. 3.18a [24].

LSFM demands precise optical alignment, requiring the excitation light sheet to remain perfectly aligned with the objective's focal plane. Even minor misalignments cause severe image degradation, including SNR reduction, blurred edges, and loss of fine details, as Fig. 3.18b. Maintaining this alignment becomes especially challenging during deep-tissue imaging, owing to intrinsic structural and chemical heterogeneity.

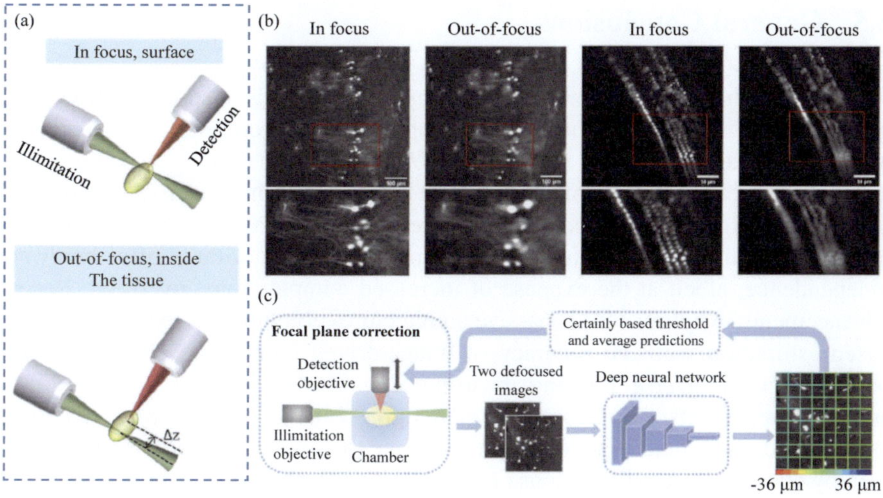

Fig. 3.18 Principle of light-sheet fluorescence microscopy and autofocus workflow

Refractive index heterogeneity within tissues introduces severe optical aberrations, including spherical aberration and micro-level focal plane drift. These refractive index-driven fluctuations lead to dynamic axial misalignment between the light sheet and focal plane during acquisition. To preserve imaging fidelity, dynamic objective adjustments are required to correct focal drift, keeping the light sheet aligned at the optimal plane and improving stability and resolution.

This section introduces a deep-learning-based autofocusing algorithm for LSFM. The method leverages two defocused image pairs as inputs, enhancing autofocus accuracy and imaging robustness, as Fig. 3.18c. During acquisition, two differently defocused frames are captured and processed through a classification network to estimate their defocus magnitude. The network was trained, validated, and tested on 421 defocus image stacks. Cross-entropy was employed as the loss function, with the Adam optimizer and a learning rate of 10^{-5}. To improve model generalization, various data augmentation strategies were applied during training, including normalization, color saturation adjustment, random cropping, and horizontal/vertical flipping. Each iteration randomly selected two frames from the same stack, spaced by Δs and differing by Δz in defocus. 128 × 128 ROIs were cropped for input, while Δz served as the reference label.

The approach reliably predicts light sheet alignment direction and demonstrates superior accuracy and convergence compared to conventional methods. With optimized dual-image inputs, the method outperforms stack-based approaches on small ROIs (~83 × 83 μm^2) and achieves comparable results on larger regions (~250 × 250 μm^2). The method incorporates an intrinsic uncertainty estimation mechanism to exclude low-quality regions, enhancing prediction accuracy and robustness while reducing the number of required images and mitigating photobleaching risk, making it suitable for rapid, stable focusing of sensitive samples.

3.5 General Conclusions

Autofocusing has evolved from manual adjustment methods to sophisticated strategies that integrate optical hardware, computational algorithms, and intelligent learning mechanisms. Active approaches, leveraging laser ranging, tunable optics, and wavefront modulation, offer rapid and precise focusing but require complex hardware configurations. In contrast, passive techniques based on image sharpness metrics, spatial–frequency analysis, and optimization algorithms offer hardware independence, albeit at the expense of increased computational overhead. Recent advancements in hybrid active–passive architectures combine the speed of ranging-based estimation with the accuracy of image-driven refinement, ensuring robust performance in diverse imaging environments. Moreover, deep learning-driven autofocus represents a transformative paradigm, achieving single-shot prediction, extended depth-of-field capability, and resilience to noise and structural complexity, thereby overcoming the limitations of traditional methods. Collectively, these developments underscore a clear trend toward compact, intelligent, and adaptive autofocus systems, which will play a pivotal role in next-generation high-resolution microscopy, whole-slide imaging, and real-time biomedical applications.

Acknowledgments This work was supported by the National Natural Science Foundation of China (Nos. 12374271, 62375065, 62405079, 62575091); Fundamental Research Funds for the Central Universities (HIT.OCEF.2024055, AUGN5750300325); The Fund of National Laboratory on Adaptive Optics, China (FNLAO-24-MS-O07); Natural Science Foundation of Heilongjiang Province of China (LH2024A009); Key Laboratory of Image Processing and Intelligent Control (Huazhong University of Science and Technology), Ministry of Education, Wuhan, China (IPIC2025-05); Key Laboratory of Advanced Medical Imaging and Intelligent Computing of Guizhou Province (AMIIC250204); the China Postdoctoral Science Foundation-CCTEG Joint Support Program (2025T011ZGMK).

References

1. Dinh VH, Hoang LP, Vu YNT, Cao XB (2023) Auto-focus methods in laser systems for use in high precision materials processing: a review. Opt Lasers Eng 167:107625
2. Hua Z, Zhang X, Tu D (2023) Autofocus methods based on laser illumination. Opt Express 31(18):29465–29479
3. Song H, Li Q, Shi Z (2024) Comparative investigation on the performance of spot positioning sensors in point-autofocus microscopy. Opt Lasers Eng 183:108491
4. Bathe-Peters M, Annibale P, Lohse MJ (2018) All-optical microscope autofocus based on an electrically tunable lens and a totally internally reflected IR laser. Opt Express 26(3):2359–2368
5. Sanz M, Trusiak M, Garcia J, Micó V (2020) Variable zoom digital in-line holographic microscopy. Opt Lasers Eng 127:105939
6. Navikas V, Descloux AC, Grussmayer KS, Marion S, Radenovic A (2021) Adaptive optics enables multimode 3D super-resolution microscopy via remote focusing. Nanophotonics 10(9):2451–2458

7. Liu Z, Zhou X, Wen X, Li YT, Liu ST (2024) Review of methods for enhancing measurement and computation speeds in computational optical imaging systems. Laser Optoelectron Prog 61(2):0211007
8. Bian ZC, Guo CF, Jiang SW, Zhu JK, Wang RH, Song PM, Zhang ZB, Hoshino K, Zheng GA (2020) Autofocusing technologies for whole slide imaging and automated microscopy. J Biophoton 13:e202000227
9. Langehanenberg P, Kemper B, Dirksen D, von Bally G (2008) Autofocusing in digital holographic phase contrast microscopy on pure phase objects for live cell imaging. Appl Opt 47(19):D176–D182
10. Choi YS, Lee SJ (2009) Three-dimensional volumetric measurement of red blood cell motion using digital holographic microscopy. Appl Opt 48(16):2983–2990
11. Guo C, Zhao Y, Tan J, Liu S, Liu Z (2018) Adaptive lens-free computational coherent imaging using autofocusing quantification with speckle illumination. Opt Express 26(11):14407–14420
12. Zhang Y, Wang H, Wu Y, Tamamitsu M, Ozcan A (2017) Edge sparsity criterion for robust holographic autofocusing. Opt Lett 42(19):3824–3827
13. Li Z, Zhou X, Wang Y, Huang G, Liu S, Gao B, Liu Z (2024) Flexible and universal autofocus based on amplitude difference of fractional Fourier transform. Opt Lasers Eng 175:107991
14. Zhou X, Xiong P, Chi D, Wen X, Ji Y, Li Y, Liu S, Jia D, Liu Z (2022) Fast autofocusing based on pixel difference with the Tanimoto coefficient between images. Opt Lett 47(15):3752–3755
15. Liu X, Wang WH, Sun Y (2007) Dynamic evaluation of autofocusing for automated microscopic analysis of blood smear and pap smear. J Microsc 227(1):15–23
16. Zhang Y, Shi Z, Wang H, Pang J (2024) An efficient autofocus method for microscope based on the improved first-order derivative Gaussian filtering operator. Opt Laser Technol 170:110225
17. Guan Y, Cui Z, Zhu J, Zhou W (2024) Accurate and fast autofocusing in off-axis digital holography based on step reduction search and particle swarm optimization. Opt Lasers Eng 181:108421
18. Zhou X, Wen X, Ji Y, Geng Y, Liu S, Liu Z (2022) Fast automatic multiple positioning for lensless coherent diffraction imaging. Opt Lasers Eng 155:107055
19. Cao J, Cheng Y, Wang P, Zhang K, Xiao Y, Li K, Peng Y, Hao Q (2017) Autofocusing imaging system based on laser ranging and a retina-like sample. Appl Opt 56(22):6222–6229
20. Pinkard H, Phillips Z, Babakhani A, Fletcher DA, Waller L (2019) Deep learning for single-shot autofocus microscopy. Optica 6(6):794–797
21. Ding H, Li F, Meng Z, Feng S, Ma J, Nie S, Yuan C (2021) Auto-focusing and quantitative phase imaging using deep learning for the incoherent illumination microscopy system. Opt Express 29(17):26385–26403
22. Wu Y, Rivenson Y, Zhang Y, Wei Z, Günaydin H, Lin X, Ozcan A (2018) Extended depth-of-field in holographic imaging using deep-learning-based autofocusing and phase recovery. Optica 5(6):704–710
23. Xin K, Jiang S, Chen X, He Y, Zhang J, Wang H, Liu H, Peng Q, Zhang Y, Ji X (2021) Low-cost whole slide imaging system with single-shot autofocusing based on color-multiplexed illumination and deep learning. Biomed Opt Express 12(9):5644–5657
24. Li C, Moatti A, Zhang X, Ghashghaei HT, Greenabum A (2021) Deep learning-based autofocus method enhances image quality in light-sheet fluorescence microscopy. Biomed Opt Express 12(8):5214–5226

Chapter 4
Optical Speckle Analysis

Guancheng Huang, Zhengjun Liu, and Yutong Li

Abstract This chapter describes optical speckle analysis. Images can be reconstructed by capturing and processing scattered light patterns. And statistical analysis, optical compensation and inverse problem solving methods are introduced to overcome optical signal attenuation and blurring. It can be used for biomedical, industrial and environmental monitoring to provide high-resolution, real-time, and non-invasive imaging.

4.1 Speckle Imaging

Optical imaging serves as a foundational modality across disciplines, such as biomedical diagnostics, environmental surveillance, industrial inspection, and security enforcement. Nevertheless, under complex turbid media, encompassing fog, atmospheric aerosols, aquatic environments, biological tissues, or particulate-laden plumes, the light propagation undergoes substantial perturbation due to stochastic interactions with inhomogeneous refractive structures [1]. These interactions precipitate multiple scattering events and severe energy attenuation, thereby disrupting the deterministic geometrical mapping and coherent wavefront integrity upon which conventional imaging systems rely. Consequently, both the spatial and temporal coherence of the light field are significantly compromised, leading to pronounced image degradation or complete failure of reconstruction.

In scattering media, photons are classified as ballistic, snake, and diffuse. Ballistic photons preserve spatial coherence but are rapidly attenuated. Snake photons undergo mild scattering with partial structural retention. Diffuse photons dominate but lack

G. Huang · Z. Liu (✉) · Y. Li
School of Physics, Harbin Institute of Technology, Harbin 150001, China
e-mail: zjliu@hit.edu.cn

G. Huang
e-mail: hgc1217048001@163.com

Y. Li
e-mail: yutong_li@hit.edu.cn

Z. Liu and Y. Li (eds.), *Intelligent Analysis of Optical Images*, Scientific Computation,
https://doi.org/10.1007/978-981-95-3720-4_4

spatial encoding, contributing mainly to noise. From the perspective of photon transport theory, speckle imaging methodologies can be dichotomized into two principal paradigms. The first capitalizes on selective acquisition of minimally scattered photons via temporal, spatial, polarization, or coherence gating, exemplified by time-gated imaging, optical coherence tomography, polarization-resolved detection, and time-correlated single-photon counting. The second paradigm embraces the modeling and exploitation of multiply scattered light through statistical inference, wavefront modulation, or inverse transmission operator reconstruction. This category includes some techniques, such as memory-effect based imaging, speckle autocorrelation analysis, transmission matrix measurement, and wavefront shaping.

Augmented by advances in computational imaging, sophisticated modeling and inversion frameworks, such as compressive sensing, Bayesian inference, and sparsity-promoting regularization, have been integrated into scattering imaging to address the ill-posedness induced by measurement incompleteness and information entanglement. Concurrently, the incorporation of programmable optical components, including spatial light modulators (SLMs), digital micromirror devices (DMDs), and single-pixel detectors (SPDs), has enabled synergistic optimization of both illumination strategies and reconstruction algorithms. Deep learning has emerged as a transformative paradigm, offering end-to-end inference pipelines for deblurring and reconstruction. While outperforming conventional methods in data-abundant regimes, such approaches remain limited by concerns over physical interpretability, generalization robustness, and theoretical transparency.

Ultimately, speckle imaging transcends algorithmic reconstruction. It inherently encompasses system-level co-design, optical hardware engineering, and signal acquisition strategies. Its ongoing evolution has engendered tangible impact across application domains, as Fig. 4.1. In medical diagnostics, optical coherence tomography and photoacoustic microscopy facilitate non-invasive visualization of subsurface tissue architecture [2]. In aquatic and atmospheric sensing, time-gated and adaptive optics systems enable underwater and aerosol-compensated imaging [3]. In intelligent vision systems, dehazing algorithms and low-light enhancement schemes improve perceptual stability and operational robustness [4]. Non-line-of-sight (NLOS) imaging further extends the frontiers of indirect vision by reconstructing hidden scenes from multiply scattered or reflected light, often leveraging time-resolved measurements, speckle correlations, or neural inference to retrieve spatial information beyond the direct line of sight [5]. The trajectory of this field increasingly converges toward compact system integration, intelligent imaging automation, and cross-domain applicability.

This chapter presents a comprehensive exposition on the fundamental mechanisms of optical scattering and analytical strategies for speckle-based imaging. It begins with an overview of light-matter interaction, elucidating the statistical characteristics of light propagation in scattering media and their implications for imaging performance. Subsequently, mainstream methodological approaches, including ballistic photon extraction, scattered light manipulation, computational inversion, and deep learning-enhanced reconstruction, are systematically examined in terms of physical models, system architectures, and application efficacy. The discussion culminates in

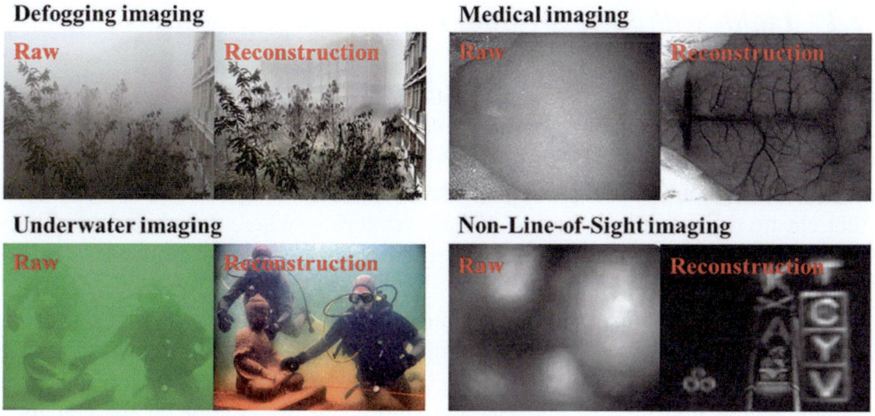

Fig. 4.1 Application of optical speckle imaging analysis

a critical analysis of representative application scenarios, highlighting prevailing technical challenges and prospective research directions. By integrating theoretical foundations, methodological frameworks, and practical implementations, this chapter aims to establish a cohesive knowledge structure for optical speckle imaging, offering a rigorous perspective to inform both academic inquiry and engineering development in complex imaging conditions.

4.2 Scattering Mechanisms and Imaging Models

The propagation of light through complex media constitutes the physical foundation of scattering-based imaging. To elucidate the intrinsic causes of image degradation in these systems, it is essential to examine light-matter interactions from the perspective of electromagnetic theory. This entails a systematic analysis of the scattering process, including its underlying mechanisms, characteristic parameters, and the spatial behavior of different photon populations. Under multiple scattering conditions, the recorded intensity distribution no longer directly corresponds to the spatial structure of the target. Instead, it embodies a complex mixture of signal components and stochastic perturbations induced by the medium. Consequently, a precise understanding of how scattering media modulate the optical field is a prerequisite for the development of effective image reconstruction, modeling, and information recovery strategies.

4.2.1 Light-Scattering Medium Interaction Mechanisms

As light propagates through a medium, its electromagnetic field interacts with the microscopic constituents of the material. When spatial variations in refractive index occur on the scale comparable to the optical wavelength, photons are recurrently scattered throughout their trajectory. These interactions are generally categorized as either elastic or inelastic scattering. Elastic scattering, where the photon frequency remains invariant while its propagation direction is altered, is of primary relevance in optical scattering imaging.

The incident electromagnetic wave induces oscillating electric dipoles within the scattering particles, effectively transforming them into secondary radiators. This process gives rise to wavefronts propagating in multiple directions, resulting in complex spatial interference patterns and significant modulation of the statistical properties of the intensity fields [6].

The microscopic nature of scattering can be quantitatively described by classical theories such as Rayleigh and Mie scattering, as Fig. 4.2. Rayleigh theory can be applied when particle diameters are substantially smaller than the incident wavelength, with scattering intensity exhibiting an inverse fourth-power dependence on wavelength. Mie theory, in contrast, accommodates scattering by particles with sizes comparable to the wavelength, making it more suitable for modeling light-tissue interactions, turbid liquids, and particulate suspensions [7]. In strongly scattering media, light typically undergoes multiple successive scattering events, leading to highly non-deterministic trajectories and the complete disruption of coherent wavefronts. This results in a dramatic loss of spatial coherence and directionality, posing substantial challenges for conventional imaging techniques.

4.2.2 Classification and Characteristics of Scattered Photons

During light propagation through a scattering medium, photons undergo varying degrees of directional perturbation and energy attenuation. Although elastic scattering conserves photon energy, it alters the direction of wave propagation and typically introduces phase retardation. By using the extent and nature of scattering events, photons can be broadly categorized into three classes, as ballistic, snake, and diffuse photons [8].

Ballistic photons traverse the medium without encountering significant scattering or with only minimal angular deviation. They retain their original directionality, phase coherence, and spatial correlation with the illumination beam, thereby conveying the most direct and high-fidelity structural information of the target. These photons arrive earliest at the detector, are spatially concentrated, and serve as the principal information carriers in high-resolution imaging.

Snake photons, having undergone a limited number of non-forward scattering events, exhibit moderate angular deviation but still propagate approximately along

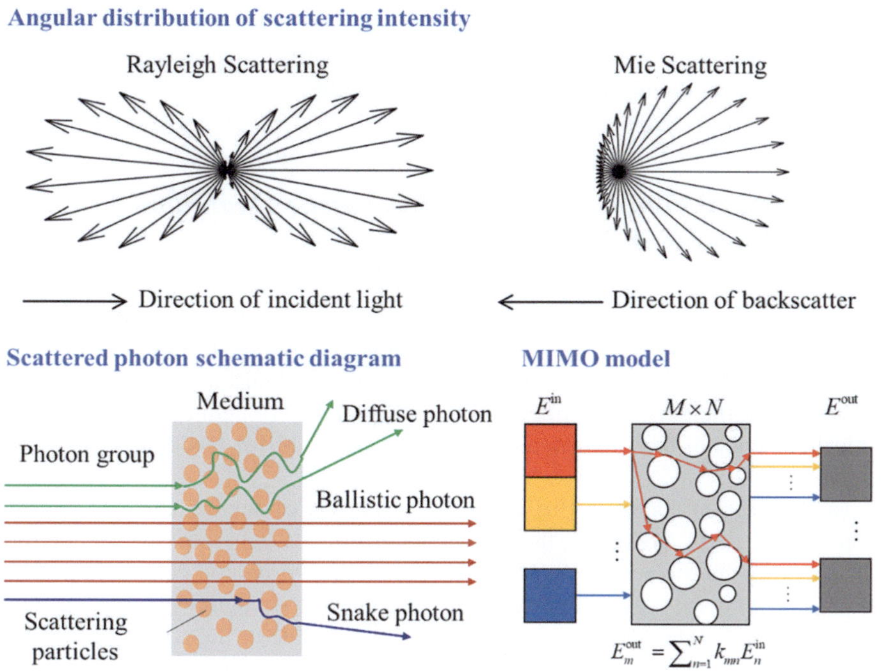

Fig. 4.2 Scattering mechanism and imaging principle

the initial incident direction. While their spatial coherence and structural correlation are partially degraded, they retain useful information content. Temporally delayed relative to ballistic photons and exhibiting broader spatial spread, snake photons are especially prevalent in peripheral tissue regions or shallow-depth imaging scenarios. Their relatively high abundance makes them an important supplementary signal source in many practical imaging modalities.

Diffuse photons (or multiply scattered photons), by contrast, experience numerous random scattering interactions. These photons lose their original direction, coherence, and polarization state, resulting in a nearly isotropic angular distribution and a randomized speckle-like intensity field. They are substantially delayed in time, exhibit strong angular divergence, and constitute the dominant source of noise and interference in scattering imaging.

In strongly scattering optical media, ballistic photons are exceedingly scarce, whereas the diffuse component dominates the detected signal. In such regimes, photons originating from a single point on the object are no longer focused into an Airy disk on the image plane but instead give rise to a chaotic, delocalized intensity pattern. This severely impairs the ability of conventional imaging systems to resolve fine structural details. Since these three photon classes exhibit markedly different responses within the imaging system, most anti-scattering strategies are fundamentally oriented toward enhancing the contribution of ballistic and snake photons while suppressing or modeling the influence of diffuse light.

4.2.3 Optical Parameters and Characteristic Quantities of Scattering Media

To quantitatively characterize the modulation imposed by scattering media on light propagation, a set of key statistical parameters is employed. These include the absorption coefficient (μ_a), scattering coefficient (μ_s), anisotropy factor (g), and extinction coefficient (μ_t). Collectively, these parameters define the macroscopic optical behavior of the medium and form the foundation for modeling light, tissue interactions [9].

The absorption coefficient μ_a denotes the fraction of energy absorbed per unit propagation length, governing the overall attenuation of light intensity. In contrast, the scattering coefficient μ_s quantifies the probability of photon deflection per unit length, and serves as the primary descriptor of scattering strength within the medium. The anisotropy factor g, defined as the expectation value of the cosine of the scattering angle, captures the angular distribution of scattering events. Its value ranges from -1 to 1: $g \approx 1$ indicates highly forward-directed scattering (characteristic of Mie scattering in biological tissues), $g \approx 0$ corresponds to isotropic scattering, and $g \approx -1$ reflects predominantly backward scattering. In biological and turbid media, g typically lies between 0.8 and 0.98, indicating a strong tendency for forward propagation even after scattering, and providing the physical basis for the formation of snake photons.

These parameters determine two critical spatial scales: the scattering mean free path ($l_s = 1/\mu_s$), representing the average distance between successive scattering events; and the transport mean free path ($l_t = 1/\mu_t$), where $\mu_t = \mu_s(1-g) + \mu_a$, which defines the characteristic distance over which the photon direction becomes randomized and its energy significantly attenuated. In highly scattering regimes where $\mu_s \gg \mu_a$, photon propagation exhibits a random walk-like behavior, necessitating the use of the radiative transfer equation or its simplified counterpart, the diffusion approximation, for accurate modeling.

4.2.4 Image Formation Mechanism in Scattering Media

In optical imaging systems, the recorded image represents not the electromagnetic field itself, but rather the spatially integrated intensity distribution after undergoing modulation by both the imaging optics and perturbation from the scattering medium. In the absence of scattering, the phase and amplitude of the incident wavefront are faithfully mapped onto the image plane via the Fourier-transform properties of lens systems, enabling direct, high-fidelity imaging. However, in the presence of a scattering medium, the incident wavefront is severely distorted, and the structural information becomes spatially mixed and diffused. The point spread function (PSF) is markedly broadened, resulting in significant deterioration of spatial resolution and image contrast.

In the frequency domain, scattering manifests as a low-pass filtering effect in the system transfer function: high spatial frequencies, which encode fine image details, are rapidly attenuated, leaving only blurred contours and low-frequency energy distributions. The addition of multiply scattered light further degrades the signal-to-noise ratio (SNR), burying relevant object information beneath stochastic background noise. Due to the inherent inhomogeneity and multi-path feedback within the scattering field, the image formation process exhibits strong nonlinearity, rendering conventional linear system models inadequate.

Under strong scattering conditions, the imaging process can be equivalently modeled as a nonlinear, multiple-input-multiple-output (MIMO) system, whose transfer characteristics vary with spatial location, incidence angle, and wavelength [10]. A MIMO is characterized by the presence of multiple independent input and output channels, enabling complex spatial interactions between the incident and detected fields. In scattering-dominant regimes, the imaging process can be analogized to a nonlinear MIMO, where the input wavefronts undergo complex multipath propagation and the outputs represent spatially mixed intensity patterns. From a system-theoretic perspective, the light-medium interaction and subsequent detection constitute a spatially variant MIMO process, wherein each input mode contributes to multiple output modes through a highly nontrivial transfer function, is expressed as

$$E_m^{\text{out}} = \sum_{n=1}^{N} k_{mn} E_n^{\text{in}}, \tag{4.1}$$

where E_n^{in} is the complex amplitude of the nth input mode representing the spatial or angular distribution of the incident field; E_m^{out} is the complex amplitude of the mth output mode after propagation through the scattering medium, and k_{mn} denotes the transmission matrix element describing the complex coupling coefficient between input mode n and output mode m. Here, N is the total number of input channels. Thus, to recover object information, it becomes imperative to model the entire imaging pipeline or to solve its inverse problem in a statistical or computational framework. This necessity underpins the theoretical foundations for subsequent optimization-based reconstruction, computational inversion, and data-driven approaches.

To analyze and compensate for scattering-induced image degradation, several mathematical models can be used to describe the relationship between light transport and imaging response. These include as follows.

Diffusion models, based on photon intensity transport, treating multiple scattering as a random walk process governed by the diffusion equation. Steady-state form is as

$$\nabla \cdot [D(\boldsymbol{r})\nabla\Phi(\boldsymbol{r})] - \mu_a(\boldsymbol{r})\Phi(\boldsymbol{r}) = -S(\boldsymbol{r}) \tag{4.2}$$

where $\Phi(\boldsymbol{r})$ is the photon flux density, $D(\boldsymbol{r})$ is the diffusion coefficient, μ_a is the absorption coefficient, and $S(\boldsymbol{r})$ denotes the source term. Here, $\boldsymbol{r}$ is the spatial position index. Diffusion models are computationally efficient and well-suited for deep tissue imaging or low-resolution applications. However, their validity breaks down in near-source regions, shallow layers, or in highly anisotropic or weakly scattering regimes, where higher-order models are required.

PSF-based convolution models assume that the scattering-degraded image is the convolution of the original object with PSF. This approximation holds in weakly scattering or near-linear regimes and is widely used in blind deconvolution and inverse filtering techniques. In strongly scattering media, PSF can be approximated by a Gaussian function under the assumption of dominant diffusion as

$$\mathrm{PSF}(\boldsymbol{r}) \approx \exp\left(-\frac{|\boldsymbol{r}|^2}{4D\tau}\right) \tag{4.3}$$

where D is the diffusion coefficient and τ is the photon dwell time, collectively determining the effective PSF width.

Transmission matrix models aim to precisely characterize the deterministic input-output relation of the field through the scattering medium. These models are particularly powerful in wavefront shaping and phase retrieval applications but require exhaustive system calibration and can be sensitive to perturbations. In cases where scattering dominates over absorption ($\mu_s \gg \mu_a$), PSF may still be approximated as Gaussian, with its spread governed by the diffusion length l_D, which depends on both scattering and absorption properties. However, in complex or spatially inhomogeneous media, simple PSF models often fail to accurately capture the field propagation, necessitating advanced numerical solvers or empirical corrections.

Advances in computational power and sensor technology have enabled the development of hybrid models, which combine physics-based formulations with data-driven learning. These models aim to bridge the gap between physical fidelity and computational efficiency by embedding physical priors into trainable architectures. Hybrid models leverage the interpretability and generalizability of physics-informed constraints while avoiding overfitting and model mismatch common in purely data-driven methods. Particularly in scenarios involving unknown or uncharacterized media, such models have demonstrated improved robustness and enhanced recovery accuracy. Compared with traditional fully physical solvers, hybrid models offer significant speed advantages, making them highly suitable for fast or real-time imaging applications.

4.3 Ballistic Photon Discrimination Techniques

In strongly scattering environments, the severe degradation of image quality is primarily attributed to the cumulative effects of multiple scattering, which disrupt the spatial and temporal coherence of the optical field. Ballistic photon discrimination techniques aim to selectively enhance photon signals that have undergone no or minimal scattering, thereby improving spatial resolution and SNR in the reconstructed image. Owing to their preserved directionality and coherence, ballistic photons are capable of transmitting critical structural information about the target scene even after traversing turbid media. Consequently, the development of efficient ballistic photon extraction strategies constitutes a foundational approach for imaging through scattering layers.

These techniques can be categorized according to the optical systems in which they operate, each exploiting distinct physical properties of light. Key classes include: Time-resolved gating, which leverages photon time-of-flight differences to isolate early-arriving ballistic components; spatial wavefront control, as implemented in confocal and multiphoton microscopy, which restricts detection to tightly focused signal paths; coherence-based interferometry, such as optical coherence tomography, which selects photons within a narrow path-length window; and polarization or spectral filtering, which distinguishes photon populations based on their polarization retention or wavelength-dependent scattering behavior.

While these methods rely on different physical mechanisms, they share the common objective of suppressing multiply scattered light prior to signal acquisition, thereby preserving the integrity of the high-fidelity ballistic component. Ballistic photon discrimination serves as a critical enabler for high-resolution imaging in complex scattering environments.

4.3.1 Image Formation Mechanism in Scattering Media

Time-gated imaging (TGI) [11–15] represents one of the most prominent and conceptually straightforward ballistic photon discrimination techniques, exploiting differences in photon time-of-flight to selectively capture minimally scattered light. By imposing a narrow temporal detection window, the system preferentially records photons arriving within a specific time interval, typically corresponding to shallow reflections or quasi-ballistic paths, while rejecting later-arriving, multiply scattered components. This allows useful structural information to be extracted from an otherwise severely degraded intensity background, as Fig. 4.3.

A pulsed light source emits temporally separated optical wave packets toward a scattering medium. Upon interaction, the light undergoes multiple scattering, resulting in two distinct echo components: early-arriving ballistic or quasi-ballistic photons (green) carrying target information, and delayed diffuse photons (orange) forming noise. Time-gated detection restricts acquisition to a narrow temporal

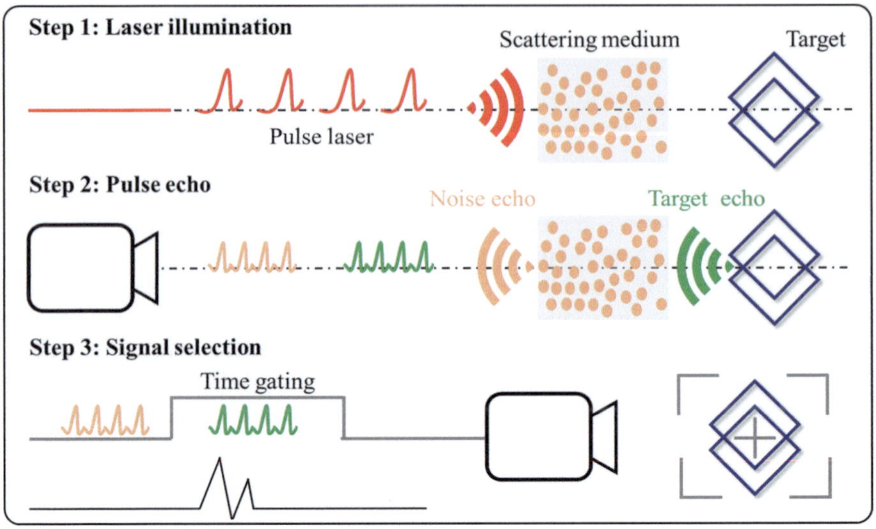

Fig. 4.3 Principle of time-gated imaging

window synchronized with the expected arrival time of the ballistic echo, thereby suppressing the contribution from multiply scattered noise photons and enhancing the fidelity of the reconstructed image.

In scattering media, photons traverse varying optical path lengths and undergo different numbers of scattering events, resulting in a broad distribution of arrival times at the detector [13]. Ballistic photons typically arrive first, followed by snake photons, while diffuse photons exhibit the longest delays. The temporal distribution of detected photon intensity $I(t)$, can thus be modeled as a mixture of distinct photon populations

$$I(t) = I_b(t) + I_s(t) + I_d(t), \tag{4.4}$$

where $I_b(t)$ corresponds to nearly unscattered (ballistic) photons, $I_s(t)$ corresponds to snake photons that experience slight deflections, and $I_d(t)$ corresponds to multiply scattered photons contributing to the background. Assuming a gate width Δt and a trigger time t_0, the gated image I_g acquired by the detector can be expressed as

$$I_g(t_0, \Delta t) = \int_{t_0}^{t_0+\Delta t} I(t)dt. \tag{4.5}$$

A narrow gate width enhances spatial resolution and contrast by suppressing late-arriving scattered light, whereas a wider window admits more photons, improving SNR at the cost of increased blurring and reduced contrast.

A typical time-gated imaging system comprises a pulsed laser source, a synchronized timing controller, a time-gated detector, and a high-bandwidth data acquisition unit. By adjusting the delay between the illumination pulse and the detector gate, photon signals corresponding to different propagation depths can be isolated, enabling depth-resolved imaging or layer-specific sampling. Common time-resolved detectors include intensified charge-coupled devices (ICCDs), streak cameras, and single-photon avalanche diode (SPAD) arrays. ICCDs operate by electronically controlling the gate timing of the microchannel plate, offering nanosecond-level resolution, and are well suited for applications such as underwater or near-infrared imaging. SPAD arrays, with picosecond-level timing precision and single-photon sensitivity, excel in low-light or high-dynamic-range scenarios, making them ideal for biomedical or autonomous vision tasks [14].

Time-gated imaging has been successfully applied in underwater laser scanning, fog-penetrating vision systems, and biological tissue imaging. Its primary advantages lie in its well-defined physical mechanism and controllable implementation, enabling selective acquisition of minimally scattered photons for enhanced image contrast and depth resolution. Nonetheless, the method also faces intrinsic limitations, including stringent requirements on the temporal resolution of detection hardware, increased system cost, and the need for precise synchronization under dynamic conditions. Ongoing advances in ultrafast photodetectors, timing electronics, and adaptive gating strategies are progressively mitigating these constraints, paving the way for broader adoption of time-gated imaging in real-time and large-scale scattering scenarios.

4.3.2 Optical Coherence Tomography

Optical coherence tomography (OCT) [16–20] is a noninvasive imaging modality based on low-coherence interferometry, enabling depth-resolved reconstruction of internal microstructures by detecting the interference between reference and sample-reflected light beams. Due to the coherence gating effect, only ballistic or quasi-ballistic photons with optical path lengths matching the reference arm within the coherence length contribute to high-contrast interference signals. Thus, OCT intrinsically functions as a ballistic photon selection system.

As Fig. 4.4, an OCT system typically consists of a broadband light source, a beam splitter, a reference arm, a sample arm, and an interferometric detector. The incident beam is divided into two paths: one directed toward a reference mirror, and the other toward the sample. The backscattered light from the sample interferes with the reference beam, and interference fringes are generated only when the optical path difference is within the coherence length of the source. The amplitude of the interference signal corresponds to the reflectivity at a specific depth within the sample.

The axial resolution δz of OCT is determined by the central wavelength λ_0 and the spectral bandwidth $\Delta\lambda$ of the source, following the relation

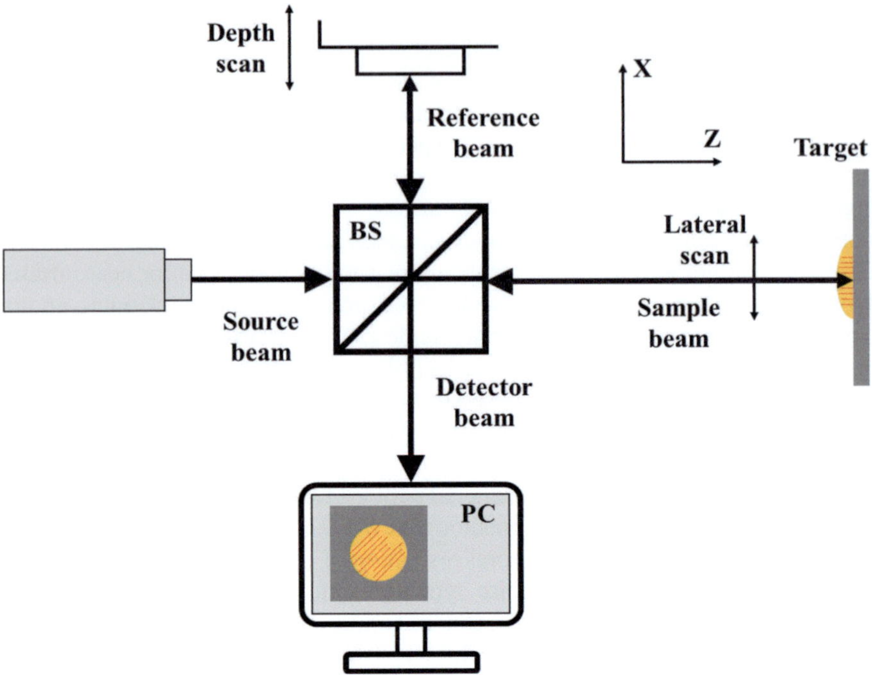

Fig. 4.4 Principle of optical coherence tomography

$$\delta z \approx \frac{2 \ln 2}{\pi} \cdot \frac{\lambda_0^2}{\Delta\lambda}. \tag{4.6}$$

A broader bandwidth yields finer resolution. Typical broadband near-infrared sources, such as superluminescent diodes, can achieve axial resolutions in the range of 1–10 μm. The lateral resolution δx is governed by the focusing characteristics of the beam and is given by

$$\delta x \approx \frac{1.22\lambda_0}{\mathrm{NA}}, \tag{4.7}$$

where NA is the numerical aperture of the focusing optics. Lateral resolution typically ranges from several to tens of micrometers.

OCT has been widely adopted in biomedical imaging, particularly in ophthalmology, dermatology, and neuroscience, due to its high resolution, rapid acquisition speed, and compact system design. However, its penetration depth is limited to a few millimeters, and the system is sensitive to sample motion. To enhance image contrast and depth resolution, advanced schemes such as coherence gating optimization, angular compounding, and speckle reduction algorithms have been integrated into modern OCT systems. Beyond the conventional time-domain OCT (TD-OCT), more advanced system architectures have emerged, notably spectral-domain OCT

(SD-OCT) and swept-source OCT (SS-OCT), both of which eliminate the need for mechanical depth scanning and offer enhanced imaging speed and sensitivity [18].

SD-OCT eliminates mechanical depth scanning by measuring the interference spectrum across wavelengths. A broadband light source is used, and the spectrally resolved interference signal is recorded by a line-scan camera coupled to a diffraction grating. By applying Fourier transform to the acquired spectrum, the depth-resolved reflectivity profile (A-scan) is reconstructed. Since all depths are acquired simultaneously in a single shot, SD-OCT offers dramatically faster acquisition rates (typically tens to hundreds of kHz) and improved sensitivity due to coherent signal averaging. It is widely used in ophthalmology, dermatology, and anterior tissue imaging.

SS-OCT employs a narrow-linewidth, rapidly tunable laser that sequentially scans across a range of wavelengths. At each time point, a single interference fringe is recorded by a high-speed photodetector. A complete spectral scan yields the full interferogram, which is Fourier-transformed to recover the depth profile. SS-OCT operates typically at longer central wavelengths (e.g., 1050 or 1300 nm), providing deeper penetration in highly scattering media such as skin or blood-rich tissues. Compared with SD-OCT, SS-OCT offers better phase stability, higher dynamic range, and superior performance in long-range imaging.

These advanced modalities have substantially broadened the clinical and scientific scope of OCT. Combined with algorithmic innovations such as motion correction, deep learning denoising, and adaptive optics enhancement, modern OCT systems are evolving into versatile platforms for scattering-resilient, real-time, and quantitative optical diagnostics [20]. As hardware and computational capabilities continue to advance, future OCT implementations are expected to further improve imaging depth, resolution, and specificity, facilitating broader deployment in complex scattering environments and pushing the frontiers of non-invasive optical imaging in both clinical and fundamental research applications.

4.3.3 Polarization-Gated Imaging

Polarization-gated imaging (PGI) techniques [21–25] exploit the polarization-state transformations of photons as they propagate through scattering media, offering a powerful strategy for ballistic photon enhancement. Unlike conventional intensity-based imaging, polarization-sensitive methods account not only for the amplitude of the light field, but also for the evolution and depolarization of its polarization state. As linearly polarized light traverses a turbid medium, its polarization becomes progressively randomized with increasing scattering events. In contrast, ballistic and quasi-ballistic (snake) photons tend to retain a significant portion of their original polarization. This distinction provides a useful contrast mechanism for selectively enhancing weakly scattered components [21, 22].

As linearly polarized light propagates through a turbid medium, such as biological tissue or foggy water, it undergoes direction-dependent scattering events. Each scattering interaction can alter the polarization state of the photon, depending on the angle, refractive index mismatch, and particle size. Ballistic and forward-scattered photons tend to preserve their polarization, while multiply scattered photons become progressively depolarized [23]. This polarization memory effect serves as a key physical principle for separating signal from background noise.

As Fig. 4.5, polarization gating enables differentiation between polarized (coherent) and depolarized (diffuse) components of the detected light. When imaging underwater scenes, the observed signal consists of two primary components: the direct radiance from the target object and veiling light caused by scattering from the surrounding medium. As light propagates through water, scattering by suspended particles alters its direction and polarization state. Ballistic and quasi-ballistic photons preserve their original polarization, while multiply scattered photons become depolarized. By placing a polarizing filter in front of the camera, the system can suppress the depolarized veiling light and enhance the contrast of the polarization-maintaining object radiance. This polarization-based filtering improves visibility and structural clarity in turbid aquatic environments.

The stochastic evolution of polarization can be modeled using the Stokes-Mueller formalism, where the Stokes vector $S = [I, Q, U, V]^{\mathrm{T}}$ describes the polarization state of light, and the Mueller matrix M characterizes the transformation imposed by the medium $S_{\mathrm{out}} = M \cdot S_{\mathrm{in}}$. Through appropriate analysis of the output Stokes vector, one can extract physical parameters, such as polarization contrast ratio and degree of depolarization, which are directly correlated with scattering properties and tissue microarchitecture.

Polarization contrast ratio (PCR) quantifies the intensity difference between polarization-preserving and depolarized regions, reflecting the spatial variation in polarization retention. It is typically calculated as the normalized ratio of the polarized intensity I_{p} to the total detected intensity I_{t}, and is expressed as

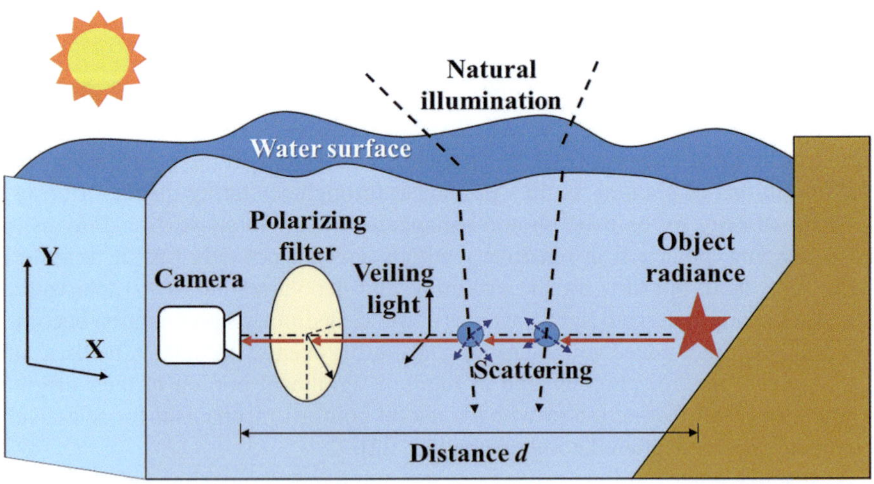

Fig. 4.5 Principle of underwater polarization imaging

$$\mathrm{PCR} = I_{\mathrm{p}} \big/ I_{\mathrm{t}}. \tag{4.8}$$

A higher polarization contrast ratio implies stronger retention of structured information and is particularly beneficial in shallow-depth imaging of tissue surfaces or rough materials. Degree of depolarization (DOD) describes the extent to which the incident polarization is scrambled due to multiple scattering events. It is commonly derived from the Stokes vector components of the scattered light. Given the total intensity I, and polarization components (Q, U, V), the DOD is calculated as

$$\mathrm{DOD} = 1 - \sqrt{Q^2 + U^2 + V^2} \Big/ I. \tag{4.9}$$

A high DOD indicates strong depolarization and, by implication, dominance of multiply scattered light. Conversely, a low DOD corresponds to preserved polarization, suggesting the presence of ballistic or snake photons.

By analyzing both PCR and DOD metrics, polarization imaging can distinguish between weakly scattered, polarization-maintaining photons and highly scattered, depolarized ones. This allows for enhanced visualization of structural features that may otherwise be obscured by scattering noise. PI has demonstrated efficacy in scenarios involving complex geometries and highly scattering media, such as biological tissues and turbid liquids [22, 24].

A typical polarization imaging system consists of a linear polarizer (at the source), a polarization analyzer (at the detector), and a polarization-resolving detector (e.g., division-of-focal-plane polarization camera or rotating-analyzer system) [25]. Applications include fog-penetrating automotive vision, marine surface target detection, and low-light imaging enhancement. While polarization imaging offers clear advantages in terms of shallow penetration contrast and scene robustness, it is inherently limited in depth performance. In highly scattering or multilayered media, polarization memory degrades rapidly, restricting the effectiveness in deep-tissue or high-optical-depth scenarios. Nevertheless, recent developments in polarization-sensitive tomography, hybrid contrast mechanisms, and data-driven polarization demodulation are actively extending the functional range of polarization imaging, making it a valuable component in multi-modal optical systems for scattering-aware sensing.

4.3.4 Photon-Counting Imaging

Photon-counting imaging (PCI) [26–30] is an ultrasensitive optical modality that reconstructs high-resolution images by discretely detecting individual photon events, making it particularly well suited for operation under low-light or strongly scattering conditions. In contrast to conventional intensity-based imaging, which continuously integrates analog light levels, PCI records both the temporal arrival and spatial location of each photon with high precision. This event-driven acquisition paradigm enables effective image formation in extremely low SNR regimes, as Fig. 4.6.

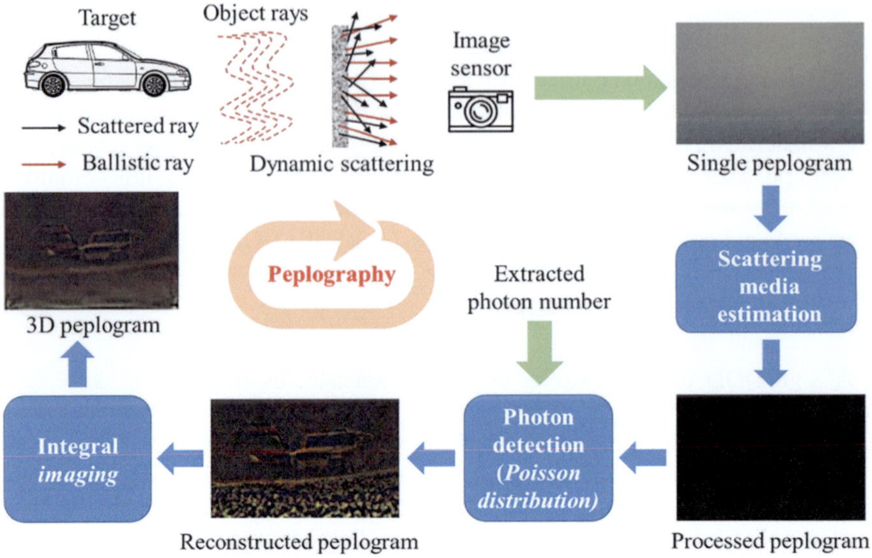

Fig. 4.6 Principle of photon-counting imaging

In PCI systems, each detected photon is individually time-stamped and spatially localized using a detector array, typically modeled as a Poisson process reflecting the quantum-statistical nature of photon arrivals. The resulting spatiotemporal data are leveraged to infer the optical properties of the medium and reconstruct latent scene information through statistical estimation and iterative inversion, such as peplographic recovery [26]. This framework enables the recovery of object structures otherwise buried in multiply scattered backgrounds.

The heart of PCI lies the statistical modeling of photon arrival time distributions [27]. In highly scattering environments, photons traverse heterogeneous and stochastic propagation paths, resulting in a wide distribution of arrival times. Ballistic photons typically arrive earlier and cluster within a narrow temporal window, while multiply scattered photons exhibit delayed arrivals and form an exponentially decaying tail. These temporal features provide a statistical basis for separating signal from scattering-induced noise. To formalize this distinction, probabilistic mixture models are often employed to describe the composite arrival-time distribution as a sum of statistically distinct components. Ballistic photons, which propagate with minimal angular deviation, are generally modeled using a Gaussian distribution. In contrast, the long-tailed and asymmetric nature of multiply scattered photons is better captured using a Gamma distribution. The combined temporal probability density function is given by

$$P(t) = \alpha \cdot \mathcal{N}(t; \mu, \sigma^2) + (1 - \alpha) \cdot \Gamma(t; k, \theta). \tag{4.10}$$

where, $\mathcal{N}(t; \mu, \sigma^2)$ denotes a Gaussian distribution modeling the arrival times of ballistic photons with mean μ and variance σ^2, while $\Gamma(t; k, \theta)$ represents a Gamma distribution characterizing the long-tailed temporal response of multiply scattered photons, with shape parameter k and scale parameter θ. The mixing coefficient $\alpha \in [0, 1]$ quantifies the relative contribution of ballistic components to the overall signal.

A specialized and increasingly prominent variant of PCI is single-photon imaging [29, 30], which is tailored for operation in the extreme photon-starved regime, where the mean photon count per spatiotemporal resolution element approaches or falls below unity. Unlike conventional photon-counting systems that can tolerate moderate photon flux and rely on aggregated measurements, single-photon imaging is fundamentally event-driven and inference-based, requiring the reconstruction of complex scene structures from highly sparse, often binary detection events. Each detected photon is treated as a probabilistic carrier of spatial, temporal, and occasionally angular information, enabling scene reconstruction even when operating near or below the quantum noise floor, conditions under which traditional sensors fail to yield meaningful outputs.

To enable robust reconstruction from sparse photon data, single-photon imaging systems commonly employ structured illumination using DMDs, SLMs, or static coded masks. Spatial patterns, typically derived from orthogonal bases such as Hadamard, Fourier, or Walsh, encode scene information into the photon acquisition process. The modulated responses, collected by single-photon detectors or SPAD arrays, are demultiplexed via inverse transforms, compressive sensing, or Bayesian inversion, enabling high-resolution imaging with sub-Nyquist measurements. In addition, single-photon imaging often leverages the time-of-flight domain, assigning picosecond-scale timestamps to individual photons. This temporal encoding supports depth-resolved reconstruction without mechanical scanning and enables advanced modalities such as transient imaging and NLOS vision. These applications require modeling the spatiotemporal impulse response of the system and scene, often addressed via deconvolution or probabilistic inference.

As photon-counting hardware continues to advance, through innovations such as high-fill-factor SPAD arrays, integrated time-to-digital converters (TDCs), low-jitter superconducting nanowire detectors (SNSPDs), and event-driven CMOS architectures, PCI is increasingly being integrated with computational optics and learn-based modeling. These synergies are paving the way toward high-resolution, scattering-resilient imaging platforms capable of operating in extreme environments, including deep tissue, underwater scenes, planetary atmospheres, and security-critical NLOS contexts.

4.3.5 *Trends in Ballistic Photon Extraction Techniques*

Advances in ballistic photon extraction are increasingly oriented toward overcoming the fundamental limitations imposed by hardware constraints, scattering complexity, and dynamic scene variability. A key trajectory involves the integration of ultrafast time-resolved detection systems with higher temporal resolution, lower jitter, and scalable architectures. Emerging SPAD arrays and time-gated intensified CMOS sensors offer picosecond-level precision while maintaining photon sensitivity, enabling finer discrimination between ballistic and multiply scattered photons across broader depth ranges.

In parallel, multi-contrast strategies are gaining traction, wherein temporal gating is complemented by polarization-resolved, coherence-gated, or spectral selection mechanisms. Such hybrid frameworks allow multidimensional filtering of optical signals, effectively enhancing the signal-to-background ratio in heterogeneous scattering environments. For instance, combining time gating with polarization filtering not only improves shallow-layer contrast, but also increases robustness to depolarizing background noise.

On the computational front, data-driven models and physics-informed deep learning have emerged as powerful tools for reconstructing ballistic content from highly degraded measurements. These methods are capable of learning complex mappings between photon arrival distributions and underlying scene structure, often outperforming purely analytical inversion approaches under extreme scattering conditions. Furthermore, real-time reconstruction algorithms based on GPU-accelerated Bayesian inference and compressive sensing are reducing latency and broadening the applicability of ballistic imaging in in vivo and field-deployable scenarios.

Another important issue is the development of compact, integrated systems for portable ballistic imaging. Advances in photonic integration, ultrafast laser miniaturization, and FPGA-controlled synchronization are enabling the design of lightweight and energy-efficient systems suited for clinical diagnostics, environmental monitoring, and navigation in visually degraded environments.

Collectively, these trends are a transition from isolated optical strategies toward cohesive, computationally enhanced platforms capable of robust operation across a range of scattering regimes. Ballistic photon extraction is evolving into a critical enabler for next-generation scattering-resilient imaging systems, offering not only enhanced resolution and contrast, but also deeper penetration, higher frame rates, and broader modality integration.

4.4 Scattering Light Manipulation Techniques

In most practical scenarios, the proportion of ballistic photons diminishes rapidly with increasing medium thickness or scattering strength, thereby severely limiting the applicability of ballistic-only imaging techniques in deep-tissue or high-turbidity environments. Consequently, when ballistic photons are insufficient to support reliable reconstruction, a critical alternative is to model and exploit the information embedded within multiply scattered photons. Despite undergoing significant alterations in propagation paths, coherence, and directionality, scattered light retains statistical signatures that can, under appropriate conditions, provide indirect yet meaningful cues about the underlying structure. Understanding and controlling these diffuse signals constitutes a foundational challenge in optical imaging through scattering media.

Current strategies for leveraging scattered photons fall into three principal domains: (1) correlation-based imaging methods that exploit statistical features, such as the optical memory effect; (2) wavefront shaping and inverse transport modeling techniques aimed at reconstructing the distorted light field; and (3) system-level modeling and inversion frameworks that incorporate the complete scattering response. Rather than rejecting scattered light as noise, these approaches integrate it as a legitimate component of the imaging pipeline, treating it as a carrier of recoverable structural information.

4.4.1 Optical Wavefront Shaping

Optical wavefront shaping (OWS) [31–35] constitutes a foundational strategy in modern scattering optics, offering a means to regain control over light propagation in highly disordered media. By precisely tailoring the incident wavefront, it becomes possible to counteract the scrambling effects of multiple scattering and induce constructive interference at targeted spatial locations beyond the scattering layer. This enables localized focusing or image reconstruction that would otherwise be inaccessible in strongly diffusive environments.

Under typical conditions, a coherent input field, such as a plane wave or a focused beam, is rapidly degraded as it traverses a scattering medium, with its phase and amplitude distributions randomized to form a speckle pattern at the detector plane. This loss of spatial coherence renders conventional imaging approaches ineffective. However, if the scattering medium is static or exhibits only slow temporal variation, the input-output light field relationship remains deterministic [31]. This determinism underpins the theoretical reversibility of the scattering process and serves as the operational basis for wavefront shaping.

As schematically illustrated in Fig. 4.7, the original wavefront is severely distorted after passing through a turbid medium, resulting in an unstructured intensity distribution at the detector that conceals spatial information. By inserting a modulation element prior to the scattering layer, one can impose a phase correction on the incident field. When properly optimized, this modulation causes the multiply scattered components to interfere constructively at a designated target region. The resulting engineered interference effectively undoes the scattering-induced distortions, restoring spatial focusing or structured illumination beyond the diffusive barrier. At its core, wavefront shaping relies on the principle that light interference within a scattering environment can be deterministically engineered if sufficient knowledge or feedback is available. This inverse design approach exploits both the phase memory of the medium and the principle of optical reciprocity. Practical implementation demands accurate estimation or inference of the medium transmission characteristics, which can be achieved through direct transmission matrix measurement, iterative feedback

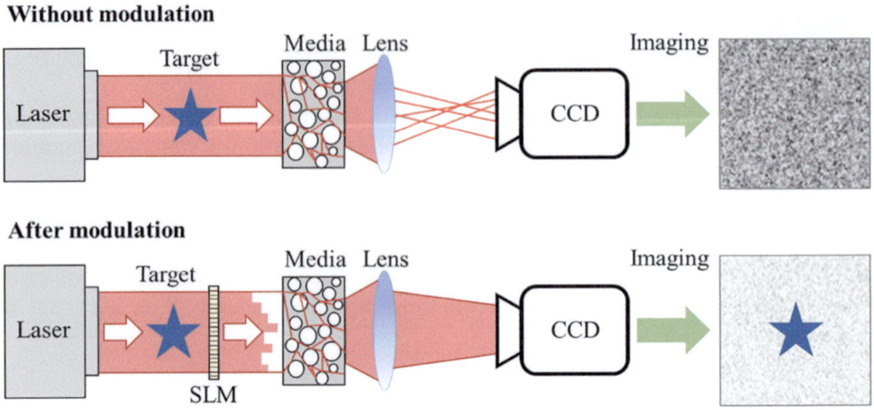

Fig. 4.7 Principle of optical wavefront shaping

optimization, or optical phase conjugation, each differing in complexity, robustness, and experimental accessibility.

The fundamental premise of wavefront shaping is that coherent interference can be deliberately engineered within a scattering medium by modulating the input phase such that multiply scattered waves constructively interfere at a desired output position. This "inverse design" of light propagation relies on the medium phase memory and optical reciprocity. The realization of OWS typically requires precise characterization and control of the light-medium interaction, which can be implemented through several strategies depending on the degree of access to the internal parameters.

To implement wavefront shaping in scattering environments, it is essential to accurately characterize and manipulate the light transmission behavior. As Fig. 4.8, three primary approaches have been established for practical realization: transmission matrix (TM)-based modulation, iterative feedback optimization, and optical phase conjugation (OPC). Each of these pathways exploits different levels of access to the internal parameters and offers trade-offs between accuracy, adaptability, and experimental complexity.

In the TM-based approach [34], light propagation through a static or quasi-static scattering medium can be approximated as a deterministic linear mapping

$$\boldsymbol{E}_{\text{out}} = \boldsymbol{T} \cdot \boldsymbol{E}_{\text{in}}, \tag{4.11}$$

where $\boldsymbol{E}_{\text{in}}$ and $\boldsymbol{E}_{\text{out}}$ represent the complex input and output field vectors, respectively, and $\boldsymbol{T}$ is the complex-valued transmission matrix of the medium. Each element T_{ij} encodes the amplitude and phase response from an input mode i to an output mode j. The implementation of this approach generally comprises two sequential stages. In the first stage, the transmission matrix of the scattering system is characterized by sequentially illuminating structured input modes, while the corresponding output

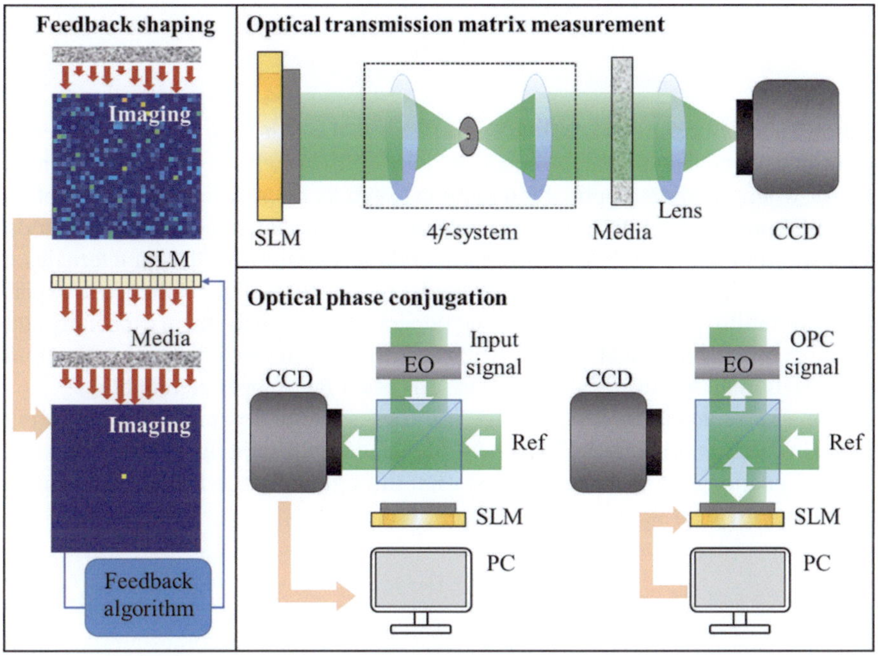

Fig. 4.8 Representative wavefront shaping implementations through scattering media

fields are recorded using interferometric or holographic techniques to reconstruct the complete transmission response. In the subsequent stage, the optimal incident wavefront is computed to generate a target output field, such as a focused spot or an arbitrary image, through either pseudoinverse computation or numerical optimization. This scheme enables versatile wavefront modulation for a broad range of applications, including single- and multi-point focusing, multi-channel beam reconstruction, and spatially controlled image projection. While system stability can be affected by environmental fluctuations, such as thermal drift or mechanical vibration, ongoing advancements in stabilization techniques and adaptive compensation strategies continue to enhance the robustness and applicability of this approach in practical imaging scenarios.

In contrast, feedback-based wavefront shaping [32, 33] circumvents the explicit measurement of the transmission matrix by relying on iterative optimization algorithms that adjust the input wavefront based solely on intensity or image feedback from the output. Algorithms such as continuous sequential modulation, genetic optimization, simulated annealing, and gradient descent are employed to iteratively maximize a target function, typically the optical intensity at a desired output location as

$$\max_{\{\phi_n\}} \left| \sum_n A_n e^{i\phi_n} \right|^2 \tag{4.12}$$

where A_n and ϕ_n denote the amplitude and phase of the n-th input mode. This method is particularly advantageous in scenarios, where direct field measurement is infeasible or where nonlinear optical interactions, such as fluorescence or photothermal responses, are involved. While the convergence rate may be influenced by the modulation speed of the spatial light modulator and the signal-to-noise characteristics of the feedback signal, feedback-based strategies demonstrate high resilience to experimental variability. Their adaptability and generalizability make them especially suitable for complex biological environments and dynamic in vivo imaging applications, where traditional wavefront shaping methods often fail to operate reliably.

OPC [35] constitutes a physically rigorous strategy for reversing scattering-induced wavefront distortions by generating the complex conjugate of the distorted field and enabling its backward propagation through the same medium. Assuming the original incident wavefront can be represented as

$$E_{\text{in}}(r) = A(r)e^{i\phi(r)} \tag{4.13}$$

where the transmitted field through the scattering medium becomes a randomized speckle pattern due to multiple phase perturbations. By synthesizing the conjugate wavefront as

$$E_{\text{OPC}}(r) = A(r)e^{-i\phi(r)} \tag{4.14}$$

and directing it back into the medium, the reciprocal phase response enables the conjugated wave to accumulate inverse phase distortions during reverse propagation. This results in constructive interference at the original source location, thereby enabling spatial refocusing or image reconstruction. Often conceptualized as time-reversal optics, this process leverages the principle of optical reciprocity to facilitate high-fidelity wavefront recovery in strongly scattering media. While OPC presents practical challenges, such as the need for precise phase measurement and system stability, it remains a powerful paradigm for wavefront correction and has been successfully implemented in various scattering imaging scenarios, including biomedical imaging, optical communication, and quantum information processing.

These wavefront shaping modalities offer complementary advantages: TM-based methods deliver high precision under stable conditions, feedback optimization enhances adaptability to uncontrolled or nonlinear environments, and OPC provides a physically intuitive and often hardware-efficient pathway for real-time correction in reciprocal systems.

4.4.2 Speckle Correlation Imaging

In highly scattering media, the optical wavefront undergoes multiple elastic scattering events, leading to the loss of its original phase and directionality. This process gives rise to speckle patterns, spatially random yet statistically deterministic interference distributions originating from coherent multipath interference. Speckle-based statistical analysis and its associated framework, correlation imaging (CI) [36–40], have become powerful tools for inferring medium characteristics, reconstructing obscured objects, and enhancing image quality under scattering conditions. CI refers to a class of imaging techniques that do not rely on lens-based conjugate mapping, but instead retrieve spatial information through statistical correlations in the light field, deviating from conventional geometric optics. The central concept is to extract structural information by analyzing spatial or temporal correlations of the detected light, with two dominant strategies being ghost imaging (GI) and speckle auto-correlation imaging.

GI reconstructs an object by correlating integrated intensity measurements collected from the object path with structured illumination patterns from a reference arm, as Fig. 4.9. The object information is indirectly encoded into the temporal fluctuations of the detected intensity signal, and image recovery is achieved via second-order correlation with known spatial modulations [37]. In scattering environments, such structured priors can be imposed via dynamic scattering configurations, where the randomness itself can encode diverse spatial constraints. Formally, the reconstructed image is given by

$$G(r) = \langle P_i(r) \cdot D_i \rangle - \langle P_i(r) \rangle \langle D_i \rangle, \tag{4.15}$$

where $P_i(r)$ denotes the i-th spatial modulation pattern, D_i is the corresponding single-pixel intensity measurement from the object path, and $\langle \cdot \rangle$ represents ensemble averaging.

Single-pixel imaging (SPI) [36, 38] further simplifies the optical configuration of ghost imaging by integrating a SLM and a single-pixel detector into a coaxial setup, as Fig. 4.9. Through deterministic encoding with orthogonal spatial patterns, SPI

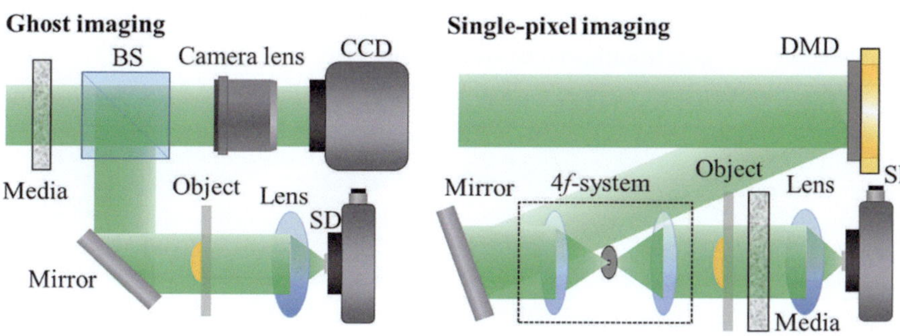

Fig. 4.9 Schematic diagram of ghost imaging and single pixel imaging

enables full recovery of the second-order spatial correlation of the target scene. The statistical characteristics of the scattering transmission matrix arise from classical random walk processes, wherein each matrix element represents the coherent summation of numerous propagation paths connecting input and output modes within the medium. Consequently, the transmission matrix can be modeled as an independent and identically distributed random matrix with Gaussian statistics. Under backscattering configurations, spatial averaging over the detector surface suppresses cross-terms between different input modes, rendering their contributions uncorrelated and uniformly weighted. This statistical decorrelation permits direct measurement of the correlation coefficients needed for image reconstruction, even in the presence of strong scattering.

In static or quasi-static scattering media, when the incident wavefront or object undergoes small angular displacements, the resulting speckle pattern in the far field experiences a corresponding shift without changing its overall structure, as Fig. 4.10. This phenomenon, known as the optical memory effect (OME), indicates that the speckle field retains partial information about the incident perturbation within a limited angular range, as Fig. 4.10. If the incident direction shifts from θ_0 to $\theta_0 + \Delta\theta$, the speckle pattern I_{out} undergoes a proportional lateral displacement Δr as

$$I_{\text{out}}(r; \theta_0 + \Delta\theta) \approx I_{\text{out}}(r + \Delta r; \theta_0). \tag{4.16}$$

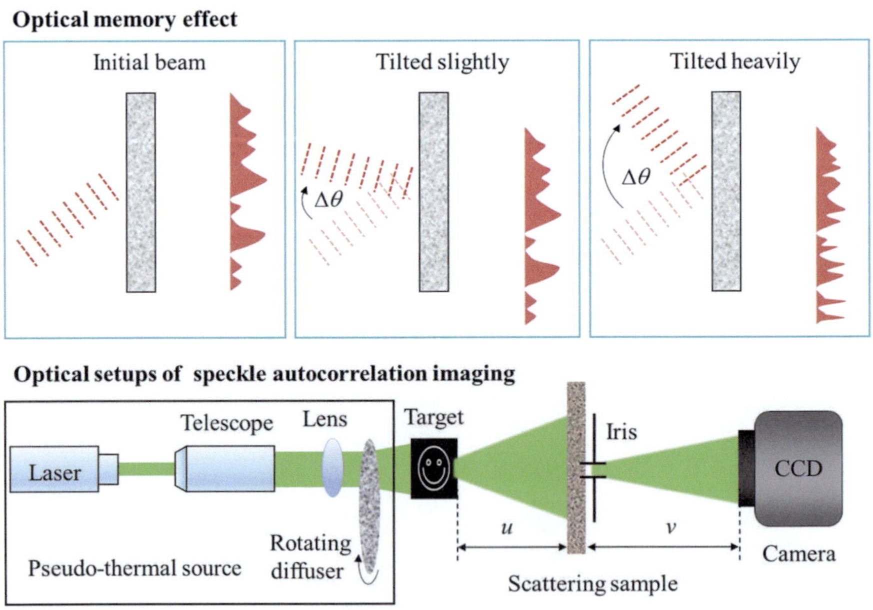

Fig. 4.10 Optical memory effect and speckle autocorrelation imaging

Speckle auto-correlation imaging (SAI) provides a lensless and calibration-free strategy for imaging through strongly scattering media by exploiting the optical memory effect and the convolutional model of static speckle formation [39]. In such media, the spatial coherence and phase structure of incident light are rapidly randomized, yielding highly disordered speckle fields. However, under quasi-static conditions, small angular or lateral shifts in the hidden object lead to corresponding lateral translations of the speckle pattern at the detector, owing to the memory effect. This spatial invariance enables the statistical correlation between input perturbations and output patterns to be exploited for image reconstruction without explicit knowledge of the transmission matrix or wavefront information, as Fig. 4.10.

Assuming the object intensity distribution $O(r)$ is convolved with point spread function $h(r)$, the recorded speckle image is modeled by convolution as

$$I(r) = O(r) * h(r). \tag{4.17}$$

Although $h(r)$ is unknown and spatially disordered, its autocorrelation is approximately shift-invariant under the memory effect, allowing the approximation

$$I(x) \star I(r) \approx O(r) \star O(r). \tag{4.18}$$

This statistical relation forms the basis for reconstructing object features via phase retrieval algorithms, such as the Gerchberg-Saxton method or the ptychographic iterative engine, which estimate both amplitude and phase in the speckle domain. These algorithms are particularly advantageous in scenarios where coherent detection, reference beams, or interferometric calibration are infeasible [40].

A representative implementation involves recording two speckle images at distinct object positions behind the scattering layer. The temporal difference between these frames suppresses static background contributions and isolates the modulation introduced by object motion. The autocorrelation of the resulting difference image provides an estimate of the object's intensity autocorrelation, which is then used as input for phase retrieval to recover the spatial structure. This memory-effect-based computational framework facilitates robust imaging in dynamic or non-line-of-sight conditions, where traditional wavefront shaping becomes impractical.

Despite its inherent limitations in field of view due to the angular extent of the memory effect, several strategies, such as regional tiling, adaptive sliding windows, and sparse-domain decompositions, have been developed to extend its applicability. Furthermore, a related technique, speckle cross-correlation imaging (SCC), leverages statistical consistency across multiple frames or object realizations to enhance noise robustness and reconstruction stability. This class of correlation-based imaging has shown promising results in scenarios involving surface-scattering media like ground glass, paper, or biological tissue, and is gaining traction in applications ranging from microscopy to security surveillance.

4.4.3 Speckle Deconvolution and Blind Reconstruction

Under the assumption of a spatially invariant PSF, imaging through scattering media can be mathematically formulated as a convolutional degradation model, where the observed speckle image results from the convolution between the object intensity distribution and PSF. In principle, this formulation enables the use of deconvolution techniques for image recovery. However, in scattering environments, PSF is typically unknown, inaccessible, or highly dynamic, rendering conventional deconvolution approaches ineffective.

This scenario gives rise to a blind reconstruction problem, in which neither PSF nor the object is known a priori [41–43]. To address this, blind deconvolution strategies aim to jointly estimate both the object and the convolution kernel from the observed data alone. This inverse problem is severely ill-posed and requires the integration of statistical priors, structural constraints, and iterative optimization algorithms to ensure solution uniqueness and stability. The generic blind deconvolution model is expressed as

$$I = O * h + \eta, \tag{4.19}$$

where I denotes the recorded speckle image, O is the unknown object, h is the unknown PSF, and η represents measurement noise. The corresponding optimization objective can be formulated as

$$\min_{O,h} \; \mathcal{L}(I, O * h) + \lambda_1 R_1(O) + \lambda_2 R_2(h), \tag{4.20}$$

where $L(\cdot)$ is a data fidelity term, typically the squared $\ell 2$-norm or a robust divergence under noise modeling, $R_1(O)$ and $R_2(h)$ denote object- and kernel-specific regularizations. Common choices for R_1 include total variation, ℓ_1-sparsity in wavelet or dictionary domains, or low-rank matrix priors for structured objects. For R_2, constraints such as non-negativity, compact support, or energy normalization (e.g., $\|h\|_1 = 1$) are used to suppress trivial or degenerate solutions.

Algorithmically, blind deconvolution is typically solved via alternating minimization, in which O and h are updated in turns while fixing the other, exploiting convexity in individual subproblems [41]. Advanced approaches further integrate Bayesian inference frameworks, where prior distributions over O and h are imposed, and posterior inference is performed via expectation-maximization, variational Bayes, or Markov Chain Monte Carlo sampling. In particular, when noise follows Gaussian or Poisson statistics, maximum a posteriori estimation leads to analytically tractable objective functions. The blind deconvolution problem can also be interpreted in a variational inference setting, where an approximate posterior over latent variables is optimized using evidence lower bounds.

From a statistical physics perspective, the scattering PSF h can be modeled as a random process arising from multi-path interference, whose spatial properties are governed by ensemble statistics such as speckle correlation width and angular memory effect [44]. In this context, stochastic priors informed by speckle autocorrelation properties may be incorporated into R_2, enabling physics-informed optimization. For high-dimensional or large-field imaging tasks, computational efficiency becomes critical. Efficient solvers exploit Fourier domain convolution, alternating direction method of multipliers frameworks, or learned proximal priors through neural networks. Recent advances in deep unrolling and plug-and-play priors have enabled real-time performance and superior generalization to unseen scattering conditions.

Importantly, blind deconvolution models exhibit inherent scale ambiguity, i.e., any solution pair $(\alpha O, h/\alpha)$ yields the same observation for any scalar $\alpha > 0$. Therefore, normalization constraints or data consistency priors are typically introduced to enforce identifiability. In scattering imaging, the blind deconvolution approach is particularly useful in non-calibrated systems, biological tissues with unknown scattering characteristics, and non-invasive setups where PSF estimation is either infeasible or unstable. Its flexibility allows integration with multi-frame fusion, sparse-view acquisition, or compressive measurement strategies, extending its applicability across diverse imaging modalities.

4.4.4 Trends in Scattering Light Manipulation Techniques

The evolution of scattering light manipulation techniques is progressively shaped by the integration of optical control, statistical modeling, and computational inference. As imaging in highly diffusive environments increasingly exceeds the capabilities of ballistic-photon-based methods, the emphasis shifts toward extracting information from multiply scattered light. Rather than treating scattering as a detrimental phenomenon, emerging approaches conceptualize it as an information-bearing process whose underlying structure can be quantitatively characterized and effectively inverted.

A central trend is the refinement of adaptive wavefront modulation strategies that incorporate physical priors and real-time feedback. Spatial light modulators and tunable metasurfaces enable precise spatiotemporal shaping of the incident field, allowing constructive interference to emerge even after multiple scattering events. When informed by statistical properties of the medium, such as angular correlations or speckle decorrelation dynamics, wavefront shaping becomes a targeted optimization process governed by coherent control rather than stochastic trial.

Parallel advances are being made in inverse modeling frameworks that jointly infer object and medium parameters from scattered measurements. Radiative transport models are increasingly integrated into differentiable solvers and physics-informed learning architectures, enabling end-to-end reconstruction pipelines that generalize across varying media without requiring explicit calibration. These hybrid formulations combine the interpretability of forward models with the adaptability of data-driven inference, offering enhanced robustness in heterogeneous or poorly characterized environments.

Hardware development is also contributing to this transformation. The incorporation of ultrafast detectors, on-chip interferometers, and photonic-integrated circuits provides precise temporal resolution and compact signal acquisition platforms. When coupled with computational frameworks based on compressive sensing, low-rank priors, or manifold learning, these systems achieve real-time, scattering-resilient imaging suitable for non-line-of-sight, deep-tissue, or minimally invasive scenarios.

Overall, the field is transitioning toward a unified paradigm in which scattered light is no longer rejected or averaged out, but actively modulated, modeled, and mined for latent structural information. This conceptual shift positions scattering-aware imaging not merely as a set of compensatory techniques, but as a foundational modality in the broader domain of computational optics.

4.5 Computational Reconstruction Method

With the continuous advancement of computational power and imaging technologies, computational optics has emerged as a vital subfield within scattering imaging. By integrating advanced mathematical modeling, numerical optimization, and efficient image processing algorithms, computational optics offers new paradigms and tools that transcend the limitations of conventional optical imaging. Through modeling the image degradation process and coupling it with numerical inversion strategies, this approach not only enables the reconstruction of high-resolution images but also facilitates noise suppression, distortion correction, and adaptive compensation in low SNR conditions.

In scattering imaging, the stochastic and complex nature of scattered light often renders traditional imaging techniques inadequate for reliable reconstruction. Computational optics addresses these challenges by formulating the forward imaging process into tractable mathematical models, thereby transforming classical optical problems into solvable computational tasks. Techniques such as inverse problem solving, compressive sensing, sparse representation, and optimization-based reconstruction have proven effective in recovering high-fidelity images from severely degraded data. Importantly, this computational framework enables the extraction of ballistic photon information and the recovery of structural details under multiple scattering and low-photon conditions, greatly extending the functional capacity of optical imaging systems in complex environments.

4.5.1 Inverse Problem Solution

In conventional imaging systems, the relationship between the object and the acquired image can often be modeled as a linear or quasi-linear process, typically represented by a convolution operation involving the object and PSF. However, in scattering environments, where light undergoes multiple random perturbations due to complex interactions with the medium, this assumption no longer holds. The image formation process becomes highly nonlinear and spatially variant, rendering traditional linear models inadequate for faithful reconstruction.

To recover the underlying object structure from such degraded observations, the image reconstruction task is reformulated as an inverse problem. This involves constructing a forward degradation model that accounts for light transport through the scattering medium, the statistical or physical properties of the medium, and the spatial response of the imaging system. The inverse problem is then solved using optimization algorithms to infer the most plausible object that could have generated the observed data [45, 46].

One classical approach to inverse problem solving is the least-squares estimation, which aims to minimize the discrepancy between the predicted and observed images under the assumed forward model. While effective in noise-free or weakly degraded scenarios, least-squares methods often struggle with the severe distortions and high noise levels characteristic of scattering-based imaging. Their performance is particularly limited in cases where the system is ill-posed or underdetermined.

To address these limitations, Bayesian inference offers a more flexible and robust framework by incorporating prior information into the reconstruction process. Instead of minimizing a deterministic loss function, Bayesian methods treat the unknown object and possibly the forward model parameters as random variables, and seek to estimate the posterior distribution given the observed data. The maximum a posteriori estimation then provides a statistically grounded solution that integrates both the fidelity to measurements and the regularization imposed by the prior. Commonly employed priors include sparsity, smoothness, or total variation, which effectively suppress noise and guide the solution toward physically plausible reconstructions. This probabilistic formulation enhances resilience to noise, compensates for model uncertainty, and enables principled handling of inverse problems in highly scattering conditions.

4.5.2 Compressive Sensing Reconstruction

Compressive sensing (CS) has emerged as a powerful theoretical and algorithmic framework in signal processing and computational imaging. Its central premise is that a signal, if sparse or compressible in some transform domain, can be accurately reconstructed from far fewer measurements than required by the Nyquist sampling criterion. This principle is particularly well-suited to scattering imaging, where the recorded data often exhibit significant sparsity when expressed in appropriate bases such as wavelets, Fourier, or curvelets.

In scattering scenarios, the target image can often be approximated by a sparse representation, allowing the use of randomized or nonuniform sampling strategies during data acquisition. By exploiting this inherent sparsity, CS enables substantial reduction in the number of measurements while preserving essential structural information. During reconstruction, CS algorithms solve an optimization problem that recovers the high-resolution image by promoting sparsity and fidelity to the measurements [47, 48].

Beyond reducing data acquisition requirements and hardware complexity, CS also enhances noise robustness and allows for efficient imaging under photon-limited or bandwidth-constrained conditions. Its advantages become even more pronounced in high-dimensional or large-scale imaging tasks, such as volumetric or hyperspectral scattering imaging, where exhaustive sampling is prohibitively expensive or technically infeasible. However, the performance of CS heavily relies on the validity of the sparsity assumption. In complex scattering environments where this assumption may not strictly hold, the reconstruction quality can deteriorate due to model mismatch.

Sparse reconstruction methods extend the capabilities of CS by integrating additional priors and advanced regularization strategies to better capture image structure. These methods typically employ regularization techniques, such as L1-norm minimization, L2-norm smoothing, or total variation constraints, to enforce sparsity while simultaneously suppressing noise and artifacts. Compared to standard CS, sparse reconstruction techniques offer improved fidelity in reconstructing fine details and sharp edges, and are better suited for structured images or those with piecewise smooth features.

By incorporating domain-specific priors and leveraging advanced optimization algorithms, sparse reconstruction has become an indispensable component of modern computational scattering imaging pipelines. It bridges the gap between underdetermined measurements and high-quality image estimation, contributing significantly to practical implementations of low-sampling, high-resolution imaging systems.

4.5.3 Reconstruction Optimization Algorithm

Optimization algorithms play a central role in computational optics, serving as indispensable tools for image recovery in scattering environments. Given the inherently nonlinear and ill-posed nature of scattering-induced image degradation, conventional analytical inversion methods often fail to yield satisfactory results. Iterative optimization techniques tailored to the specific complexities of scattering models have become essential for achieving high-fidelity reconstructions [49, 50].

A broad class of optimization algorithms has been adopted for this purpose, including gradient descent, conjugate gradient, Newton method, and quasi-Newton schemes. These algorithms iteratively refine the estimated image by minimizing a cost function that quantifies the discrepancy between the forward model prediction and the measured data. In scattering imaging, commonly used loss functions include the L2-norm for Gaussian noise models, L1-norm for robustness to outliers, and composite metrics incorporating image priors such as sparsity, smoothness, or edge preservation.

Gradient descent remains the most widely used baseline method due to its simplicity and ease of implementation. By computing the gradient of the objective function with respect to the image estimate, it incrementally updates the solution along the direction of steepest descent. In scattering applications, this technique is often embedded within iterative phase retrieval or deconvolution frameworks to progressively correct for multiple scattering effects. However, standard gradient descent suffers from slow convergence and susceptibility to local minima, especially in high-dimensional non-convex landscapes.

To address these limitations, accelerated gradient methods have been developed, notably Nesterov accelerated gradient and the Fast Iterative Shrinkage-Thresholding Algorithm. These algorithms introduce momentum terms that exploit curvature information or extrapolated updates, significantly speeding up convergence without sacrificing stability. In scattering image reconstruction, such accelerations not only reduce computation time but also improve reconstruction fidelity, particularly when integrated with regularization strategies that incorporate prior knowledge of the object structure.

By embedding optimization algorithms within model-based or data-driven frameworks, computational scattering imaging can efficiently recover complex scene information, even under severe noise or incomplete data conditions. As algorithmic development continues to evolve, especially in conjunction with parallel computation and GPU acceleration, optimization-based image reconstruction will remain a foundational component of high-performance scattering imaging systems.

4.6 General Conclusions

This chapter systematically explores the physical mechanisms, statistical properties, and computational strategies related to optical speckle in scattering environments. By reconsidering speckle not as noise but as an information carrier, diverse imaging methodologies are developed to extract structural details from multiply scattered light. Through wavefront modulation, correlation analysis, and statistical modeling, speckle fields are manipulated to enable imaging without direct line-of-sight or ballistic photon reliance. Computational frameworks, ranging from inverse problem formulations and compressed sensing to deep learning, further enhance reconstruction quality under low signal-to-noise and strong scattering conditions. Together, these advances establish optical speckle as a central tool in scattering-resilient imaging, bridging physical optics and data-driven computation for robust scene recovery in complex media.

Acknowledgments This work was supported by the National Natural Science Foundation of China (Nos. 12374271, 62375065, 62405079, 62575091); Fundamental Research Funds for the Central Universities (HIT.OCEF.2024055, HIT.DIJJ.2025037, AUGN5750300325); The Fund of National Laboratory on Adaptive Optics, China (FNLAO-24-MS-O07); Natural Science Foundation of Heilongjiang Province of China (LH2024A009); Key Laboratory of Image Processing and Intelligent Control (Huazhong University of Science and Technology), Ministry of Education, Wuhan, China (IPIC2025-05); Key Laboratory of Advanced Medical Imaging and Intelligent Computing of Guizhou Province (AMIIC250204); the China Postdoctoral Science Foundation-CCTEG Joint Support Program (2025T011ZGMK).

References

1. Yoon S, Kim M, Jang M, Choi Y, Choi W, Kang S, Choi W (2020) Deep optical imaging within complex scattering media. Nat Rev Phys 2(3):141–158
2. Sdobnov A, Piavchenko G, Bykov A, Meglinski I (2024) Advances in dynamic light scattering imaging of blood flow. Laser Photon Rev 18(2):2300494
3. Wu H, Zhao M, Xu W (2020) Underwater de-scattering imaging by laser field synchronous scanning. Opt Lasers Eng 126:105871
4. Singh D, Kumar V (2019) A comprehensive review of computational dehazing techniques. Arch Comput Methods Eng 26(5):1395–1413
5. Faccio D, Velten A, Wetzstein G (2020) Non-line-of-sight imaging. Nat Rev Phys 2(6):318–327
6. Bertolotti J, Katz O (2022) Imaging in complex media. Nat Phys 18(9):1008–1017
7. Dorodnyy A, Smajic J, Leuthold J (2023) Mie scattering for photonic devices. Laser Photon Rev 17(9):2300055
8. Radford J, Lyons A, Tonolini F, Faccio D (2020) Role of late photons in diffuse optical imaging. Opt Express 28(20):29486–29495
9. Re R, Spinelli L, Martelli F, Di Sieno L, Bargigia I, Amendola C, Maffeis G, Torricelli A (2025) A review on time domain diffuse optics: principles and applications on human biological tissues. La Rivista del Nuovo Cimento 1–83
10. Popoff SM, Lerosey G, Carminati R, Fink M, Boccara AC, Gigan S (2010) Measuring the transmission matrix in optics: an approach to the study and control of light propagation in disordered media. Phys Rev Lett 104(10):100601
11. Dahan M, Laurence T, Pinaud F, Chemla DS, Alivisatos AP, Sauer M, Weiss S (2001) Time-gated biological imaging by use of colloidal quantum dots. Opt Lett 26(11):825–827
12. Duncan MD, Mahon R, Tankersley LL, Reintjes J (1991) Time-gated imaging through scattering media using stimulated Raman amplification. Opt Lett 16(23):1868–1870
13. Kadobianskyi M, Papadopoulos IN, Chaigne T, Horstmeyer R, Judkewitz B (2018) Scattering correlations of time-gated light. Optica 5(4):389–394
14. Di Sieno L, Bettega G, Berger M, Hamou C, Aribert M, Mora AD, Puszka A, Grateau H, Contini D, Hervé L, Coll JL, Dinten JM, Pifferi A, Planat-Chrétien A (2016) Toward noninvasive assessment of flap viability with time-resolved diffuse optical tomography: a preclinical test on rats. J Biomed Opt 21(2):025004
15. Baek WJ, Park J, Gao L (2024) Depth-resolved imaging through scattering media using time-gated light field tomography. Opt Lett 49(22):6581–6584
16. Huang D, Swanson EA, Lin CP, Schuman JS, Stinson WG, Chang W, Hee MR, Flotte T, Gregory K, Puliafito CA, Fujimoto JG (1991) Optical coherence tomography. Science 254(5035):1178–1181
17. Tearney GJ, Brezinski ME, Bouma BE, Boppart SA, Pitris C, Southern JF, Fujimoto JG (1997) In vivo endoscopic optical biopsy with optical coherence tomography. Science 276(5321):2037–2039
18. Bouma BE, de Boer JF, Huang D, Jang IK, Yonetsu T, Leggett CL, Leitgeb R, Sampson DD, Suter M, Vakoc BJ, Villiger M, Wojtkowski M (2022) Optical coherence tomography. Nat Rev Methods Primers 2(1):79
19. Zhang X, Li Z, Nan N, Wang X (2022) Denoising algorithm of OCT images via sparse representation based on noise estimation and global dictionary. Opt Express 30(4):5788–5802
20. Akca BI, Považay B, Alex A, Wörhoff K, De Ridder RM, Drexler W, Pollnau M (2013) Miniature spectrometer and beam splitter for an optical coherence tomography on a silicon chip. Opt Express 21(14):16648–16656
21. Li X, Han Y, Wang H, Liu T, Chen SC, Hu H (2022) Polarimetric imaging through scattering media: a review. Front Phys 10:815296
22. Gao C, Zhao M, Cao F, Wang Z, Lu D, Hu Y, Dou J, Dai J (2023) Underwater polarization de-scattering imaging based on orthogonal polarization decomposition with low-pass filtering. Opt Lasers Eng 170:107796

23. Lv S, Zhang W, Zhang M, Zia A, Man T, Zhou H, Wan Y (2025) Parameter optimization based polarization difference fusion imaging method for underwater turbid medium. Opt Laser Technol 180:111465
24. Seow KLC, Török P, Foreman MR (2020) Single pixel polarimetric imaging through scattering media. Opt Lett 45(20):5740–5743
25. Raveendran S, Patil MD, Birajdar GK (2021) Underwater image enhancement: a comprehensive review, recent trends, challenges and applications. Artif Intell Rev 54:5413–5467
26. Cho M, Javidi B (2016) Peplography: a passive 3D photon counting imaging through scattering media. Opt Lett 41(22):5401–5404
27. He F, Liu R, Tian X (2023) Scanning-based photon-limited imaging through scattering media. Opt Lasers Eng 161:107388
28. Li ZP, Ye JT, Huang X, Jiang PY, Cao Y, Hong Y, Yu C, Zhang J, Zhang Q, Peng CZ, Xu F, Pan JW (2021) Single-photon imaging over 200 km. Optica 8(3):344–349
29. Tobin R, Halimi A, McCarthy A, Laurenzis M, Christnacher F, Buller GS (2019) Three-dimensional single-photon imaging through obscurants. Opt Express 27(4):4590–4611
30. Lee J, Ingle A, Chacko JV, Eliceiri KW, Gupta M (2023) CASPI: collaborative photon processing for active single-photon imaging. Nat Commun 14(1):3158
31. Cao H, Mosk AP, Rotter S (2022) Shaping the propagation of light in complex media. Nat Phys 18(9):994–1007
32. Yeminy T, Katz O (2021) Guidestar-free image-guided wavefront shaping. Sci Adv 7(21):eabf5364
33. Haim O, Boger-Lombard J, Katz O (2025) Image-guided computational holographic wavefront shaping. Nat Photon 19(1):44–53
34. Fan W, Chen Z, Yakovlev VV, Pu J (2021) High-fidelity image reconstruction through multimode fiber via polarization-enhanced parametric speckle imaging. Laser Photon Rev 15(5):2000376
35. Baek YS, de Aguiar HB, Gigan S (2023) Phase conjugation with spatially incoherent light in complex media. Nat Photon 17(12):1114–1119
36. Tajahuerce E, Durán V, Clemente P, Irles E, Soldevila F, Andrés P, Lancis J (2014) Image transmission through dynamic scattering media by single-pixel photodetection. Opt Express 22(14):16945–16955
37. Paniagua-Diaz AM, Starshynov I, Fayard N, Goetschy A, Pierrat R, Carminati R, Bertolotti J (2019) Blind ghost imaging. Optica 6(4):460–464
38. Zhao X, Nie X, Yi Z, Peng T, Scully MO (2022) Imaging through scattering media via spatial-temporal encoded pattern illumination. Photon Res 10(7):1689–1694
39. Wang D, Sahoo SK, Zhu X, Adamo G, Dang C (2021) Non-invasive super-resolution imaging through dynamic scattering media. Nat Commun 12(1):3150
40. Shi Y, Sheng W, Fu Y, Liu Y (2023) Overlapping speckle correlation algorithm for high-resolution imaging and tracking of objects in unknown scattering media. Nat Commun 14(1):7742
41. Lane RG (1992) Blind deconvolution of speckle images. J Opt Soc Am A 9(9):1508–1514
42. Antipa N, Kuo G, Heckel R, Mildenhall B, Bostan E, Ng R, Waller L (2017) DiffuserCam: lensless single-exposure 3D imaging. Optica 5(1):1–9
43. Pan L, Shen Y, Qi J, Shi J, Feng X (2023) Single photon single pixel imaging into thick scattering medium. Opt Express 31(9):13943–13958
44. Bao G, Li P, Lin J, Triki F (2015) Inverse scattering problems with multi-frequencies. Inverse Probl 31(9):093001
45. Borden B (2001) Mathematical problems in radar inverse scattering. Inverse Probl 18(1):R1
46. Colton D, Kress R (2006) Using fundamental solutions in inverse scattering. Inverse Probl 22(3):R49
47. Durán V, Soldevila F, Irles E, Clemente P, Tajahuerce E, Andrés P, Lancis J (2015) Compressive imaging in scattering media. Opt Express 23(11):14424–14433
48. Anselmi N, Oliveri G, Hannan MA, Salucci M, Massa A (2017) Color compressive sensing imaging of arbitrary-shaped scatterers. IEEE Trans Microw Theory Tech 65(6):1986–1999

49. Yang CX, Zhang J, Tong MS (2021) A hybrid quantum-behaved particle swarm optimization algorithm for solving inverse scattering problems. IEEE Trans Antennas Propag 69(9):5861–5869
50. Woo CM, Zhao Q, Zhong T, Li H, Yu Z, Lai P (2022) Optimal efficiency of focusing diffused light through scattering media with iterative wavefront shaping. APL Photon 7(4):046109

Chapter 5
Moving Images for Velocity Measurement

Longhui Fu, Zhengjun Liu, and Yutong Li

Abstract The chapter introduces a method for measuring velocity in moving images. The velocity calculation encompasses the trajectories of target motion within dynamic image sequences, extracting motion information from temporally continuous inter-frame variations. Key technologies involved include feature extraction, speed modeling, and noise suppression.

5.1 Introduction

Classic speed measurement methods include inertial, magnetic, and capacitive sensors, as well as lasers and radars. These techniques are often invasive and come with disadvantages such as high costs and bulky equipment. Measuring the speed of an object using video or image sequences captured by imaging devices has emerged as a new trend, known as "moving images velocity measurement". This technology can be categorized into two schemes based on the implementation methods: Particle Image Velocimetry (PIV) and Blurred Image Velocimetry (BIV). Due to the differences in their underlying principles, these two methods exhibit variations in application, performance, efficiency, and development.

L. Fu · Z. Liu · Y. Li (✉)
School of Physics, Harbin Institute of Technology, Harbin, China
e-mail: yutong_li@hit.edu.cn

L. Fu
e-mail: 1224572137@qq.com

Z. Liu
e-mail: zjliu@hit.edu.cn

Z. Liu and Y. Li (eds.), *Intelligent Analysis of Optical Images*, Scientific Computation,
https://doi.org/10.1007/978-981-95-3720-4_5

5.2 Particle Image Velocimetry

PIV is a method for measuring speed that utilizes cameras. This technique is grounded in the optical flow method, which captures sequences of images at high frame rates and calculates the projections of the same particle points from consecutive frames. The velocity is determined based on the dynamic information represented by the minute spatial displacements observed in these projections. PIV technology is particularly well-suited for non-contact measurements of fluids, including fog and water flow.

5.2.1 Basic Principle

PIV is a technique grounded in flow visualization technology. It involves injecting tracer particles into the fluid and capturing rapid sequential images using imaging equipment. This process allows for the correlation analysis of the tracer particle positions, thereby providing valuable in-field motion information [1–4]. The two-dimensional projected velocity of particles within a fluid is defined as follows:

$$v_x = \frac{\mathrm{d}x(t)}{\mathrm{d}t} \approx \frac{x(t+\Delta t) - x(t)}{\Delta t}, \; v_y = \frac{\mathrm{d}y(t)}{\mathrm{d}t} \approx \frac{y(t+\Delta t) - y(t)}{\Delta t} \tag{5.1}$$

where $x(t)$ and $y(t)$ represent the positions of the particles in the image at time t, and their positions in the next frame of the image are $x(t+\Delta t)$ and $y(t+\Delta t)$ respectively.

Particles are propelled within a flow field. When the particles are sufficiently lightweight and exhibit adequate followability to the flow, their movement can accurately reflect the velocity of the flow field. High-energy lasers are employed to illuminate the particles dispersed throughout the flow field, allowing visualization of the flow dynamics by observing the particles' movement. PIV technology utilizes a camera to capture the motion of the particles and applies image processing algorithms to calculate the displacement of particles between two consecutive frames, thereby determining the velocity of the flow field. Figure 5.1 presents a typical schematic diagram of a PIV system, which includes:

(1) Tracer Particle Generator: This device is used to emit tiny particles that can trace and move within the flow field.
(2) Lighting Equipment: Generally, this refers to a high-energy pulsed laser that can achieve an extremely short interval between two frames. This type of laser is capable of controlling the interval between two emissions at the sub-microsecond level. The emitted sheet-shaped laser can illuminate a specific plane within the flow field, facilitating the measurement of a two-dimensional flow field.

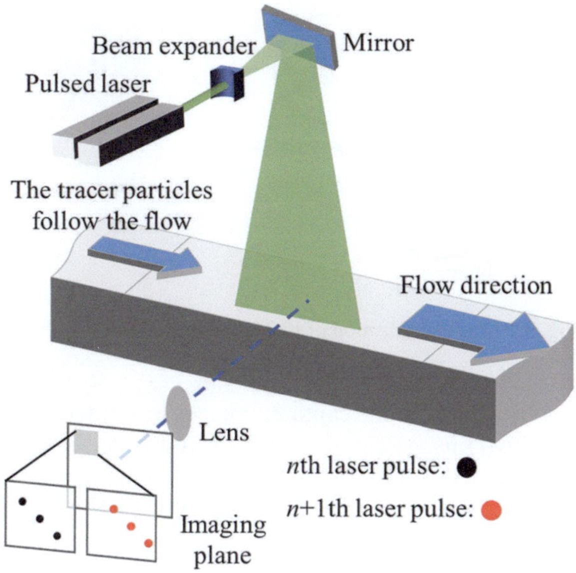

Fig. 5.1 Schematic diagram of the PIV system

(3) Recording Equipment: Typically, this involves a high-speed camera or a camera specifically designed for PIV technology, capable of rapidly capturing two frames of images.
(4) The synchronizer facilitates the simultaneous activation of both the camera and the laser, ensuring that the camera effectively captures the particles scattered by the laser.
(5) Calibration: A standard for assessing particle movement.

PIV can be categorized into two primary methods: Particle tracking velocimetry (PTV) and Particle correlation velocimetry (PCV). PTV focuses on tracking individual particles, as Fig. 5.2a [5]. This method generally assumes that the local velocity gradient is comparable to the particle motion and that the displacement of particles between images does not exceed the average interval, making it effective in sparse fields. In contrast, PCV is more suitable for dense fields, as Fig. 5.2b. Experimental results indicate that while PTV yields more accurate calculations, its matching efficiency is relatively low. Conversely, although the overall accuracy of PCV is lower, it provides a more comprehensive coverage of the entire field.

5.2.2 *The Development of PIV*

(1) Relaxation filtering technology based on PIV: Unlike classical PIV, which captures two projections of the same particle, this technology captures only the motion-blurred image of a single particle projection. A linear spatial filter

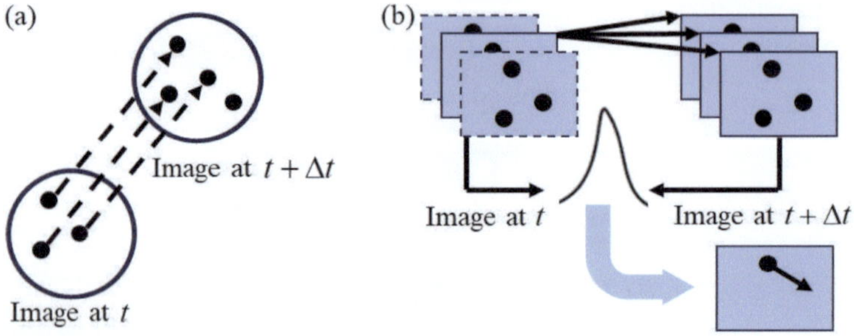

Fig. 5.2 The classification of PIV technology: **a** PTV, **b** PCV

is applied to the captured images to extract the persistent states of motion blur. Two images are generated using the detected persistent positions as input for the PIV algorithm.

The angle between [−45°, 45°] [−45°, 45°] is sufficient for filtering motion blur. To detect these consistently blurred positions, two spatial filters are applied to the input image as

$$S_X^+ = \begin{bmatrix} 1 & 2 & -i & -2 & -1 \\ 4 & 8 & -i & -8 & -4 \\ 6 & 12 & 0 & -12 & -6 \\ 4 & 8 & -i & -8 & -4 \\ 1 & 2 & -i & -2 & -1 \end{bmatrix} \tag{5.2}$$

$$S_X^- = \begin{bmatrix} -1 & -2 & -i & 2 & 1 \\ -4 & -8 & -i & 8 & 4 \\ -6 & -12 & 0 & 12 & 6 \\ -4 & -8 & -i & 8 & 4 \\ -1 & -2 & -i & 2 & 1 \end{bmatrix} \tag{5.3}$$

The two equations above represent the kernels of spatial filters S_X^+ and S_X^-, which consist of vertical Sobel operators modified by the parameter i. This parameter facilitates the suppression of the vertical component of particle trajectories that are not entirely horizontal. Figure 5.3 illustrates the effects of motion blur at various angles. In contrast to the Sobel result, which is effective only for horizontal blur, the result for filter S_X also accounts for blur separation at a 45° angle. The suppression factor i serves as a measure of the sensitivity of filter, and its optimal value is contingent upon the maximum occurrence angle, the modulation transfer function, and the luminance of both the particles and the background.

The relaxation filtering technology based on PIV must effectively suppress the vertical components of motion blur in the scene, maximize the number of

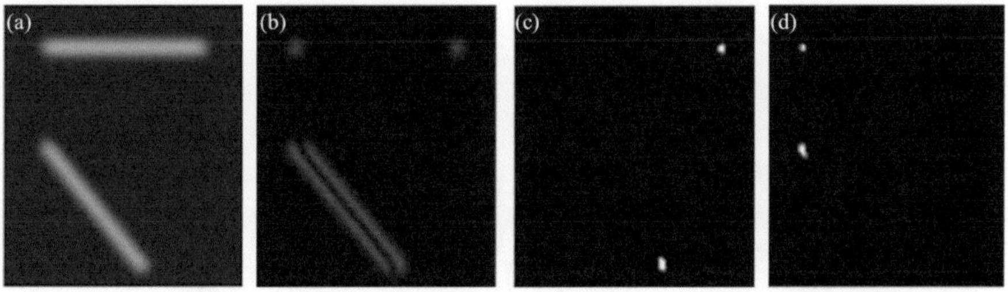

Fig. 5.3 Motion blur effects from various angles: **a** motion blur at 0° and 45° angles, **b** the result of vertical sobel, **c** the result of S_X^+, **d** the result of S_X^-

persistent positions, and select the smallest possible suppression factor based on the worst-case scenario.

(2) Cell segmentation PIV technology (CSPIV) [6]: This method consists of three steps. Firstly, the inquiry field is divided into units based on the matching rate and statistical correlation of point measurements; Secondly, the single-point matching method is applied to track the smaller particle swarm within the unit, and the best velocity estimation is found through the competition process among these sub-units. The motion of the final winning unit is used to represent the velocity vector field.

This technology identifies a broad correlation region between the first and second frames, which is approximately assessed using the inquiry window and is referred to as the "parent space". Its size is related to the maximum dimensions of the inquiry window and encompasses the initial unit set. During the experimental process, the parent space can be considered the relevant area. To enhance accuracy, normalized covariance correlation was applied to each cell in the two images as

$$r(k,l) = \frac{\sum_{i=0}^{M}\sum_{j=0}^{N}\left[f(i,j) - f_m\right]\left[g(i+k,j+l) - g_m\right]}{\sqrt{\sum_{i=0}^{M}\sum_{j=0}^{N}\left[f(i,j) - f_m\right]^2 \sum_{i=0}^{M}\sum_{j=0}^{N}\left[f(i,j) - g_m\right]^2}} \tag{5.4}$$

The relevant results can be represented as a matrix, CorrVal, of size $l \times c$ as

$$\mathrm{CorrVal}(\mathrm{MotherSpace}(\Delta t)) = \begin{bmatrix} r_{11} & r_{12} & \cdots & r_{1c} \\ r_{21} & r_{22} & \cdots & r_{2c} \\ r_{31} & r_{32} & \cdots & r_{3c} \\ \vdots & \vdots & \vdots & \vdots \\ r_{l1} & r_{l2} & \cdots & r_{lc} \end{bmatrix} \tag{5.5}$$

where f represents the gray value distribution of the evaluation sample from the first frame image, confined within a rectangular inquiry window of size $M \times N$. The parameter f_m is the average gray value of the evaluation area of the first

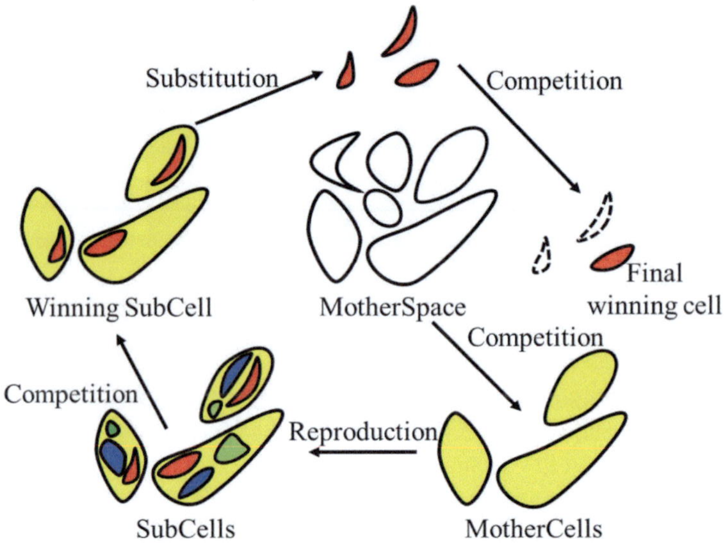

Fig. 5.4 The model of CSPIV

frame image. The parameters g and g_m represent the gray value distribution and the average gray value of the second frame image, respectively.

In a given area, there are typically multiple cells, collectively referred to as Cell_i ($i = 1, 2, 3, \ldots, M_c$), where M_c represents the total number of cells. Utilizing the gray-scale distribution and coordinate information from the particle image, Eq. (5.5) is employed to select the cells. Initially, all cells in the mother space are selected based on a threshold fitness value $T_{\text{threshold}}$, which can either be a fixed value or an adaptive value depending on the specific mother space. If the value of r_{ij} is greater than or equal to $T_{\text{threshold}}$, the corresponding cells are retained. The retained cells then continue to compete until only one cell remains, as Fig. 5.4.

Cells selected from the mother space by $T_{\text{threshold}}$ belong to the set labeled MotherCell_i ($i = 1, 2, 3, \ldots, N$). A mother cell is a matching set that contains both correct and incorrect matching points. A mother cell is generated in the mother space and selected based on a threshold.

When the initial competition concludes, the mother cells are regarded as being in a balanced state. To identify the most compatible mother cell among all candidates, it is essential to understand the internal matching information of the mother cells and to segment each internal element within the matching region. The point matching function ContriRatio(·) quantifies the degree of correlation between each point as

$$\text{ContriRatio}(x) = \frac{2f_x g_x}{f_x^2 + g_x^2} \tag{5.6}$$

where f_x and g_x represent the gray values of the two corresponding points at position x. Function ContriRatio(·) begins to be calculated when the mother cell initiates division. Each mother cell divides into two daughter cells as

$$\text{MotherCell}_i = \{\text{SubCell}_{1i}, \text{SubCell}_{2i}\}, \quad i \in \{1, 2, \ldots, N\} \tag{5.7}$$

$$\text{SubCell}_{1i} = \{x_{1i}, x_{2i}, x_{3i}, \ldots, x_{ni}\} = A_i, \quad i \in \{1, 2, \ldots, N\} \tag{5.8}$$

$$\text{SubCell}_{1i} = \{x_{1i}, x_{2i}, x_{3i}, \ldots, x_{ni}\} = A_i, \quad i \in \{1, 2, \ldots, N\} \tag{5.9}$$

where SubCell_{1i} and SubCell_{2i} represent the segmentation results of the i th mother cell, while x_{ji} and y_{ji} denote the coordinates of the j th point associated with subcells 1 and 2, respectively. The selection of elements in subcellular environments is as follows

$$\begin{cases} \frac{\text{ContriRatio}(x_{ji})}{\text{CorrVal}(\text{MotherCell}_i)} \geq 1, \ j = 1, 2, 3, \ldots, n \\ \frac{\text{ContriRatio}\left(y_j^i\right)}{\text{CorrVal}(\text{MotherCell}_i)} < 1, \ j = 1, 2, 3, \ldots, m \end{cases} \tag{5.10}$$

(3) 3D PIV technology (3D-PIV): A three-dimensional particle image velocimetry technology based on correlation image sensors (CIS), captures the trajectory of tracer particles over time. This is achieved by analyzing the phase distribution related to a reference signal and the time correlation between the output intensity of the incident light and the reference signal.

By utilizing a CIS with an intensity-modulated light source, the information to be extracted is encoded in the amplitude and phase of the light intensity, which can be demodulated at each pixel of the image sensor. By employing spatially modulated light sources to illuminate particles and utilizing CIS imaging technology with three reference signals, both the temporal and positional information of the particles can be embedded into each pixel of the particle trajectory. This approach enables the estimation of flow field velocity for a single target within a single frame.

To achieve three-dimensional particle image velocimetry, it is essential not only to measure the time of the trajectory but also to utilize CIS to estimate the position in the depth direction. A spatio-temporal phase modulation light source is introduced to determine the depth position of the trajectory. Spatio-temporal phase modulation is accomplished by projecting light through a rotating disk that contains a radial lattice pattern, as Fig. 5.5.

Let the angular frequency of the disk be denoted as ω_0, the number of grid stripes as N, and the stripe frequency as $N\omega_0$. Consequently, the light intensity of the projected pattern can be expressed as follows:

$$I(t, \theta) = A\cos(N\omega_0 t + N\theta) + B \tag{5.11}$$

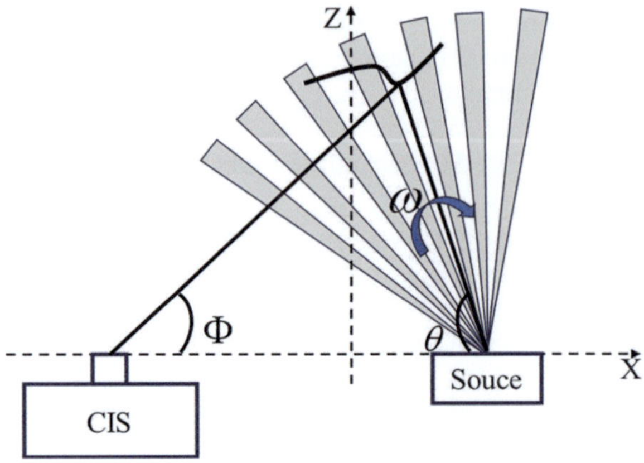

Fig. 5.5 The structure of spatio-temporal phase modulation

where A represents the amplitude of the projected pattern, B denotes the component of ambient light, and θ indicates the projection angle of the light source.

The position in the depth direction is estimated by demodulating the projection angle and utilizing the triangulation method. When imaging moving particles illuminated by spatiotemporally modulated light, the resulting image can be expressed as follows

$$\begin{aligned} g_\omega(x, y) &= \int_{-T/2}^{T/2} \frac{A}{2} e^{\mathrm{i}((N\omega_0-\omega)t+N\theta)} f'(x, y, t)\mathrm{d}t \\ &+ \int_{-T/2}^{T/2} \frac{A}{2} e^{-\mathrm{i}((N\omega_0-\omega)t+N\theta)} f'(x, y, t)\mathrm{d}t \\ &+ \int_{-T/2}^{T/2} Bf'(x, y, t)e^{\mathrm{i}\omega t}\mathrm{d}t \end{aligned} \tag{5.12}$$

where ω is the reference signal frequency input to the CIS. This indicates that the projection angle θ is encoded along the trajectory in phase $N\theta$. When demodulating the projection angle, the issue of phase uncertainty arises. The relevant images exhibit multiple identical phases, each generated by a different projection angle. The phase uncertainty problem is addressed by utilizing the frequencies of two reference signals and two distinct lattice patterns. A mixed pattern rotating disk is employed, featuring two different numbers of lattice stripes, denoted as N and M, respectively. The frequencies of the stripes are represented as $N\omega_0$ and $M\omega_0$. When the frequency difference between the two is sufficiently large, the mutual influence within each fringe can be mitigated. Consequently, two correlated images with the same phase corresponding to two angles of information can be obtained. Each difference image of the relevant phase corresponds to an image projected with $N - M$ lattice fringes. If there is only one fringe in the difference image of the phase, the phase uncertainty problem is resolved. Therefore, this method can estimate the position of the trajectory at each pixel, simultaneously obtain the time and position along the trajectory for each pixel, and estimate the three components of particle velocity at the pixel level.

PIV is widely utilized across various fields due to its non-contact measurement of fluid flow velocity. However, its effectiveness in handling motion-blurred images is limited, and its accuracy is constrained by the large number of grid points required and the high demand for data.

5.3 Blurred Image Velocimetry

Motion blur can be utilized to estimate and describe the kinematic properties of an object within a single image. When the imaging device and the subject being captured move relative to one another, a mismatch between the exposure time and the target's movement speed often results in a blurred effect. This blurring reduces the fidelity and resolution of the image, which is detrimental to extracting image information. However, these blurred images contain valuable kinematic data. By analyzing this information, one can calculate the direction and speed of the movement of target. Methods that facilitate the determination of target speed in this manner are referred to as BIV.

5.3.1 Basic Model of Motion Blur

Assuming the imaging distance is fixed and other blurring effects and aberrations are disregarded, motion blur [7] can be characterized as

$$g(x, y) = \iint I(\alpha, \beta)h(x - \alpha, y - \beta)\mathrm{d}\alpha\mathrm{d}\beta + n(x, y) \tag{5.13}$$

where $I(\alpha, \beta)$ represents the original image, $g(x, y)$ denotes the degraded image, $h(x - \alpha, y - \beta)$ refers to the point spread function (PSF), and $n(x, y)$ signifies the additive noise introduced during the blurring process.

When the PSF is spatially invariant, there is

$$g(x, y) = I(x, y) * h(x, y) + n(x, y) \tag{5.14}$$

If the direction and length of the movement can be determined, the PSF can be reconstructed when the exposure time is known. When the camera moves a certain distance L within the exposure time T, each point in the captured image frame is mapped onto several pixels of the degraded image, resulting in a blurred image along the direction of movement. The length L of the PSF is equal to the distance that the pixel shifts due to motion, and the angle θ represents the direction of motion relative to the x axis.

The angle and length parameters of the blurred kernel are respectively associated with the direction of motion and the amount of displacement. Both parameters should be used to reconstruct the image prior to down-sampling. While this is feasible under ideal conditions, it becomes more challenging to achieve with images captured in real-world scenarios due to the presence of additive noise, which interferes with the inverse filtering results of the Fourier transform, as

$$h(x, y) = \begin{cases} \frac{1}{L}, & 0 \le |x| \le L\cos\theta \text{ and } y = L\sin\theta \\ 0, & \text{otherwise} \end{cases} \tag{5.15}$$

The model described above is applicable only to global linear motion blur and cannot be directly utilized for local blur. In the case of local blurring, the background remains stationary while the target moves, resulting in blurring that occurs not only within the target but also at the boundary between the moving object and the stationary background. Consequently, the blurred image is characterized as follows

$$g(x, y) = \frac{1}{R}\int_0^R f(x - \rho\cos\theta, y - \rho\sin\theta)\mathrm{d}\rho \tag{5.16}$$

where x is in a state of global motion blur. The function $g(x, y)$ is also written as

$$g(x, y) = \frac{1}{R}\left\{(R - R')f_b(x, y) + \int_0^{R'} f(x - \rho\cos\theta, y - \rho\sin\theta)\mathrm{d}\rho\right\} \tag{5.17}$$

where $R' < R, f_b(x, y)$ is the unknown background at point (x, y). If (x, y) is not in the local motion blur, then $g(x, y)$ is equivalent to $f(x, y)$.

If the direction of motion θ is known, then Eq. (5.16) can be simplified to a one-dimensional form

$$g(x) = \frac{1}{R}\int_0^R f(x - \rho)\mathrm{d}\rho \tag{5.18}$$

Similarly, Eq. (5.17) can be simplified as

$$g(x) = \frac{1}{R}\left\{(R - R')f_b(R') + \int_0^{R'} f(x - \rho)\mathrm{d}\rho\right\} \tag{5.19}$$

Identifying the direction of motion is of vital importance. Consider horizontal linear motion in discrete cases without loss of generality as

$$(K + 1)g[i] = \sum_{j=0}^{i} f[j] + (K - i)f_b[i], \text{ for } i < K \tag{5.20}$$

and

$$(K + 1)g[i] = \sum_{j=0}^{\infty} f[i - j], \text{ for } i \geq K \tag{5.21}$$

where g represents the blurred image, f denotes the original image, and $f_b[i]$ is the value of the unknown background pixel at i. The value of $g[i]$ is determined by the average values on the right sides of Eqs. (5.20) and (5.21).

5.3.2 *The Estimation of Motion-Blur Parameter*

(1) Virtual Pixel Method: Since the proposed motion blur model will be utilized for speed detection, it is essential to estimate the blur parameters from the model. These parameters include the direction of motion, the "length" of the blurred motion, and the position of the object prior to the occurrence of motion blur. To estimate the motion blur parameters, the moving target is segmented from the static background, and the blur parameters are calculated for the segmented blurred target area. The first step involves identifying the direction of movement relative to the horizontal axis of the image. Due to the effects of movement, the image resolution primarily decreases in the direction of motion. The positive direction relative to the horizontal axis is defined by the following formula

$$\Delta f(i, j)_{[\theta \text{ degrees}]} = f(i', j') - f(i, j) \tag{5.22}$$

where $f(i', j')$ is a virtual pixel that forms an angle θ with pixel $f(i, j)$, and its intensity can be approximated using bilinear interpolation with adjacent pixels. The approximate derivative, $\Delta f(i, j)$, of the image in this direction can be characterized as follows

$$\begin{aligned}\Delta f(i, j)_{[\theta \text{ degrees}]} = &[1 - \tan(\theta)]f(i + 1, j) \\ &+ \tan(\theta)f(i + 1, j + 1) - f(i, j)\end{aligned} \tag{5.23}$$

Therefore, the direction of motion relative to the horizontal axis of the image θ can be determined by calculating the sum of the image derivatives as

$$\sum_{i=1}^{M-1}\sum_{j=1}^{N-1}\left|\Delta f(i,j)_{[\theta\ \mathrm{degrees}]}\right| \tag{5.24}$$

where M and N represent the number of rows and columns in the image, respectively.

Assuming that the estimated direction of motion forms an angle θ with the horizontal axis, the image can be corrected to produce a new representation of the object moving along the horizontal direction. The motion deblurring models (5.20) and (5.21) are employed to generate sequences of deblurred images, each containing varying amounts of blurred pixels. By calculating the modified sum, the Laplacian focusing metric assesses the blurred regions of each image. The number of blurred pixels associated with the focused image corresponds to the motion length.

(2) Cepstral analysis method: By taking advantage of the frequency-domain characteristics of motion blur to estimate the direction of motion. This is achieved by calculating the logarithmic power spectrum of local image blocks. The velocity is then estimated through cepstral analysis, and the overall estimation is finalized by detecting the period of the sinc function, which is related to the length of the motion blur.

Ignore the noise term and apply the Fourier transform to both sides of Eq. (5.14) to derive its frequency-domain representation as

$$G(u, v) = I(u, v)H(u, v) \tag{5.25}$$

where $G(u, v)$ represents the spectrum of the blurred image, $I(u, v)$ denotes the spectrum of the original image, and $H(u, v)$ signifies the system transfer function. By taking the absolute values on both sides of Eq. (5.25) and then applying the logarithm, we obtain

$$\lg|G(u, v)| = \lg|I(u, v)| + \lg|H(u, v)| \tag{5.26}$$

Then, perform the inverse Fourier transform on Eq. (5.26) to obtain the real number cepstrum as

$$F^{-1}(\lg|G(u, v)|) = F^{-1}(\lg|I(u, v)|) + F^{-1}(\lg|H(u, v)|) \tag{5.27}$$

Let the inverted independent variable be q. Then, Eq. (5.27) can be simplified as follows

$$C_g(q) = C_I(q) + C_H(q) \tag{5.28}$$

When ignoring the noise term, the blurred image in the cepstrum is created by superimposing the cepstrum of the original image with the system transfer function. Figure 5.6 presents the cepstrum of the blurred image.

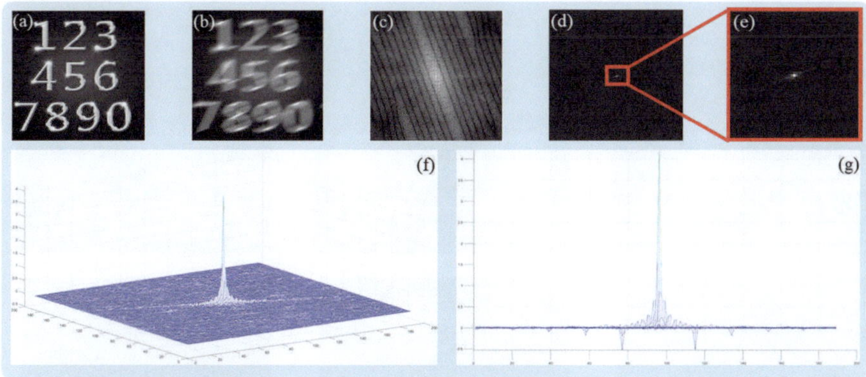

Fig. 5.6 The Cepstrum results of blurred images: **a** original image, **b** blurred image, **c** spectrum of blurred images, **d** cepstrum of blurred images, **e** cepstrum amplification display, **f** three-dimensional display of cepstrum, **g** front view of the cepstrum

Figure 5.6f, g illustrate that the three-dimensional cepstrum of the blurred image exhibits a symmetrical structure. Empirical observations indicate that the angle between the line connecting the two primary negative peaks of the cepstral graph and the axis signifies the direction of motion, while the distance between these peaks represents half the length of the motion.

(3) Discrete Cosine Transform method (DCT): This type of method holds that in Eq. (5.25), the zeros of $G(u, v)$ are related to the zeros of $I(u, v)$ and $H(u, v)$. The parameter values of PSF can be determined by identifying the zero-crossing points of the frequency response of the blurred system. Motion blur exhibits its zero points along the direction perpendicular to the blur and is spaced at regular intervals of $1/L$. This technology has low computational complexity; however, its performance deteriorates when the signal-to-noise ratio falls below 42 dB [8, 9]. Figure 5.7 illustrates the DCT visualization results of the blurred image.

As Eq. (5.15), uniform blurring can be modeled as a function of amplitude $1/L$ and length L. The motion blur length L can be determined by identifying the trough point where the sinc function equals zero in the cosine spectrum:

$$\mathrm{sinc}(\omega L) = 0 \tag{5.29}$$

Therefore, the length of motion blur, denoted as L, can be characterized as

$$L = \frac{\pi n}{\omega_n} \tag{5.30}$$

where, $n = 1, 2, 3$. Eq. (5.30) indicates that the blurred length is inversely proportional to frequency. Due to its superior energy compression characteristics, the cosine transform is more effective for tracking.

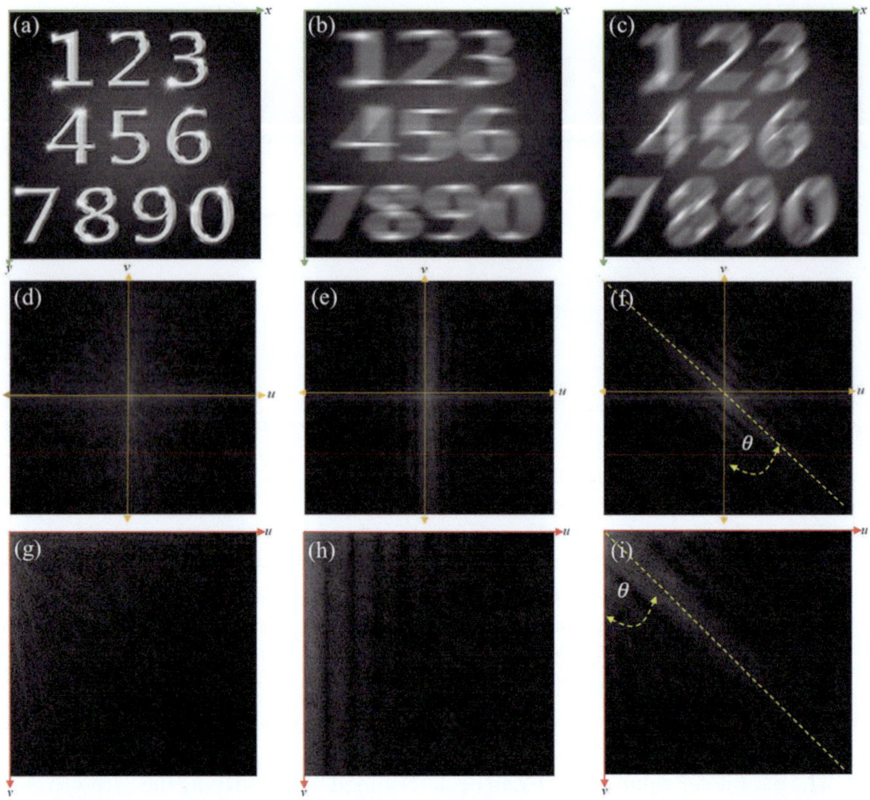

Fig. 5.7 Angle estimation of blurred images at different motion blur angles: **a** original image, **b** blurred image at 0°, **c** blurred image at 45°, **d** the DFT of (**a**), **e** the DFT of (**b**), **f** the DFT of (**c**), **g** the DCT of (**a**), **h** the DCT of (**b**), **i** the DCT of (**c**)

Due to the modulation of motion, the DFT spectrum of motion-blurred images exhibits anisotropy. This spectrum displays distinct characteristics in the direction perpendicular to the motion. By applying windows to mitigate the abrupt changes at the image edges, one can estimate the relative positions of the dark and bright areas within the ripple pattern to determine the angle of motion blur. DCT ripples are more easily identifiable and trackable in the frequency domain, as Fig. 5.7.

(4) Radon Transform Method and Hough Transform Method: Both methods are sensitive to line features in images. Based on the previous analysis and the observation of the spectrum of the blurred image, it is evident that the spectrum is modulated by motion and exhibits pronounced stripe features. These stripe features are associated with the blur parameters. Consequently, it is appropriate to employ the Radon transform and Hough transform to analyze the stripe features and extract the blur parameters. This section will use the Radon transform as an example for illustration.

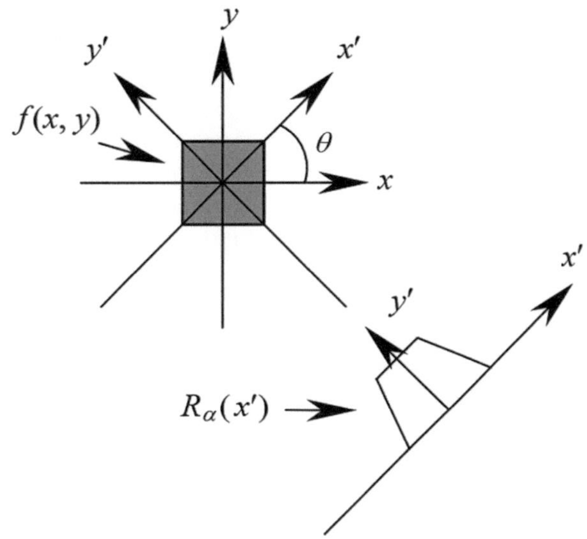

Fig. 5.8 The diagram of Radon transformation

Figure 5.8 illustrates the geometric relationship of the Radon transform. The Radon transform executes linear integration on the image matrix in a specified direction. By applying the Radon transformation, the lines present in the image are mapped to the corresponding line parameter domain, resulting in a peak at the position corresponding to each line in the image within this domain. Following the Radon transformation, parallel stripes oriented in the same direction will exhibit peaks at the corresponding angular positions in the line parameter field, with the coordinates of these peak points representing the angles of the stripes.

The Radon transform along the angle θ direction is defined as

$$R_\alpha(x') = \int_{-\infty}^{+\infty} f(x' \cos\alpha - y' \sin\alpha, x' \sin\alpha + y' \cos\alpha)\mathrm{d}y' \tag{5.31}$$

After determining the angle, a blurred kernel function model based on the sinc function can be constructed using the estimated angle.

$$K(x, y) = c \cdot \mathrm{sinc}(Lu \cos\theta + Lv \sin\theta) \tag{5.32}$$

where c is a constant. The mathematical expressions for the spatial frequencies u and v are $u = x/\lambda f$ and $v = y/\lambda f$, respectively. f represents the focal length of the lens, while L denotes the blur scale to be determined.

For determining the magnitude of the movement speed of target, the distance d between the center point of the Fourier spectrum of the motion-blurred image and the line $ua/M + vb/N = 1$ can be analyzed. According to the distance formula between the center point and the line, the following relationship can be established as

$$d = \left[\left(\frac{a}{M} \right)^2 + \left(\frac{b}{N} \right)^2 \right]^{-\frac{1}{2}} \tag{5.33}$$

According to the frequency-domain characteristics of motion-blurred images, it can be determined that the fringe spacing D on both sides of the spectral center point is twice that of the other fringe spacings d. If the blur scale of a motion-blurred image is L, then $a = L \cos \theta$, $b = L \sin \theta$. Let $M/N = \sigma$, then D is

$$D = \frac{2M}{L} \left[(\cos \theta)^2 + (\sigma \sin \theta)^2 \right]^{-\frac{1}{2}} \tag{5.34}$$

According to Eq. (5.34), the scale of motion blur is inversely proportional to the spacing D of the spectral graph stripes and is influenced by the direction of motion. Additionally, it is also related to the size of the image pixels.

By applying the Radon transform and projecting in the direction of motion blur, the width of the main lobe of the projection curve corresponds to the distance between the dark stripes on either side of the spectral center point. Consequently, a secondary Radon transform is conducted on the spectral graph in the direction of angle θ, aligned with the motion blur, to obtain the projection curve. The coordinates of two extreme points, which are symmetrical to the center point, are extracted to estimate the distance D from the center point to the dark stripes on both sides. The motion blur scale and blur direction are then estimated based on the spectral characteristics of the motion-blurred image. Given that parameters such as object distance and exposure time are known, the speed of movement of target can be calculated.

5.3.3 *Velocity Estimation*

The schematic diagram for calculating the relative velocity of a moving object from a single-frame, linear motion-blurred image is presented in Fig. 5.9. Let the angle between the vertical plane and the direction of motion be denoted as θ, and let the displacement of the object during the exposure time T be represented as d. Based on the object-image similarity relationship illustrated in Fig. 5.9, we have

$$\frac{f}{p + l} = \frac{d \sin \theta}{m} \tag{5.35}$$

$$\frac{d \cos \theta - m}{l} = \frac{s}{f} \tag{5.36}$$

where l represents the length of the target's movement on the image plane, n is the vertical distance from the camera lens to the object's starting point, and f is the focal length of the camera. To eliminate the variable m, substitute Eq. (5.35) into (5.36) and solve for d. This yields the following result

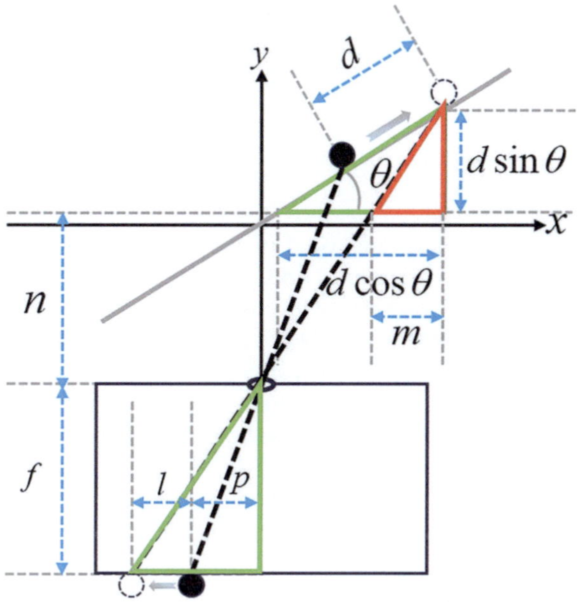

Fig. 5.9 The top view of the imaging model for estimating the motion length of the point target

$$d = \frac{nl}{f\cos(\theta) - (p + l)\sin(\theta)} \tag{5.37}$$

Suppose the rate of motion of an object is constant, then there is

$$v = \frac{d}{T} \tag{5.38}$$

$$v = \frac{nl}{T\left[f\cos(\theta) + (p + l)\sin(\theta)\right]} \tag{5.39}$$

Substituting the camera pixel size S_x into Eq. (5.39), which can be obtained during the camera calibration process, yields the following result

$$v = \frac{nLS_x}{T\left[f\cos(\theta) - S_x(P + L)\sin(\theta)\right]} \tag{5.40}$$

where P and L represent the distances from the vertical axis, both measured in pixels. If the image plane is parallel to the direction of motion, then $\theta = 0$. At this point, Eq. (5.40) can be simplified to

$$v = k\frac{L}{T} \tag{5.41}$$

where $k = nS_x/f$ is the scaling factor that can be determined using the camera calibration method of Zhang [10].

5.3.4 Reconstruction of Blurred Images

After estimating the speed of the blurred image, it is essential to reconstruct an image with high clarity and fidelity. Once the blur parameters have been determined, employing image reconstruction algorithms to restore blurred images is often a necessary step in the field of animated image speed measurement.

The essence of image quality degradation due to motion blur lies in the interaction of noise and convolution operators with the target image. Deblurring technology must establish a degradation model and restore the image by reversing the degradation process. This requires the use of blur priors and deblurring techniques, as well as quality standards for restoration to assess clarity.

Early image deblurring methods primarily involved filtering techniques, with notable examples including inverse filtering [11] and Wiener filtering. Subsequently, the constrained least squares method and regularization constraint method were introduced for deblurring. With advancements in artificial intelligence, deep learning-based deblurring has emerged as the preferred approach. Convolutional neural networks [12] and generative adversarial networks [13] have been effectively utilized in the field of image deblurring.

Blur can be categorized into local blur and global blur. Local blur arises from the stillness of the camera equipment combined with the movement of objects within the frame. In contrast, global blur occurs when an object remains stationary while the camera equipment is in motion. The factors contributing to local blur are more intricate, making its restoration more challenging.

Blind deblurring targets blurred images. In situations where the blur priors are unknown, a comprehensive degradation process model is established to investigate the relationship between blur priors and motion states. Additionally, methods are developed to mitigate or eliminate the effects of blurring. In contrast, non-blind deblurring reconstructs images using known blur priors [14]. Blind deblurring focuses on estimating the unknown blur priors. Although this approach is complex, it enables effective image reconstruction and speed measurement, and the algorithm offers multiple functionalities. In practical applications, scenarios involving unknown blur priors are more common, making blind deblurring reconstruction particularly valuable.

(1) Classic Algorithm: Inverse filtering is a well-established non-blind deblurring technique that operates in both the frequency and spatial domains. Due to the inherent challenges associated with deconvolution, this method is highly sensitive to noise. Even minor disturbances can result in unstable restoration, making it suitable primarily for deblurring images that are largely free of noise. In an ideal noise-free scenario, as described in Eq. (5.25), there is

$$F(u,v) = \frac{G(u,v)}{H(u,v)} \tag{5.42}$$

By applying a Fourier transform to the degraded image $g(x, y)$, the transformed image $G(u, v)$ is obtained. The system transfer function $H(u, v)$ is derived using the identified blurred parameters. The frequency-domain inverse filtering result $F(u, v)$ is calculated using Eq. (5.42), and subsequently, the reconstructed image $f(x, y)$ is obtained through the inverse Fourier transform.

In practical applications, the noise spectrum $N(u, v)$ often cannot be overlooked. When $|H(u, v)|$ is relatively small, particularly near zero in the frequency domain, inverse filtering can lead to significant amplification of the noise:

$$F(u,v) = \frac{G(u,v)}{H(u,v)} + \frac{N(u,v)}{H(u,v)} \tag{5.43}$$

where $G(u, v)/H(u, v)$ represents the principal term, while $N(u, v)/H(u, v)$ denotes the noise amplification term. This amplification contributes to both the noise restoration results and the ringing phenomenon observed in the reconstructed image. The mathematical principle of inverse filtering is straightforward, characterized by rapid calculation speed and the absence of iterative processes. This makes it suitable for quick implementation without the need for complex parameter adjustments. However, its sensitivity to noise and lack of robustness limit its range of applications.

Wiener filtering is a classic linear filtering technique that is grounded in the minimum mean square error criterion. This method operates under the assumption that both the image and the noise are random variables. It aims to estimate $\hat{f}$, the original image f, in order to minimize the mean square error between the two. The measurement of this error is represented by the following formula

$$e^2 = E\left\{\left(f - \hat{f}\right)^2\right\} \tag{5.44}$$

where $E\{\cdot\}$ represents the expected value. Assuming that the noise and the image are uncorrelated, with one of them having a zero mean, the estimated gray level of the image can be expressed as a linear function of the gray level of the degraded image. Under these conditions, the minimum value of the error function in Eq. (5.44) in the frequency domain is determined by the following equation

$$\begin{aligned}\hat{F}(u,v) &= \left[\frac{H^*(u,v)S_f(u,v)}{S_f(u,v)|H(u,v)|^2 + S_\eta(u,v)}\right]G(u,v)\\ &= \left[\frac{H^*(u,v)}{|H(u,v)|^2 + S_\eta(u,v)/S_f(u,v)}\right]G(u,v)\end{aligned}$$

$$= \left[\frac{1}{H(u,v)} \frac{|H(u,v)|^2}{|H(u,v)|^2 + S_\eta(u,v)/S_f(u,v)} \right] G(u,v) \tag{5.45}$$

Equation (5.45) represents the Wiener filter. The filter, which consists of the terms within the square brackets, is commonly referred to as the least mean square error filter or the least squares filter. In the absence of noise, the noise power spectrum vanishes, resulting in a simplification of the Wiener filter to inverse filtering.

When dealing with white noise, the spectrum $|N(u,v)|^2$ is constant and can be simplified using Eq. (5.45). However, the undegraded power spectrum is often unknown in many scenarios. When these quantities cannot be estimated, the following formula is frequently employed to approximate Eq. (5.45) as

$$\hat{F}(u,v) = \left[\frac{1}{H(u,v)} \frac{|H(u,v)|^2}{|H(u,v)|^2 + K} \right] G(u,v) \tag{5.46}$$

where K is a specific constant added to all terms of $|H(u,v)|^2$. Similarly, Wiener filtering is straightforward to implement, minimizes mean square error, exhibits stable performance, and does not require iteration. However, its practical application is limited due to the necessity of knowing the statistical characteristics of both signals and noise, as well as the challenge of accurately estimating the noise power spectrum. If there is a significant deviation in noise estimation, a ringing phenomenon may occur, as Fig. 5.10.

Constrained least squares filtering is a regularized image restoration technique. By incorporating constraint conditions, such as image smoothness and sparsity, into the optimization objective, this method effectively suppresses noise and mitigates the issues associated with inverse filtering.

For the image degradation model Eq. (5.25), the objective of constrained least squares filtering is to restore $F(u,v)$ from the observed image $G(u,v)$. Simultaneously, it is essential to maintain both data fidelity and regularization of the solution, ensuring that the error between the restoration result and the observed image is minimized, while the restored image adheres to the prior constraints.

Constrained l east squares filtering minimizes the following objective function as

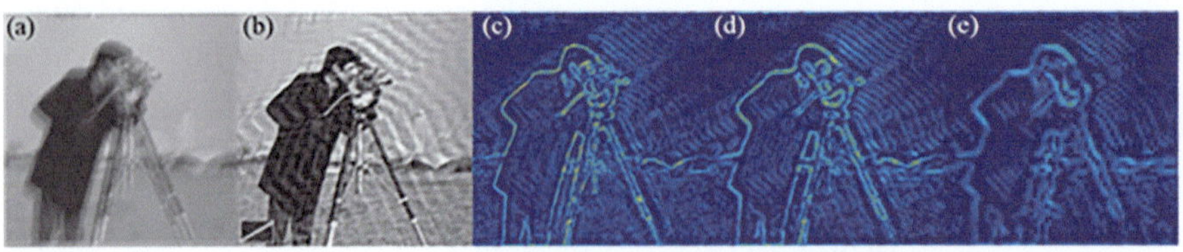

Fig. 5.10 Wiener filter reconstruction image results [15]: **a** synthesized blurred image, **b** the reconstruction result, when the prior is insufficient, contains annular artifacts, **c–e** additional parameters are included to emphasize the ring artifacts

$$\min_{F(u,v)}\left\{\|G(u,v)-H(u,v)F(u,v)\|^2+\lambda\|L(u,v)F(u,v)\|^2\right\} \tag{5.47}$$

the first term is the fidelity term, which ensures that the difference in the frequency domain between the restored image and the observed image is minimized. The second term is the regularization term, which imposes constraints on the solution through $L(u, v)$. The parameter λ serves as the regularization parameter, balancing the data fidelity against the regularization of the solution.

The optimal solution can be achieved by differentiating the frequency-domain solutions of Eq. (5.47) and setting the derivative equal to zero:

$$F(u,v)=\frac{\overline{H(u,v)}G(u,v)}{|H(u,v)|^2+\lambda|L(u,v)|^2} \tag{5.48}$$

where $|H(u,v)|^2$ represents the intensity of degradation, while $\lambda|L(u,v)|^2$ denotes the known high-frequency noise. In the context of prior constraints for the regularized $L(u, v)$ solution, the Laplacian operator, which facilitates smoothing, is typically selected as

$$L(u,v)=-\pi^2\left(u^2+v^2\right) \tag{5.49}$$

The Wiener filter can be generalized to obtain the geometric mean filter

$$\hat{F}(u,v)=\left[\frac{H^*(u,v)}{|H(u,v)|^2}\right]^{\alpha}\left[\frac{H^*(u,v)}{|H(u,v)|^2+\beta\left[\frac{S_\eta(u,v)}{S_f(u,v)}\right]}\right]^{1-\alpha}G(u,v) \tag{5.50}$$

where α and β are positive real constants. When $\alpha = 1$, the filter degenerates into an inverse filter. When $\alpha = 0$, this filter becomes a parametric Wiener filter. The parametric Wiener filter is restored to the standard Wiener filter when $\beta = 1$. If $\alpha = 1/2$, the filter becomes the product of two quantities raised to the same power, which represents the geometric mean. If $\beta = 1$ at this time, this filter is also referred to as a spectral equalization filter.

To efficiently handle blurred images, the Richardson-Lucy (RL) reconstruction method, which is grounded in Bayesian theory [16, 17], assumes that the noise follows a Poisson distribution. This method removes the blur through iterative processes utilizing the maximum likelihood estimation technique. The RL algorithm is an expectation-maximization method based on maximum likelihood estimation, with the objective of maximizing the following likelihood functions

$$P(g|f)=\prod_i\frac{(Hf)_i^{g_i}e^{-(Hf)_i}}{g_i!} \tag{5.51}$$

By solving this maximum likelihood problem, the RL algorithm employs the following iterative formula

$$f^{(k+1)} = f^{(k)} \cdot \left[h \star \left(\frac{g}{f^{(k)} * h} \right) \frac{1}{\sum h} \right] \tag{5.52}$$

where $f^{(k)}$ represents the image estimation at the k th iteration, $f^{(k+1)}$ represents the image estimation after the $k+1$ th iteration, g denotes the actual observed image, h signifies the PSF, $\star$ represents the correlation operation, and $*$ denotes the convolution operation.

The core steps of the algorithm are as follows: (1) First, convolve the currently estimated image $f^{(k)}$ with the PSF to obtain a blurred image, (2) Divide the actual observed image g point by point by the results from the previous step to obtain the error image, which reflects the discrepancy between the current estimation and the observation, (3) Perform correlation operations on the error image and the PSF to propagate the error information into the image space, (4) Multiply the current estimate $f^{(k)}$ by this correction factor to obtain the next estimate $f^{(k+1)}$.

The RL algorithm is well-suited for modeling Poisson noise and can significantly enhance image clarity without requiring explicit regularization. However, its convergence speed is relatively slow, necessitating careful selection of the number of iterations to avoid overfitting to noise. The non-blind deblurring algorithm based on RL effectively mitigates the ringing artifacts associated with the RL algorithm by incorporating gradient regularization, thereby restoring more image details [18]. Some researchers have developed a blurred model of projected motion paths to simulate the spatial transformation of motion performance, optimizing the RL algorithm and integrating regularization priors to improve deblurring outcomes [19]. Figure 5.11 illustrates the reconstruction results of both the RL algorithm and the enhanced algorithm.

Fig. 5.11 The deblurring results of the RL algorithm [19] indicate that the image dimensions are 700 × 525 pixels, and the size of the blurring kernel is 41 × 41 pixels. The following components are presented: **a** the blurred image and the blurring kernel, **b** the original image, **c** the results of the RL algorithm, and **d**–**f** the results of the improved algorithm

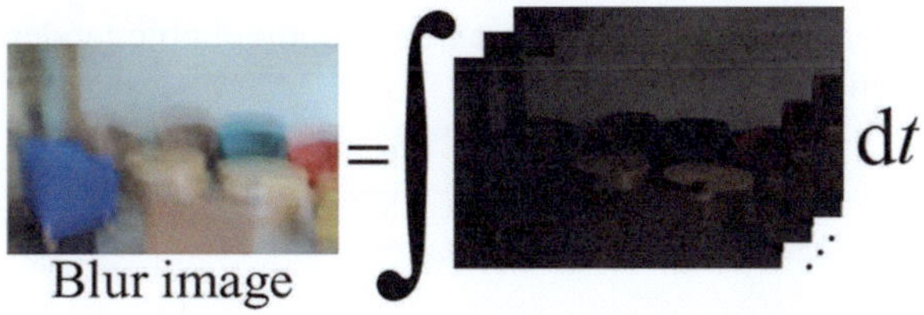

Fig. 5.12 Integration of image scenes under projection motion blur model

(2) Projection motion blur model [19]: This method represents the degraded image as an ensemble of clear scenes subjected to a series of planar projection transformations, as Fig. 5.12. One of the key advantages of this model is its enhanced suitability for capturing the self-motion of the camera compared to traditional kernel-based PSF parameterization. The conventional kernel-based PSF approach necessitates dividing the image into uniformly blurred regions, with each pixel assigned a distinct blurred kernel. In contrast, the projection motion blur model can be utilized to extend the traditional pixel-domain RL deblurring algorithm. This enhanced RL algorithm is referred to as the Projective Motion RL algorithm, which is capable of integrating multiple priority-based regularizations.

During the image capture process, the pixel intensity of an image is determined by the amount of light received by the imaging sensor during the exposure time as

$$I(x) = \int_0^T \delta I(x, t)\mathrm{d}t \approx \sum_{i=1}^{N} \Delta I(x, t_i) \tag{5.53}$$

where $I(x)$ represents the image recorded after exposure, while $\delta I(x, t)$ and its discrete form $\Delta I(x, t_i)$ denote the images captured by the sensor within an infinitesimal time interval dt. $[0, T]$ is the total exposure time, and x is a 3×1 vector indicating the uniform pixel coordinates. N is the discrete sampling rate during the exposure time. Assuming that N is sufficiently large, the difference between the continuous integral and the discrete integral can be disregarded.

When there is no relative motion between the camera and the scene, it is assumed that the influence of sensor noise is minimal, $\Delta I(x, t_1) \cong \Delta I(x, t_2) \cong \cdots \cong \Delta I(x, t_N)$, and $I(x) \cong N\Delta I(x, t_0) \triangleq I_0(x)$ is a clear image. When relative motion occurs, $I(x)$ becomes the sum of multiple disordered images $\Delta I(x, t_i)$. In the case of a static distant scene, relative motion induces a planar projection transformation in the image plane. For instance, $\Delta I(x, t_i) = \Delta I(h_i x, t_{i-1})$, where h_i is a 3×3 non-singular matrix defined by homography. Assuming that all instances of h_i are known, $\Delta I(x, t_i)$ can be expressed in terms of $I_0(x)$ using the following formula

$$\Delta I(x, t_i) = \Delta I\left(\prod_{j=1}^{i} h_j x, t_0\right) = \frac{1}{N} I_0(H_i x) \tag{5.54}$$

where $H_i = \prod_{j=1}^{i} h_j$ also exhibits homophobic tendencies. Therefore, the proposed motion model is

$$B(y) = \sum_{i=1}^{N} \Delta I(x, t_i) = \frac{1}{N} \sum_{i=1}^{N} I_0(H_i x) \tag{5.55}$$

where $B(y)$ represents the motion-blurred image, while $I_0(x)$ denotes the clear image to be estimated. According to the model, the blurred image is the average of multiple clear images, with each clear image being a planar projection transformation of $I_0(x)$.

Figure 5.13 illustrates the differences between the projected motion model and the traditional representation. The traditional spatially invariant PSF representation is a specific instance of the projected motion model, where each h_i is translated. For two-dimensional translational motion, traditional kernel-based models provide a more concise representation of motion blur compared to projective motion models. In contrast, the projected motion model is more compact and intuitive when dealing with in-plane and out-of-plane rotations.

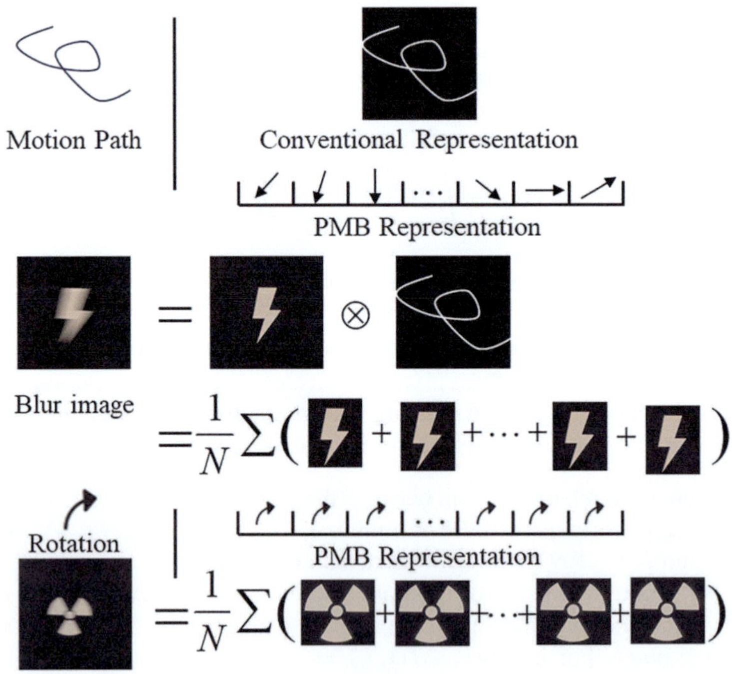

Fig. 5.13 Comparison between the projected motion model and the traditional model: given the motion path, the traditional model employs a rectangular kernel to represent the PSF. The projected motion model utilizes a series of transformation matrices. For rotational motion, the projected motion model encodes rotation through homography sequences, whereas traditional methods necessitate the storage of pixel-level PSF data

As mentioned earlier, the RL algorithm uses the Poisson distribution to model the maximum likelihood solution of the likelihood probability

$$P(B,k|I) = \prod_{x \in I} \frac{g(x)^{B(x)} e^{-g(x)}}{B(x)!} \tag{5.56}$$

$$g(x) = \sum_{y \in k} I(y)k(x-y) \tag{5.57}$$

where B is the observed blurred image, k is the PSF, I is the clear image to be estimated, and $g(x)$ is the pixel convolution process at x. Equation (5.56) assumes that the likelihood probability is independent of each other for every x.

The sufficient condition for the maximum value of I is the Kuhn-Tucker condition, where all x satisfies

$$I(x) \frac{\partial}{\partial I(x)} \ln\left(\prod_{x \in I} \frac{g(x)^{B(x)} e^{-g(x)}}{B(x)!} \right) = 0 \tag{5.58}$$

and

$$\frac{\partial}{\partial I(x)} \ln\left(\prod_{x \in I} \frac{g(x)^{B(x)} e^{-g(x)}}{B(x)!} \right) \leq 0, \text{ if } I(x) = 0 \tag{5.59}$$

To obtain the iterative update rules of the RL algorithm, the conditions represented by Eq. (5.37) are used for all $x \in I$ as

$$I(x) \frac{\partial}{\partial I(x)} \ln\left(\prod_{x \in I} \frac{g(x)^{B(x)} e^{-g(x)}}{B(x)!} \right) = 0 \tag{5.60}$$

$$I(x) \sum_{x \in I} \frac{\partial}{\partial I(x)} (B(x)\ln(g(x)) - g(x) - \ln(B(x)!)) = 0 \tag{5.61}$$

$$I(x) \sum_{x \in I} \frac{B(x)}{g(x)} \frac{\partial}{\partial I(x)} g(x) - I(x) \sum_{x \in I} \frac{\partial}{\partial I(x)} g(x) = 0 \tag{5.62}$$

$$I(x) \sum_{y \in k} \frac{B(y)}{g(y)} k(y-x) - I(x) \sum_{y \in k} k(y-x) = 0 \tag{5.63}$$

Since $\sum_{y \in k} K(y) = 1$, $\sum_{y \in k} K(y-x) = 1$. After adding the iterative index t, it can be obtained as

$$I^{t+1}(x) = I^t(x) \sum_{y \in k} \frac{B(y)}{\sum_{z \in k} I^t(z)k(y-z)} k(y-x) \tag{5.64}$$

The RL algorithm was obtained by performing convolution operations on the entire image

$$I^{t+1} = I^t \times \tilde{k} \otimes \frac{B}{k \otimes I^t} \tag{5.65}$$

where $\tilde{k}$ is the transpose of k. Equation (5.45) indicates that $B''^t = k \otimes I^t$ represents a prediction of the blurred image based on the current estimate I^t of the clear image and the given point spread function k. Therefore, B/B''^t is the residual between the actual blurred image B and the predicted blurred image B''^t.

Using the basic RL algorithm, the projection RL algorithm is derived. According to the projection motion blur model defined in Eq. (5.54), integrate the motion path at each pixel position y and define the spatially varying motion PSF:

$$B(y) = \frac{1}{N}\sum_{i=1}^{N} I(H_i x) = \sum_{x \in k_y} I(x) k_y(x) \tag{5.66}$$

and

$$k_y(x) = \begin{cases} \frac{1}{N}, & if\ x = H_i^{-1} y \\ 0, & \text{otherwise} \end{cases} \tag{5.67}$$

Since $x = H_i^{-1} y$ does not correspond to discrete integer pixel coordinates, bicubic interpolation can be used to estimate the pixel value of point x.

For the RL algorithm, replacing Eq. (5.57) with Eq. (5.66) results in

$$I^{t+1}(x) = I^t(x) \sum_{y \in k_x} \frac{B(y)}{\sum_{z \in k_y} I^t(z) k_y(z)} \tilde{k}_x(y) \tag{5.68}$$

This is the general method for removing motion blur through spatial variation using the RL algorithm. The motion path in $\tilde{k}_x$ is in the opposite direction of that in k_x.

According to Eq. (5.65), the motion paths in k_x can be described by homophobic sequences $(H_1 \ldots H_n)$. Consequently, the motion paths of $\tilde{k}_x$ can be categorized. The motion path of $\tilde{k}_x$ constitutes an original homophobic sequence, where each matrix is inverted and operates in reverse order $(H_n^{-1} \ldots H_1^{-1})$. For each point on the motion path defined by $H_1 \ldots H_n$, the sequence $H_n^{-1} \ldots H_1^{-1}$ integrates the error along the motion path through the inverse transformation, thereby yielding the iterative update rule for the projected motion blur model:

$$I^{t+1} = I^t \times \frac{1}{N}\sum_{i=1}^{N} E^t\left(H_i^{-1} x\right) \tag{5.69}$$

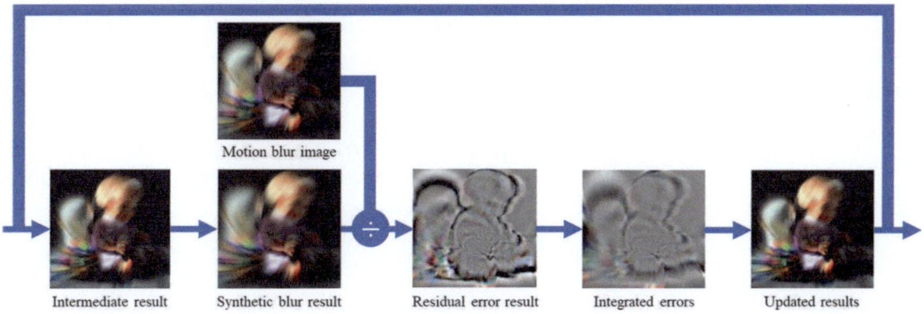

Fig. 5.14 Overview of the projection motion RL algorithm

$$E^t(x) = \frac{B(x)}{\frac{1}{N}\sum_{i=1}^{N} I^t(H_i x)} \tag{5.70}$$

Although the motion PSF of each pixel was calculated during the derivation process, Eq. (5.70) processes the image as a whole and does not necessitate the explicit reconstruction of the motion PSF for each pixel. This approach is analogous to global convolution and related processes found in conventional RL algorithms. Essentially, a series of forward projection movements and their correlations substitute for the convolution and correlation operators used in traditional RL algorithms. Figure 5.14 illustrates the projection motion RL algorithm.

The previously derived result represents the maximum likelihood estimation of the Poisson noise model, with no regularization applied to the solution. However, incorporating specific image priors can significantly reduce deconvolution artifacts. Let $P(I)$ denote the prior probability of the restored image; thus, the objective function for optimization becomes

$$\begin{aligned} \arg\max_I P(I|B,k) &= \arg\ \max_I P(B,k|I)P(I) \\ &= \arg\ \min_I - L(B,k|I) - L(I) \\ &= \arg\ \min_I \sum_{x\in I} (g(x) - B(x)\ln(g(x)) + \lambda R(I(x))) \end{aligned} \tag{5.71}$$

where, $L(\cdot)$ and $R(I) = -1/\lambda(L(I))$ are regularization terms, and λ represents the relative weight between the fidelity term and the regularization term. To obtain the necessary conditions for optimization, calculate the first derivative of Eq. (5.70) with respect to $I(x)$ and set it to zero.

$$\sum_{y\in k} \frac{B(y)}{\sum_{z\in k} I^t(z)k(y-z)} k(y-x) = 1 - \lambda\nabla R(I(x)) \tag{5.72}$$

where $\nabla R(I(x))$ is the first derivative of $R(I(x))$ with respect to $I(x)$. To make the RL algorithm converge, it is necessary for $I^{t+1}/I^t = 1$ when $t \to \infty$. Therefore, the regularized RL algorithm becomes

$$I^{t+1}(x) = \frac{I^t(x)}{1 - \lambda \nabla R(I^t(x))} \sum_{y \in k} \frac{B(y)k(y-x)}{\sum_{z \in k} I^t(z)k(y-z)} \tag{5.73}$$

and

$$I^{t+1} = \frac{I^t}{1 - \lambda \nabla R(I^t)} \times \tilde{k} \otimes \frac{B}{k \otimes I^t} \tag{5.74}$$

For the projection motion RL algorithm, the regularized form of the update rule can be derived from Eq. (5.72) as

$$I^{t+1}(x) = \frac{I^t(x)}{1 - \lambda \nabla R(I^t)} \times \frac{1}{N} \sum_{i=1}^{N} E^t\left(H_i^{-1}x\right) \tag{5.75}$$

The commonly used regularization methods in motion deblurring include total variation regularization, Laplacian regularization, and bilateral regularization. The purpose of introducing total variation regularization is to minimize the gradient magnitude in the deblurred image during the deconvolution process:

$$R_{\mathrm{TV}}(I) = \int \sqrt{\|\nabla I(x)\|^2} \mathrm{d}x \tag{5.76}$$

where $\nabla I(x)$ is the first derivative of $I(x)$. Substituting this regularization into Eq. (5.74) yields the following result:

$$I^{t+1}(x) = \frac{I^t(x)}{1 - \lambda \nabla R_{\mathrm{TV}}(I)} \times \frac{1}{N} \sum_{i=1}^{N} E^t\left(H_i^{-1}x\right) \tag{5.77}$$

where, $\nabla R_{TV}(I) = -\nabla I^t / \left|\nabla I^t\right|$.

Laplacian regularization, often referred to as sparse regularization, is characterized by a double-tailed distribution modeled by the Laplacian distribution in the histogram of gradient magnitudes as

$$R_L(I) = \exp\left(-\frac{1}{\eta}|\nabla I|^d\right) \tag{5.78}$$

where, d controls the shape distribution, while $\eta = 0.005$ represents the variance of the image noise. This regularization method was incorporated into the projected motion RL algorithm, leading to a new set of update rules as

$$I^{t+1}(x) = \frac{I^t(x)}{1 - \lambda \nabla R_L(I)} \times \frac{1}{N} \sum_{i=1}^{N} E^t\left(H_i^{-1} x\right) \tag{5.79}$$

where,

$$\nabla R_L(I) = -\frac{1}{\eta} \exp\left(-\frac{1}{\eta} |\nabla I|^d\right) |\nabla I|^{d-1} \nabla^2 I \tag{5.80}$$

To suppress visual artifacts while preserving sharp edges, bilateral regularization has been proposed.

$$R_B(I) = \sum_x \sum_{y \in N(x)} g_1\left(\|x - y\|^2\right)\left[1 - g_2\left(\|I(x) - I(y)\|^2\right)\right] \tag{5.81}$$

where, $N(x)$ is a local neighborhood of x, and $g_1(\cdot)$ and $g_2(\cdot)$ are two Gaussian functions with a mean and variance of zero. Under this regularization, the new iterative form is

$$I^{t+1}(x) = \frac{I^t(x)}{1 - \lambda \nabla R_B(I)} \times \frac{1}{N} \sum_{i=1}^{N} E^t\left(H_i^{-1} x\right) \tag{5.82}$$

where,

$$\nabla R_B(I) = \sum_{y \in N(x)} \left(I_y^d - I_y^d D_y\right) \tag{5.83}$$

$$I_y^d(x) = g_1\left(\|x - y\|^2\right) g_2\left(\|I(x) - I(y)\|^2\right) \frac{(I(x) - I(y))}{\sigma_r} \tag{5.84}$$

Due to its reliance on a Gaussian distribution and a larger neighborhood, bilateral regularization typically yields smoother results.

(3) L_0-regularization prior based on intensity and gradient: This prior knowledge pertains to the distinctive characteristics of text images and does not necessitate any heuristic edge selection methods. In clear images, the text and background areas typically exhibit uniform intensity values. This phenomenon occurs because a blurred image can be viewed as the weighted sum of several adjacent pixels from a clear image, causing the intensity distribution to compress from both ends toward the center, which results in a decrease in the number of pure black pixels. This characteristic is applicable to general images but is more pronounced in text images. For image x, regularization can be employed to describe this property as

$$P_t(x) = \|x\|_0 \tag{5.85}$$

This metric can be utilized to differentiate between clear images and blurry ones. The gradient prior is effective in suppressing artifacts and can be applied to image deblurring. The intensity values of clear images are close to dichroism, and the non-zero values of the pixel gradient are sparser compared to those of blurred images. The image gradient can be modeled using the L_0-class regularized prior $P_t(\nabla x)$.

Using the above-mentioned regularization priors regarding intensity and gradient, the priors for deblurring text images are defined by the following formula

$$P(x) = \sigma P_t(x) + P_t(\nabla x) \tag{5.86}$$

where σ is the balanced weight of the two priors.

Given the blurred image y, the latent image x, and the blur kernel k are estimated based on the prior $P(x)$ using a regularization formula as

$$\min_{x,k} \|x * k - y\|_2^2 + \gamma \|k\|_2^2 + \lambda P(x) \tag{5.87}$$

The first item pertains to the image data, while the other two items represent the constraints of the blurred kernel and the latent image, with their respective weights denoted as γ and λ.

By alternately solving the latent image x, as

$$\min_x \|x * k - y\|_2^2 + \lambda P(x) \tag{5.88}$$

and blur kernel k,

$$\min_k \|x * k - y\|_2^2 + \gamma \|k\|_2^2 \tag{5.89}$$

the deblurring process can be modeled as an optimization process.

Calculate Eq. (5.88) using the alternating minimization method. Introduce the auxiliary variables u and v in relation to x, which correspond to the image gradients in the horizontal and vertical directions, respectively. Equation (5.88) can be expressed as follows

$$\min_{x,u,g} \|x * k - y\|_2^2 + \beta \|x - u\|_2^2 + \mu \|\nabla x - g\|_2^2 + \lambda\big(\sigma \|u\|_0 + \|g\|_{00}\big) \tag{5.90}$$

where σ represents the balanced weight of the two priors, while β and u are the penalty parameters. As β and u approach infinity, the solution to Eq. (5.90) converges to Eq. (5.91). Equation (5.90) can be solved by alternately minimizing x, u and g while keeping the other variables fixed. The initial values of u and g are set to zero. In each iteration, x is calculated using the following formula

$$\min_x \|x * k - y\|_2^2 + \beta \|x - u\|_2^2 + \mu \|\nabla x - g\|_2^2 \tag{5.91}$$

The closed solution to this least squares minimization problem is

$$x = \mathcal{F}^{-1}\left(\frac{\overline{\mathcal{F}(k)}\mathcal{F}(y) + \beta\mathcal{F}(u) + \mu F_G}{\overline{\mathcal{F}(k)}\mathcal{F}(k) + \beta + \mu\left(\sum_{i\in\{h,v\}} \overline{\mathcal{F}(\nabla_i)}\mathcal{F}(\nabla_i)\right)}\right) \tag{5.92}$$

where, $\mathcal{F}(\cdot)$ and $\mathcal{F}^{-1}(\cdot)$ represent the Fourier transform and the inverse Fourier transform, respectively, while $\overline{\mathcal{F}(\cdot)}$ denotes the complex conjugate operator. Additionally, ∇_h and ∇_v of $F_G = \overline{\mathcal{F}(\nabla_h)}\mathcal{F}(g_h) + \overline{\mathcal{F}(\nabla_v)}\mathcal{F}(g_v)$ represent the horizontal and vertical differential operators, respectively.

Given x, calculate u and g respectively through the following two formulas

$$\min_u \beta\|x - u\|_2^2 + \lambda\sigma\|u\|_0 \tag{5.93a}$$

$$\min_g \mu\|\nabla x - g\|_2^2 + \lambda\|g\|_0 \tag{5.93b}$$

Eq. (5.92) is a pixel-by-pixel problem. Therefore, the solutions of u and g are

$$u = \begin{cases} x, & |x|^2 \geqslant \frac{\lambda\sigma}{\beta} \\ 0, & \text{otherwise} \end{cases} \tag{5.94}$$

$$g = \begin{cases} \nabla x, & |\nabla x|^2 \geqslant \frac{\lambda}{\mu} \\ 0, & \text{otherwise} \end{cases} \tag{5.95}$$

The rationale for proposing the subproblem u in the outer loop is that if subproblem Eq. (5.93b) is introduced in the outer loop, the algorithm of the inner loop becomes equivalent to solving the following minimization problem by incorporating the auxiliary variable u as

$$\min_x \|x * k - y\|_2^2 + \lambda\sigma\|x\|_0 + \mu\|\nabla x - g\|_2^2 \tag{5.96}$$

Similarly, presenting a sub-problem Eq. (5.93a) in the outer loop is equivalent to introducing an auxiliary variable g into the inner loop to solve the following minimization problem

$$\min_x \|x * k - y\|_2^2 + \beta\|x - u\|_2^2 + \lambda\|\nabla x\|_0 \tag{5.97}$$

Because the intensity prior relies on the parallax of independent pixels rather than adjacent ones, it introduces significant noise and artifacts during image restoration. In contrast, the gradient prior is based on the differences between adjacent pixels, which promotes a smoother result with fewer artifacts in the restored image. Since the internal loop involves image restoration, addressing sub-problems in the outer loop can help mitigate the artifacts caused by the intensity prior.

Given x, Eq. (5.89) transforms into a least squares minimization problem, and the closed-form solution can be computed using the Fourier transform. Since gradient-based estimation has been shown to be more accurate, the blur kernel k can be estimated using the following formula

$$\min_k \|\nabla x * k - \nabla y\|_2^2 + \gamma \|k\|_2^2 \tag{5.98}$$

After obtaining k, set any negative elements to 0 and normalize the vector so that the sum of its elements equals 1. Conduct the kernel estimation process in a coarse-to-fine manner using an image pyramid.

In light of the characteristics of the results generated by the two aforementioned priors, a non-blind deconvolution method that preserves details is employed to mitigate ringing artifacts. The specific steps are outlined as follows.

(1) Estimate the latent image I_l using the Laplace prior.
(2) Estimate the latent I_0 using the method outlined in Eq. (5.88), and introduce gradient regularization $P_t(\nabla x)$ only at this stage.
(3) Calculate the difference graph between I_l and I_0, and apply bilateral filtering to the result.
(4) Subtract the filtered difference map from I_l to eliminate artifacts and obtain the restored image.

5.4 General Conclusions

This chapter introduces methods for measuring velocity in moving images, mainly including particle image velocity and blur image velocity measurement technology. PIV is applicable for fluid measurements, while blur Image Velocimetry is suitable for estimating velocity in single-frame images. Each method is tailored for different application scenarios. As related technologies continue to advance, the performance of both techniques in practical applications is expected to improve further.

Acknowledgments This work was supported by the National Natural Science Foundation of China (12374271, 62375065, 62405079, 62575091); Fundamental Research Funds for the Central Universities (HIT.OCEF.2024055, AUGN5750300325); The Fund of National Laboratory on Adaptive Optics, China (FNLAO-24-MS-O07); Natural Science Foundation of Heilongjiang Province of China (LH2024A009); Key Laboratory of Image Processing and Intelligent Control (Huazhong University of Science and Technology), Ministry of Education, Wuhan, China (IPIC2025-05); Key Laboratory of Advanced Medical Imaging and Intelligent Computing of Guizhou Province (AMIIC250204); the China Postdoctoral Science Foundation-CCTEG Joint Support Program (2025T011ZGMK).

References

1. Adrian RJ (1991) Particle imaging techniques for experimental fluid mechanics. Annu Rev Fluid Mech 23:261–304
2. Gao D, Wang Y, Shen G (2001) DPIV technique and its application in flow field measurement. Hydraul Pneum Seals 21:30–33
3. Wang C, Lin J, Yamamoto F (2001) An algorithm for two-dimensional PIV images. J Hydrodyn 16(4):399–404
4. Gao C, Cao Y, Guo Y (2004) Particle image velocimetry techniques used in blood flow field analysis. Opto-Electron Eng 31(8):37–40
5. Li M, Du H, Zhang Q, Wang J (2008) Improved particle image velocimetry through cell segmentation and competitive survival. IEEE Trans Instrum Meas 57(6):1221–1229
6. Pan J, Ren W, Hu Z, Yang M (2018) Learning to deblur images with exemplars. IEEE Trans Pattern Anal Mach Intell 41(6):1412–1425
7. Cannon M (1976) Blind deconvolution of spatially invariant image blurs with phase. IEEE Trans Acoust Speech Signal Process (ASSP) 24(1):58–63
8. Kundur D, Hatzinakos D (1996) Blind image deconvolution. IEEE Signal Process Mag 13(3):43–64
9. Yoshida Y, Horiike K, Fujita K (1993) Parameter estimation of uniform image blur using DCT. IEICE Trans Fundam Electron Commun Comput Sci E76-A(7):1154–1157
10. Zhang Z (2000) A flexible new technique for camera calibration. IEEE Trans Pattern Anal Mach Intell 22(11):1330–1334
11. Twogood R, Sommer F (2007) Digital image processing. IEEE Trans Nucl Sci 29(3):1075–1086
12. Hu X, Zhang Z, Shan C (2021) Meta-USR: a unified super-resolution network for multiple degradation parameters. IEEE Trans Neural Netw Learn Syst 32(9):4151–4165
13. Shao W, Liu Y, Ye Y (2020) DeblurGAN plus: revisiting blind motion deblurring using conditional adversarial networks. Signal Process 168:107738
14. Tang S, Gong W, Li W (2013) Non-blind image deblurring method by local and nonlocal total variation models. Signal Process 94:339–349
15. Mosleh A, Elmi S, Zargari F (2018) Explicit ringing removal in image deblurring. IEEE Trans Image Process 27(2):580–593
16. Richardson Z (1972) Bayesian-based iterative method of image restoration. J Opt Soc Am A 62(1):55–59
17. Lucy B (1974) An iterative technique for the rectification of observed distributions. J Astrophys Astron 79:745–754
18. Yang L, Huang H, Lai H (2014) A novel gradient attenuation Richardson-Lucy algorithm for image motion deblurring. Signal Process 103:399–414
19. Tai Y, Tan P, Brown M (2011) Richardson-Lucy deblurring for scenes under a projective motion path. IEEE Trans Pattern Anal Mach Intell 33(8):1603–1618

References

Chapter 6
Image Stitching

Xiaoying Chen, Zhengjun Liu, and Yutong Li

Abstract Image stitching is the process of integrating multiple images with overlapping regions into a unified wide-angle or panoramic image. It involves techniques such as image registration and image fusion to ensure seamless transition between images. With the advancement of algorithms, image stitching has been widely used in the fields of virtual reality, medical imaging and remote sensing.

6.1 Overview of Image Stitching

6.1.1 Definition and Key Steps

Image stitching is one of the key technologies in the field of computer vision, aiming to integrate multiple images with overlapping regions into a seamless, high-resolution panoramic or wide-angle image. The core of image stitching lies in the two key steps of image registration and image fusion, as Fig. 6.1.

In the initial stage of registration, it is necessary to accurately detect the feature points of the image to be spliced. Subsequently, through a series of complex calculations, the inter-image geometric transformation relationship and deformation model are established to complete the image registration. Image registration is a key step to ensure that overlapping regions can be seamlessly fused. Image fusion can effectively eliminate the splicing traces between images and realize seamless splicing.

The template matching method used in the early days, by searching for the most similar template in the overlapping region of an image with another image to achieve

X. Chen · Z. Liu · Y. Li (✉)
School of Physics, Harbin Institute of Technology, Harbin, China
e-mail: yutong_li@hit.edu.cn

X. Chen
e-mail: xychen@hit.stu.edu.cn

Z. Liu
e-mail: zjliu@hit.edu.cn

Z. Liu and Y. Li (eds.), *Intelligent Analysis of Optical Images*, Scientific Computation,
https://doi.org/10.1007/978-981-95-3720-4_6

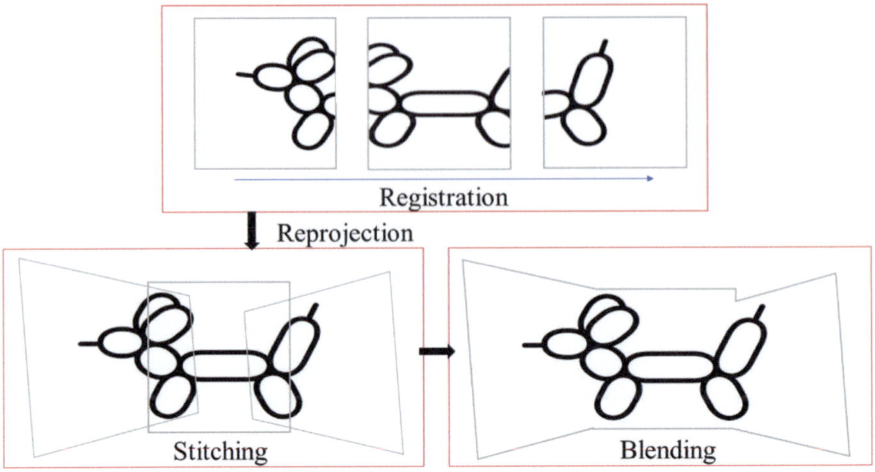

Fig. 6.1 Key steps in image stitching

image registration and splicing, but this method is computationally intensive, inefficient, and sensitive to factors such as illumination and changes in viewing angle of the image.

With the continuous progress of technology, feature-based methods have gradually become mainstream, such as Harris corner detection, scale-invariant feature transform (SIFT), speeded up robust features (SURF) and other algorithms. These algorithms significantly improve the accuracy and robustness of image stitching by extracting the feature points in the image and performing image registration based on the matching relationship of the feature points.

In recent years, local adaptive registration algorithms (e.g., as-projective-as-possible (APAP) [1], adaptive-as-natural-as-possible (AANAP) [2]) have been continuously improved, which has enhanced the quality of image stitching, and the fusion of deep learning technology and image stitching at the same time has brought new development opportunities for the field. The integration of deep learning technology and image splicing has brought new development opportunities in the field, making the processing ability of image splicing in complex scenes significantly enhanced.

6.1.2 Classification of Splicing Methods

According to the different splicing scenes and actual needs, image splicing methods can be roughly divided into three categories (Fig. 6.2). Among them, feature-based splicing method is one of the most commonly used methods due to its advantages of small computing power and fast splicing speed. Due to some uncertainty in the

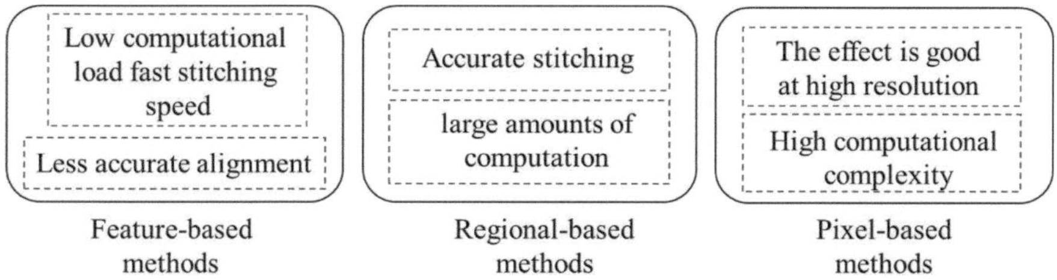

Fig. 6.2 Classification of image splicing methods

selection and matching of feature points, it may lead to the problem of inaccurate registration.

Compared with feature-based stitching methods, region-based methods utilize image region block similarity metrics (e.g., mutual information) to improve the registration accuracy, which is especially suitable for texture-complex scenes, such as medical images, but the computational complexity is high.

In addition, pixel-based methods directly compute pixel-level similarity, which presents optimal performance in areas such as high-resolution satellite remote sensing, and its ultra-high computational complexity relies on hardware acceleration. The three types of methods form a complementary relationship between computational efficiency, registration accuracy and applicable scenes, and the actual selection needs to be considered in combination with resolution requirements, real-time constraints and scene characteristics.

6.1.3 Core Objectives

The core objectives of image splicing include the following three levels:

(1) Geometrically consistent registration: Eliminating translation, rotation, scale scaling and even perspective distortion between images through precise registration algorithms to ensure that pixel points in overlapping regions strictly correspond in spatial coordinates. This process relies on mathematical transformation models, such as the homography matrix for modeling projective transformations of planar scenes, or thin-plate spline (TPS) for describing non-rigid deformations.
(2) Visual seamless fusion: to solve the problems of splicing seams and ghosting in overlapping regions due to lighting differences, inconsistent camera parameters or dynamic changes in the scene, and to achieve a natural transition of the synthesis effect through pixel-level fusion (e.g., weighted averaging, multi-scale pyramid fusion) or seam optimization (e.g., graph cut to find the minimum energy path).
(3) Scene semantic preservation: Preserve the structural features (e.g., straight lines, texture) and semantic information (e.g., object contours) of the original image

during the synthesis process, avoiding the introduction of visual distortion due to geometric deformation or fusion operations, and ensuring that the splicing results are in line with human perceptual habits.

6.2 Image Registration Technology

6.2.1 Model Registration Classification

The core of image stitching technology is image registration. Image registration is the process of integrating different images acquired for the same region at different times, from different viewpoints and with different sensors. It can be unimodal or multimodal, rigid or non-rigid. In rigid registration, image registration is achieved by global transformation models (e.g., projection transformation, affine transformation transformation) and is suitable for flat scene stitching. In non-rigid registration, local deformation fields (e.g., thin plate spline, elastic transformation) need to be constructed to deal with non-planar deformation, and the computational complexity is significantly increased. Panoramic stitching usually adopts unimodal rigid registration to balance efficiency and accuracy.

The mainstream registration methods are divided into two branches according to the driving elements, one branch is the region-based image registration, and the global search method is mainly used for image registration in the early stage, which is extremely computationally intensive. The template matching method is simple and easy to implement compared to the global search method, but both can only solve the translation problem. The extended phase correlation method combines the phase correlation method and the logarithmic polar coordinate transform to solve the rotation and scale scaling problems, but the method is more complicated to compute. Another branch is feature-based image registration, including corner points, edges, speckles and so on. The classical Harris corner detection algorithm, which takes the corner points in the image with rotational and translational invariance as the feature points for image registration. The SIFT algorithm, which is well adapted to translations, rotations, scale scaling and uneven illumination, is a hot research topic at present.

6.2.2 Global Transformation Model

Suppose x_i and x_i' denote multiple pairs of feature points matched in overlapping regions in two images. It is hoped to find a projection transformation matrix or homography transformation matrix H, to correspond to the relationship between x_i and x_i' as

$$x_i' \sim Hx_i \tag{6.1}$$

In the chi-square coordinate system, the vectors x'_i and Hx_i are not the same in scale, but have the same direction, so this linear relationship can be expressed as

$$x'_i \times Hx_i = 0 \tag{6.2}$$

Let $x'_i = (x'_i, y'_i, \omega'_i)$, then Eq. (6.2) is satisfied as follows

$$x'_i \times Hx_i = \begin{pmatrix} y'_i h^{3T} x_i - \omega'_i h^{2T} x_i \\ \omega'_i h^{1T} x_i - x'_i h^{3T} x_i \\ x'_i h^{2T} x_i - y'_i h^{3T} x_i \end{pmatrix} = 0 \tag{6.3}$$

Also with $h^{jT} x_i = x_i^T h^j, \quad j = 1, 2, 3$, the Eq. (6.3) is written in the form of a matrix as

$$\begin{bmatrix} 0^T & -\omega'_i x_i^T & y'_i x_i^T \\ \omega'_i x_i^T & 0^T & -x'_i x_i^T \\ -y'_i x_i^T & x'_i x_i^T & 0^T \end{bmatrix} \begin{bmatrix} h^1 \\ h^2 \\ h^3 \end{bmatrix} = 0 \tag{6.4}$$

Only two rows in Eq. (6.4) are linearly independent, so the above equation is equivalent to solving the least squares solution of the following linear problem when the number of matched pairs of points ≥ 4 as

$$Ah = 0 \quad \text{s.t.} \ \|h\| = 1 \tag{6.5}$$

where: for n points, A is a matrix of $2n \times 9$. Eq. (6.5) can be solved by singular value decomposition (SVD) $A = UDV^{\mathrm{T}}$, h is the last column of the updated V.

In practical applications for image stitching, the Random Sample Agreement (RANSAC) algorithm (described later) is usually used first to remove some mismatched feature point pairs, whose mis-matching is one of the causes of in-point bias. Stitched images obtained using the global transform method usually have more serious ghosting phenomenon, so it is rarely applied directly in image stitching at present. Figures 6.3 show the global transform spliced images of a temple, respectively.

6.2.3 Local Adaptive Registration Methods

To solve the problem of splicing distortion caused by parallax, three types of strategies have been developed by the current mainstream methods: (1) seam-driven splicing, (2) content-holding deformation, and (3) local adaptive transformation. The main-stream local plane transformation methods are prioritized below. Figure 6.4 illustrates

Fig. 6.3 Global transform stitching of a temple

the three mainstream methods and the two technical approaches of the local adaptive registration method among them.

The proposed local plane transform method further breaks through the limitations of the global transform model's geometric expressiveness in complex scenes, and its core idea is to realize more accurate local registration by establishing spatially adaptive mapping functions and estimating independent plane transform relations at different locations of the image. The method is based on the assumption that the scene can be decomposed into multiple approximate planar regions, and the specific implementation is divided into two types of technical paths.

Local homography methods (e.g., APAP, AANAP methods), the APAP method estimates continuously varying homography matrices for each pixel position by shift-weighted DLT $H(x)$, which adapts to medium parallax while maintaining the projection property, but the extrapolated region is prone to degradation to global homography resulting in boundary distortion; mesh optimized morphing (shape-preserving half-projective, SPHP [3]), discretizes the image into a mesh and jointly optimizes the vertex displacements, explicitly balancing the registration accuracy and geometric fidelity through the energy function (registration term+ smoothing term+ shape constraints), which is suitable for low parallax scenarios but may trigger mesh collapsing at sudden depth changes.

The intrinsic limitation is that both of them rely on the segmented plane assumption, which makes it difficult to deal with depth discontinuities caused by non-planar structures or large parallaxes, and therefore they are mainly suitable for aerial photography of the ground, building elevations, and other scenes with explicit planar geometries.

Again assuming that the multi-pair matching feature points of the two images are x_i and x_i', and that for a particular location x_*, the local one-shoulder transformation matrix is H_*, then there is

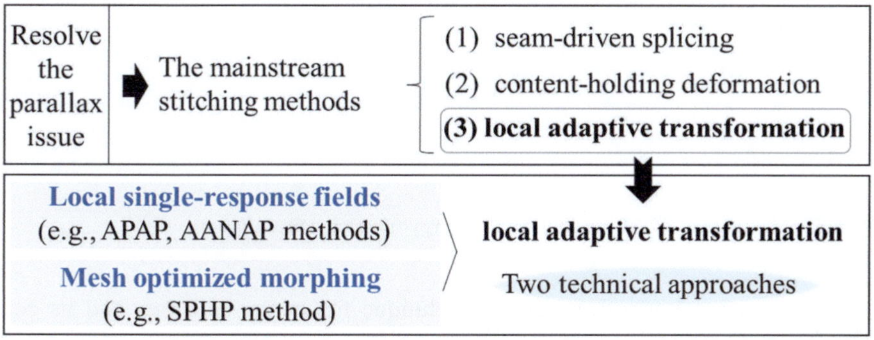

Fig. 6.4 Two technical paths of local adaptive registration methods

$$x'_* \sim H_* x_* \tag{6.6}$$

As shown in Fig. 6.5 (a) and (b), the APAP algorithm further divides the image into a dense grid and then uses Moving DLT to compute the local homography matrix of the grid H_*, adding a Gaussian weight to H_*

$$\omega_*^i = \exp\left(-\|x_* - x_i\|^2/\sigma^2\right) \tag{6.7}$$

where: σ is the scale factor, indicating that the weight of x_i generated at a distance x_* farther away is smaller.

Student's t-weighting was used in the AANAP method.

$$\omega_*^i = \left(1 + \frac{\|x_* - x_i\|}{\nu}\right)^{\frac{-(\gamma+1)}{2}} \tag{6.8}$$

where the parameter ν is used to adjust the rate of decay of the weights with respect to distance. Since the tail of the student's t-weighting distribution decays more slowly, when x_* is far away from the matching points, all the matching points are given similar weighting, which improves the robustness of the algorithm.

From Eqs. (6.7) and (6.8), the weights of neighboring x_* are close, and thus the local transformation matrices are also similar. Therefore, it is too wasteful to estimate the local homography for all x_* in the image, for this reason APAP divides the image into grids of $C_1 \times C_2$, each with a center point of x_*, and pixels in the same grid use the same local transform matrix H_*. In this way, only $C_1 \times C_2$ matrices H_* need to be computed, which significantly reduces the computational complexity. In addition, the AANAP method reduces the perspective distortion that occurs in APAP by gradually converting the local homography transforms to global similar transforms from the overlapping regions outward. The gridded processing of local planar transforms estimates the corresponding planar transform relations for different regions, which can effectively improve the registration accuracy and stitching effect.

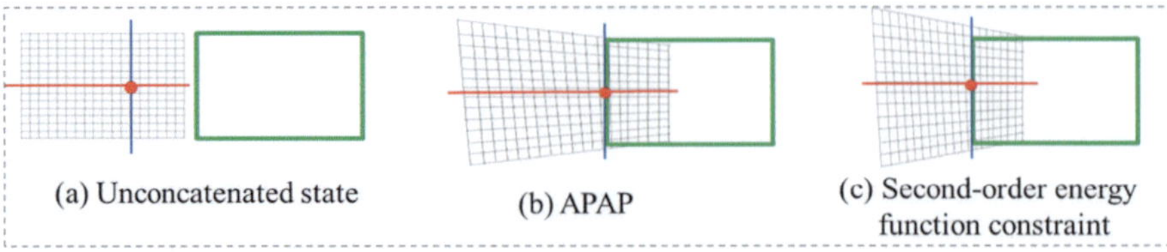

Fig. 6.5 Schematic of APAP algorithm with energy function constraints

SPHP is a local adaptive warping technique for image stitching, and its core objective is to achieve accurate registration in overlapping regions while maintaining geometric naturalness in non-overlapping regions. Its mechanism is realized by segmentation transformation strategy and smooth transition function. The registration accuracy is balanced with shape preservation through energy function minimization optimization as

$$E = E_{\text{align}} + \alpha E_{\text{smooth}} + \beta E_{\text{shape}} \tag{6.9}$$

where E_{align} is the feature point registration error (projection transformation part). E_{smooth} is the neighboring pixel transform continuity constraint (mesh vertex displacement difference). E_{shape} is the similar transformation bias penalty term (e.g. $\|\nabla f - S\|^2$) (Fig. 6.5(c)).

6.3 Image Transformation and Deformation Optimization

6.3.1 Projection Transformation and Deformation Modeling

The dual homography model (DHW) method [4], essentially belongs to a parameterized segmented projective transform method. The mapping relationship between two images is represented by a mixture of two plane transforms, according to which a better registration accuracy is obtained, denoted as

$$H_{ij} = \omega_{ij} H_{background} + (1 - \omega_{ij}) H_{foreground} \tag{6.10}$$

where $H_{background}$ and $H_{foreground}$ denote the homography matrices in the background and foreground planes, respectively; ω_{ij} is the weight of each pixel, which is used to control the contribution of each homography matrix at pixel location (i, j).

DHW targets a specific scene structure, and the model assumes that the scene to be stitched can be approximated as two dominant planes: a ground plane and a far plane (e.g., sky or background) (Fig. 6.6).

Under this assumption, DHW utilizes two different homography transforms to align the image regions belonging to these two planes respectively, constituting a

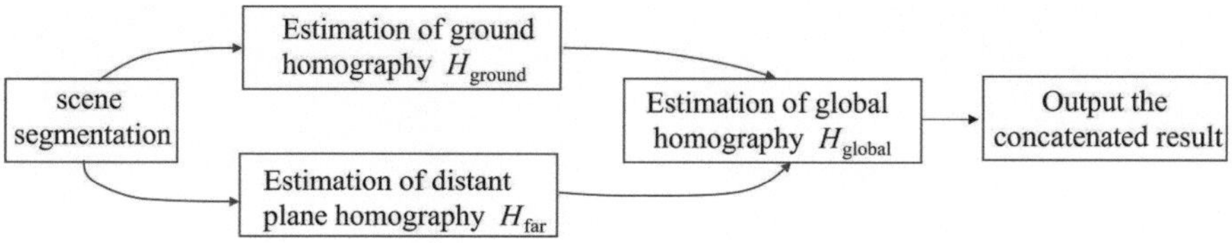

Fig. 6.6 Core mechanisms and technical positioning of DHW

special case of segmented projective transform. Compared to methods relying on a single global homography transform, this dual-transform strategy provides higher flexibility and is better able to deal with parallax or depth discontinuities caused by the two dominant planes.

When the scene structure conforms to its biplane assumption, the DHW model is able to produce good stitching results. However, the core limitation of the model is its strong dependence on the a priori structure of the scene: it is difficult to generalize the model directly to arbitrary scenes containing complex geometries (e.g., multiple significant depth layers or non-planar objects).

The key challenges are how to pre-determine the number of dominant planes required for the scene, and how to efficiently estimate the homography transformation parameters corresponding to each plane. These challenges limit the robustness and utility of the DHW model in generic scenes, prompting the development of subsequent, more generalized multiplanar or segmented transformation methods (e.g., multi-homography warping, MHW). In Fig. 6.7, the target figure is the foreground moving object, while the others are the background. Using the concept of DHW, the two can be separated and processed separately.

Smoothly varying affine (SVA) stitching is an image registration method designed to overcome the limitations of global affine transformations. The core of the method is to construct a spatially smooth varying affine transform field (SVA field). The basic principle is to first initialize the registration function using the global affine transform, and subsequently, under the constraint of maintaining the global affine consistency of the overall transformed field (i.e., affine regularization is imposed), adaptive deformations are allowed to take place in local regions to minimize the registration error among feature points through an optimization framework similar to Expectation Maximization (EM) (Fig. 6.7).

The advantage of the SVA method is that it balances flexibility with extrapolation properties: the local affine deformation allows it to adapt to a certain degree of parallax variation, while its parametric affine nature inherits the advantages of affine transformations when dealing with extrapolation tasks such as image boundaries, occluded regions, etc., which are crucial for generating reasonable stitching results.

However, the method has inherent limitations determined by its affine modeling nature. The affine transformation itself lacks the degrees of freedom required to realize a full perspective transformation (e.g., its strict maintenance of parallelism), resulting in the inability of the SVA field to guarantee global projectivity, which

Fig. 6.7 Schematic of DHW stitching

fundamentally limits its accuracy in scenarios requiring precise perspective registration. More importantly, in extrapolated regions (e.g., image periphery or occlusion boundaries) that lack sufficient matching point data to guide local deformations, affine regularization forces the transformation to degenerate into a global affine one, which often produces significant, perspective-inconsistent geometrical distortions (e.g., unnatural stretches or tilts). While SVA establishes constraints modeling imperfect matches to the true perspective geometry, the APAP approach overcomes its inherent shortcomings by It significantly improves the registration accuracy by directly estimating the local perspective parameters.

When the views of I and I' are not different by rotation alone, or are not planar scenes, the use of basic projective distortions will inevitably produce ghosting effects during the registration process. To solve this problem, the idea is to apply a position-dependent homography transform for each position as

$$x'_* = H_* x_* \tag{6.11}$$

where H_* is estimated from the weighting problem as

$$h_* = \arg\min_h \sum_{i=1}^{N} \left\| w_*^i a_i h \right\|^2 \tag{6.12}$$

constrained by $\|h\| = 1$. The change in scalar weight $\{w_*^i\}_{i=1}^N$, is computed from x^* and x_i as

$$w_*^i = \exp\left(-\frac{\|x_* - x_i\|^2}{\sigma^2}\right) \tag{6.13}$$

where σ is a scale parameter and x_i is the coordinate of the ith point in the source image I to a point in $\{x_i, x_i'\}$.

Intuitively, the projective transformation H_* is able to respect the local structure around x^*, since Eq. (6.13) assigns a higher weight to the data close to x^*. In addition, the transformation H_* changes smoothly as x^* moves continuously within its definitional domain I. This produces an overall transformation that flexibly adapts to the data while maintaining the projective properties as much as possible. We refer to this estimation process as "moving DLT".

The problem in Eq. (6.12) can be written in the form of a matrix as

$$h_* = \arg\min_h \|W_* A h\|^2 \tag{6.14}$$

where the weight matrix $W_* \in \mathbb{R}^{2N \times 2N}$ is composed as follows

$$W_* = \mathrm{diag}\left(\left[\omega_*^1, \omega_*^1, \ldots, \omega_*^N, \omega_*^N\right]\right) \tag{6.15}$$

And the diag() function creates a diagonal matrix from the given vectors. This is a Weighted Singular Value Decomposition (WSVD) problem whose solution is the minimum right singular vector of $W_* A$.

The above equation may be unstable when many of the weights are not significant, such as when the data is sparse or in extrapolated regions x_*. To prevent numerical problems in the estimation process, the weights are added to a small value $\gamma \in [0\ 1]$ as

$$w_*^i = \max\left(\exp\left(-\|x_* - x_i\|^2/\sigma^2\right), \gamma\right) \tag{6.16}$$

This also helps to regularize the distortions, where a larger value of γ reduces the complexity of the distortion; in fact, the distortion reduces to a global projected distortion when $\gamma \to 1$.

Conceptually, the moving DLT can be considered as a projected version of the MLS. In the context of distorting points in 2D image processing, the MLS estimates for each of the x^* affine transformations defined by the matrix $F_* \in \mathbb{R}^{2\times 3}$ as

$$x_*' = F_* \begin{bmatrix} x_* \\ 1 \end{bmatrix} \tag{6.17}$$

$$F_* = \arg\min_F \sum_{i=1}^{N} \left\| w_*^i \left(F \begin{bmatrix} x_i \\ 1 \end{bmatrix} - x_i' \right) \right\|^2 \tag{6.18}$$

Including non-smooth weights $\{w_*^i\}_{i=1}^N$ produces a flexible distortion, but this distortion is ultimately only as affine as possible. In addition, further limiting the overall flexibility of the distortion avoids undesired shearing of the shape.

Based on the moving DLT method for estimating local homography, a method for linearizing it in non-overlapping regions is described below. Explain how to compute the global similarity transform between the reference image and the target image. Since there may be multiple similarity transforms, the one with the smallest rotation angle will be automatically selected as the best candidate in the following. Finally, the details of the AANAP transform, which is constructed by combining the homography or its linearized version with the global similarity transform over the entire image, will be presented.

Let the target and reference images be denoted as I and I', respectively. For a pair of matching points $p = [x, y]^T$ and $p' = [x', y']^T$ between the images I and I', the homography transform $p' = h(p)$ can be expressed as follows

$$h_x(p) = \frac{h_1 x + h_2 y + h_3}{h_7 x + h_8 y + h_9} \tag{6.19}$$

$$h_y(p) = \frac{h_4 x + h_5 y + h_6}{h_7 x + h_8 y + h_9} \tag{6.20}$$

The chi-square coordinates $\hat{p} = [x, y, 1]^T$ and $\hat{p}' = \left[x', y', 1\right]^T$ can be represented by the normalized chi-square transformation matrix $H \in \mathbb{R}^{3\times3}$, where H can be scaled, as

$$\hat{p} \sim H\hat{p} \tag{6.21}$$

The column vectors of the matrix H are $h_1 = [h_1, h_4, h_7]^T$, $h_2 = [h_2, h_5, h_8]^T$ and $h_3 = [h_3, h_6, h_9]^T$. Cross-multiplying both sides of Eq. (6.21) yields

$$o_{3\times1} = \hat{p}' \times H\hat{p} \tag{6.22}$$

This can be rewritten as follows

$$o_{3\times1} = \begin{bmatrix} o_{3\times1} & -\hat{p}^T & y'\hat{p}^T \\ \hat{p}^T & o_{3\times1} & -x\hat{p}^T \\ -y\hat{p}^T & x'\hat{p}^T & o_{3\times1} \end{bmatrix} \begin{bmatrix} h_1 \\ h_2 \\ h_3 \end{bmatrix} \tag{6.23}$$

Denote the 9×1 vector in Eq. (6.23) as h. Since only two rows of the 3×9 matrix in Eq. (6.23) are linearly independent, for a set of N matches $\left\{\hat{P}_i\right\}_{i=1}^N$, and $\left\{\hat{P}_i'\right\}_{i=1}^N$, we can use Eq. (6.24) to estimate h as

$$h = \arg\min_{h} \sum_{i=1}^{N} \left\| \begin{bmatrix} a_{i,1} \\ a_{i,2} \end{bmatrix} h \right\|^2 = \arg\min_{h} \|Ah\|^2 \tag{6.24}$$

where $a_{i,1}$ and $a_{i,2}$ correspond to the two rows of the matrix of Eq. (6.23), respectively. The constraint $\|h\|^2 = 1$ will also be added since the homography transformation has only 8° of freedom.

The mobile DLT framework is introduced to estimate the the local homography by adding localizability constraint weights to the objective function of Eq. (6.24). The localized homographyness located at p_j is estimated as

$$h_j = \arg\min_{h_j} \sum_{i=1}^{N} \omega_{i,j} \left\| h \begin{bmatrix} a_{i,1} \\ a_{i,2} \end{bmatrix} \right\|^2 \tag{6.25}$$

where $W_j = \text{diag}\left(\left[\omega_{1,j}, \omega_{1,j}, \ldots, \omega_{N,j}, \omega_{N,j}\right]\right)$. Eq. (6.25) can be written as $h_j = \arg\min_{h} \left\| W_j A h \right\|^2$. The weights are generated by an offset Gaussian function that assumes that pixels near p_j have higher values and pixels farther away have equal values, i.e., $\omega_{i,j} = \max\left(\exp\left(-\left\|p_i - p_j\right\|^2/\sigma^2\right), \gamma\right)$. The parameter $\gamma \in [0\ 1]$ is an offset used to prevent numerical problems. Note that the local homography can only be calculated in the region where the target image overlaps with the reference image (Fig. 6.8).

For each pixel in the non-overlapping region, its transformation is computed by a weighted linear combination of the local homography in the overlapping region. Here it becomes important to choose a suitable offset to avoid extrapolation artifacts. Setting $\gamma = 0$ can lead to a "wave" effect due to the isotropy of the Gaussian weighting, whereas choosing the right offset gives good results. Even so, perspective distortion in non-overlapping regions is still noticeable when using APAP (Fig. 6.9).

In AANAP, an unbiased moving direct linear transform (DLT) is used in the overlapping region to estimate the local homography and extrapolates it to the non-overlapping region by homography linearization. The extrapolation weighting scheme of AANAP has less dependence on the choice of parameters. Extrapolation of the homography transform to non-overlapping regions produces an extreme and unnatural scaling effect. To understand the reason for this effect, consider a one-dimensional perspective transform as

$$x' = ax + bx/c + d \tag{6.26}$$

If the parameter $\{a, b, c, d\}$ is estimated using a set of correspondent points, including points beyond the range of available correspondent points, then the relationship between x and x' will be nonlinear. This can lead to severe perspective distortion in two dimensions. However, this distortion can be minimized by the linearization transform.

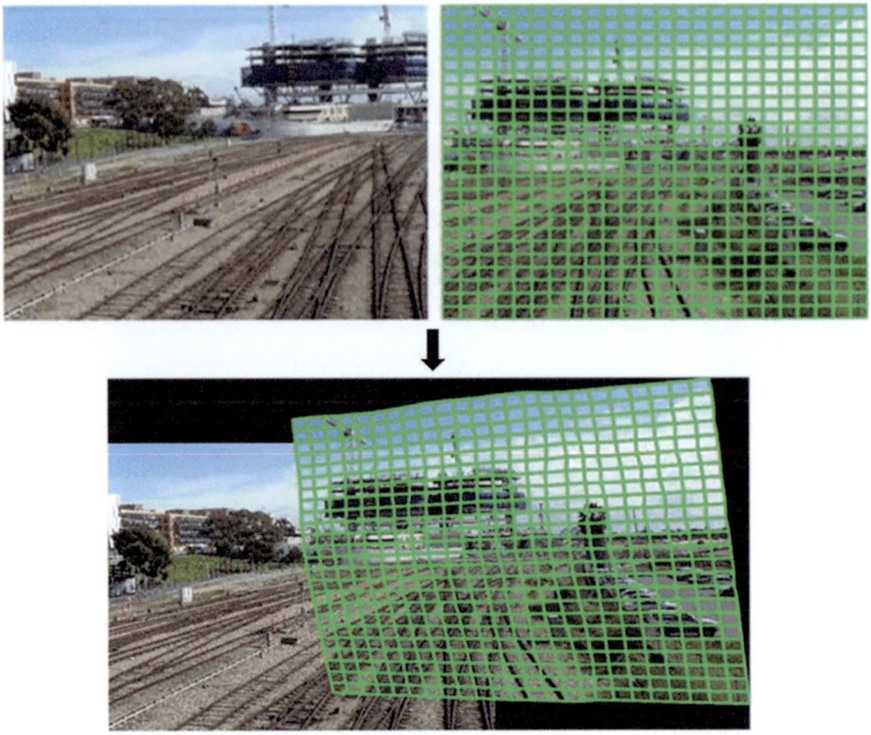

Fig. 6.8 DLT-based APAP stitching model

Fig. 6.9 DLT-based APAP temple stitching

For images, the linearization of the homography at an arbitrary point q near the reference point p can be understood by considering the Taylor series of the homography transformation $h(q)$ at that point, where $h : \mathbb{R}^2 \rightarrow \mathbb{R}^2$ as

$$h(q) = h(p) + J_h(p)(q - p) + o(\|q - p\|) \tag{6.27}$$

where $J_h(p)$ is the Jacobi matrix of the homography transformation h at the point p. The first two terms in Eq. (6.26) provide the optimal linearization for $h(q)$. Since, if h is differentiable at p, then $J_h(p)$ is invertible, the linearization of the homography is an affine transformation.

It is not straightforward to compute the linearization at an arbitrary point q in the non-overlapping region because the boundary between the overlapping region and the non-overlapping region may contain more than one point due to the fact that it is not known where the Jacobi matrix should be computed. Therefore, anchor points on the boundary are considered for linearization and the weighted average of the transformations is computed.

For a set of R anchor points $\{p_i\}_{i=1}^R$ at the boundary, the weighted combination of the linearization takes the following form for the case where there may be different local homography as

$$h^L(q) = \sum_{i=1}^{R} \alpha_i [h(p_i) + J_h(p_i)(q - p_i)] \tag{6.28}$$

Assume that α_i is a function of $\|q - p_i\|$, in particular we consider Gaussian weighting, where $\alpha_i = \exp\left(-\|q - p_i\|^2 / \sigma^2\right)$, or student t-distribution weighting, where $\alpha_i = \left(1 + \|q - p_i\|^2 / \nu\right)^{-(\nu+1)/2}$. Student t-distribution weighting is robust because the tails of this distribution decay more slowly than a Gaussian distribution, so that all anchors are given similar weights when q is far from the anchor, and so this weighting is robust. However, if Gaussian weighting is chosen, the tails should be flattened at the offset parameter γ to avoid "fluctuation" effects.

Extrapolating non-overlapping regions using linearized homography produces less perspective distortion than the results using APAP. This result is similar to the splicing results obtained using the dual-homography warping method. However, there is no need to estimate the two homography in the far plane and the ground plane using our method. AANAP is able to adapt to more complex scenes and is a generalized method compared to the dual-homography method. It does not require the case-specific parameter γ and is therefore less dependent on the choice of parameters.

6.3.2 Deformation Control and Regularization

In response to the serious perspective distortion problem induced by the APAP method when extrapolating projection transformations in non-overlapping regions,

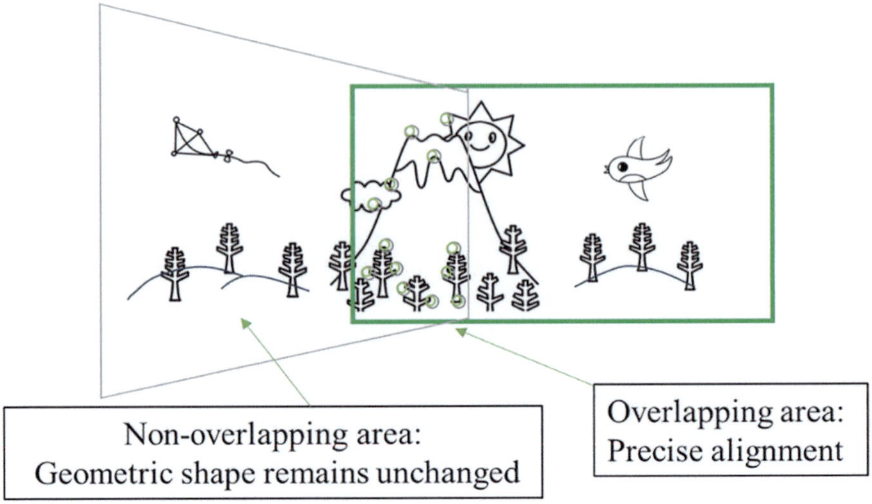

Fig. 6.10 Schematic diagram of SPHP algorithm

the SPHP transform is proposed to improve the geometric fidelity in non-overlapping regions. The SPHP algorithm is a mesh optimization class of algorithms.

The core idea of the method is to analytically decompose the projection transformation along the rotated coordinate axes and design a spatially asymptotic transformation model: In the overlapping region of the image, SPHP maintains the projection transformation consistent with APAP to achieve accurate registration; while in the non-overlapping region, the transformation property is asymptotically transitioned to similarity transformation, as Fig. 6.10. This continuous transition from projection to similarity effectively constrains the scale and angular distortion of the non-overlapping regions, which significantly improves the visual naturalness of the local shapes, and is a local adaptive warping technique for image stitching.

SPHP divides the image into two regions and applies different transformations. The total deformation field $f(x)$ can be expressed as

$$f(x) = (1 - \lambda(x)) * H(x) + \lambda(x) * S(x) \tag{6.29}$$

where $H(x)$ is a local homography transform, $S(x)$ is a similar transform, $\lambda(x)$ is a transition function (usually a radial basis function) which is 0 in the overlapping region and 1 in the non-overlapping region.

The overlapping region uses a projection transformation (homography matrix) to realize the exact registration of the feature points as

$$H(x) = \arg\min_{h} \sum_{i} w_i \|y_i - Hx_i\|^2 \tag{6.30}$$

where, w_i is the feature point weights, x_i and y_i are the matching point pairs.

The non-overlapping regions are forced to degenerate into similar transforms, retaining only translation, rotation and uniform scaling to avoid perspective distortion. The similarity transform equation as

$$S(x) = s \cdot R \cdot x + t \tag{6.31}$$

where s is the scaling factor. R is the rotation matrix. t is the translation vector.

The smooth transition of the transformation is realized by the transition region smoothing function, which is generally realized by the radial basis function as

$$\lambda(x) = 1 - \exp\left(-\frac{\|x - x_0\|^2}{\sigma^2}\right) \tag{6.32}$$

where x_0 is the boundary point of the overlap region, and σ is the attenuation coefficient to control the transition speed. The attenuation coefficient σ controls the width of the transition region: the larger the value, the smoother the transition, the smaller the value, the sharper the boundary. Typical value is $\sigma \in [5, 20]$.

To calculate the specific steps, firstly, use RANSAC to estimate the global homographyness matrix H_{global} as the initial value of the projection transformation. Secondly, based on the boundary feature points in the overlapping region, the similarity transformation parameters are solved by the least squares method as

$$\min_{s,R,t} \sum_j \left\| y_j - \left(s \cdot R \cdot x_j + t\right) \right\|^2 \tag{6.33}$$

where x_j is the boundary point.

Finally, the registration accuracy and shape preservation are balanced by energy function minimization optimization by Eq. (6.9). The weights α and β balance registration accuracy and shape preservation: the larger α is, the smoother the deformation field is, and the larger β is, the stronger the shape preservation is. Usually $\alpha = 1.0$, $\beta = 0.5$.

SPHP balances registration accuracy and shape preservation in a lightweight framework through projection, similar hybrid transformation and radial transition function. Its core value is to solve the extrapolation distortion problem of traditional projection transformations, which is suitable for low parallax scenes such as aerial ground and building façade. However, the adaptability to complex depth structures is insufficient and needs to be further optimized in combination with depth-aware models such as Depth-Aware SPHP.

For multi-planar scenes or complex registration problems with significant non-rigid deformations, parametric models often fail due to insufficient degrees of freedom. Such scenarios rely on nonparametric deformation models, the core idea of which is to directly fit a continuous transformation field through a dense correspondence of points without the need to preset the analytical form of the transformation. PS, as a representative of such models, solves for the transformation function by minimizing the following objective function as

$$E(f) = \sum_{i=1}^{N} \|y_i - f(x_i)\|^2 + \lambda \iint \left[\left(\frac{\partial^2 f}{\partial x^2}\right)^2 + 2\left(\frac{\partial^2 f}{\partial x \partial y}\right)^2 + \left(\frac{\partial^2 f}{\partial y^2}\right)^2 \right] \mathrm{d}x\mathrm{d}y \tag{6.34}$$

The first term is the fitting error (x_i and y_i are the pair of corresponding points), which ensures the registration accuracy. The second term is the bending energy regularity term, which enforces the global smoothness of the transform and suppresses the physically implausible violent deformations; and λ is the trade-off parameter, which controls the balance between the smoothness and the strength of the registration.

The TPS has a high degree of freedom to accurately model arbitrary local deformations (e.g., surface deformation, non-planar parallax), and theoretically completely eliminates feature point registration errors. The bending energy term simulates the deformation of an elastic thin plate, and its minimization guarantees the smoothness and differential continuity of the transformation (C^2 continuity). The regular term constraints allow the deformation in the non-overlapping region to decay naturally to the affine transformation, avoiding the extrapolation distortion problem of APAP.

Its inherent defects are: first, overfitting sensitivity, susceptibility to noise at low λ values, and high-frequency oscillatory distortion (e.g., the "tent effect") in the sparse point set region; second, $O(N^3)$ complexity restricts the application of large-scale matching points (the efficiency plummets at $N > 10^3$); and third, localized geometrical distortions, which are not able to maintain the rotational/scaling invariance, leading to angular shifts (straightness) in the non-overlapping region. angular offset (right angle to obtuse angle) or texture stretching in overlapping regions; and fourth, occlusion region failure, where regions without match point constraints rely on regular term derivation, which may trigger unphysical displacements (e.g., background fracture).

Given two overlapping images I_p and I_q and their matching points $p_i = (x_i, y_i)^{\mathrm{T}}$, $q_i = (u_i, v_i)^{\mathrm{T}}$, $i = 1, \ldots, n$. The projection of the computed I_p in I_q is denoted by $p_i' = \left(x_i', y_i'\right)^{\mathrm{T}}$. The parallax error on p_i' is represented by the projection deviation $g_i = p_i' - q_i = (g_i, h_i)^T$. For input images with parallax, $g_i \neq 0$.

Elastic morphing technique was introduced to solve the medical image registration problem with multiple morphing methods with different properties to meet specific application requirements. Radial Basis Function (RBF) is a powerful tool for image deformation given some anchor points. The TPS with a simple RBF type is chosen here to characterize image morphing, which excels in both registration quality and efficiency.

Specifically, the deformation in the image I_q can be denoted as $\mathbf{g}(x, y) = (g(x, y), h(x, y))$, where $g(x, y)$ and $h(x, y)$ denote the deformation in the x and y directions, respectively. For simplicity, we use $g(x, y)$ instead of $\mathbf{g}(x, y)$ in the following analysis, since the computations of $g(x, y)$ and $h(x, y)$ are identical and independent of each other. The energy function of the optimal deformation consists of two terms: the registration term J_D and the smoothing term J_S, which are defined as follows

$$J_D = \sum_{i=1}^{n} \left[g(x_i', y_i') - g_i\right]^2 \tag{6.35}$$

$$J_S = \iint_{(x,y)\in\Omega} \left|\nabla^2 g\right|^2 \mathrm{dxdy} \tag{6.36}$$

The overall energy function to be minimized is

$$J_\lambda = J_D + \lambda J_S \tag{6.37}$$

According to the basic theory of TPS, the optimal solution to minimize Eq. (6.37) can be expressed as

$$g(x, y) = \sum_{i=1}^{n} w_i \phi_i(\mathbf{x}) + \alpha_1 x + \alpha_2 y + \alpha_3 \tag{6.38}$$

where $\phi_i(\mathbf{x}) = \left|x - p_i'\right|^2 ln\left|x - p_i'\right|$ is the radial basis function. The coefficients $w = (\omega_1, \ldots, \omega_n)^{\mathrm{T}}$ and $\mathbf{a} = (\alpha_1, \alpha_2, \alpha_3)^{\mathrm{T}}$ can be calculated by solving the following system of linear equations

$$\begin{pmatrix} K + \lambda C_{2,2} I & P \\ P^T & 0_{3\times 3} \end{pmatrix} \begin{pmatrix} w \\ a \end{pmatrix} = \begin{pmatrix} f \\ 0_{3\times 1} \end{pmatrix} \tag{6.39}$$

where $C_{2,2} = 8\pi$ is a constant, $K = \varphi_i p_j' \in \mathbb{R}_{n\times n}$, $\mathbf{P} = \left(\hat{p}_1', \ldots, \hat{p}_n'\right)^T \in \mathbb{R}_{n\times 3}$, $f = (g_1, \ldots, g_n)$. Eq. (6.38) has a unique solution.

The solution of this transformation is obtained when the number of matches $n \geq 4$. By substituting w and a into Eq. (6.39), the transformation at any position $(x, y)^{\mathrm{T}}$ in the source image can be obtained. In order to speed up the computation, a uniform grid of $C_x \times C_y$ cells is set up on the image plane. The deformation is first computed at the grid nodes and then linearly interpolated to all other pixels.

Based on the calculated deformation $g(x, y)$, I_q can be warped to remove the misregistration with I_p. The distorted image is denoted as $\tilde{I}_q$. For any point $q_0 = (x_0, y_0)^{\mathrm{T}}$ in I_q and its corresponding point $q = (x, y)^{\mathrm{T}}$ in $\tilde{I}_q$, we have

$$\begin{cases} x_0 = x - g(x, y) \\ y_0 = y - h(x, y) \end{cases} \tag{6.40}$$

The twist operation is realized by calling the inverse mapping with bilinear interpolation. Seamless panoramas are synthesized by re-projecting and fusing the aligned images on the panoramic image plane.

6.3.3 Outliers Removal

The processing of natural images in the feature detection and matching phase of the image stitching process faces inherent uncertainties: positional errors (localization sub-pixel offsets) and matching errors triggered by texture repetitions, lighting variations, local occlusions, etc. are almost inevitable.

These outliers, if not effectively suppressed, lead to serious deviations from the true geometric relationships in subsequent deformation model estimates. In order to achieve accurate registration, a robust estimation mechanism becomes necessary, the mainstream method adopts the Randomized Sampling Consistency Algorithm (RANSAC) as the core tool for outlier removal.

The core idea lies in: fitting the transformation hypothesis by iteratively randomly sampling the minimum set of points, and assessing the quality of the model based on the number of interior points (matching points conforming to the model), and ultimately selecting the transformation with the highest proportion of interior points as the optimal solution. It is worth noting that, from the perspective of computational efficiency optimization, global transformations (e.g., homography matrices, which require only 4 pairs of matching points to solve) are often preferred as the minimum solver for RANSAC due to their few parameters and fast solution.

The strategy first uses RANSAC to estimate a global consistency transform (e.g., global homography) to obtain an initial set of interior points while filtering out significant outliers; this clean data is then used as an input to drive the fine estimation of subsequent local distortion models (e.g., APAP, TPS). This hierarchical structure of "global coarse screening → local fine-tuning" balances robustness and computational feasibility.

There is a contradiction between the global RANSAC and the local adaptive model, which leads to two negative consequences: (1) if the projection deviation of the interior points is larger than the threshold, they may be removed incorrectly; (2) if the projection deviation of the anomalies is smaller than the threshold, they may be retained incorrectly.

Figures 6.11 and 6.12a show the result of using global RANSAC to cull outliers, where some important matches on the ground are incorrectly removed while a few outliers are retained. In this case, it is difficult to achieve accurate registration. Figure 6.12b illustrates the result of APAP stitching, which contains obvious artifacts.

For the problem of local outlier removal in image registration, there are fewer related studies. In the wide-baseline image stitching method, the outlier points can be removed by a local DLT method. The homographyness of each feature point is calculated by applying DLT in the R-pixel neighborhood of each feature point. If the residual error of a match is less than a threshold γ, the match is considered as an interior point. The localized DLT method is effective for textured regions, but fails for regions with insufficient features. Specifically, if the number of matching features in the neighborhood of a location is less than four, its homography cannot be reliably estimated. Figure 6.12c illustrates the outlier removal results for a local DLT with settings $R = 50$ and $\gamma = 5$. Some of the interior points are incorrectly removed

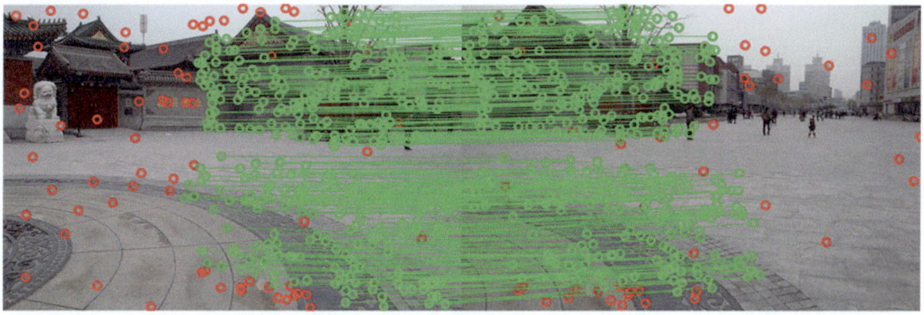

Fig. 6.11 RANSAC result graphs

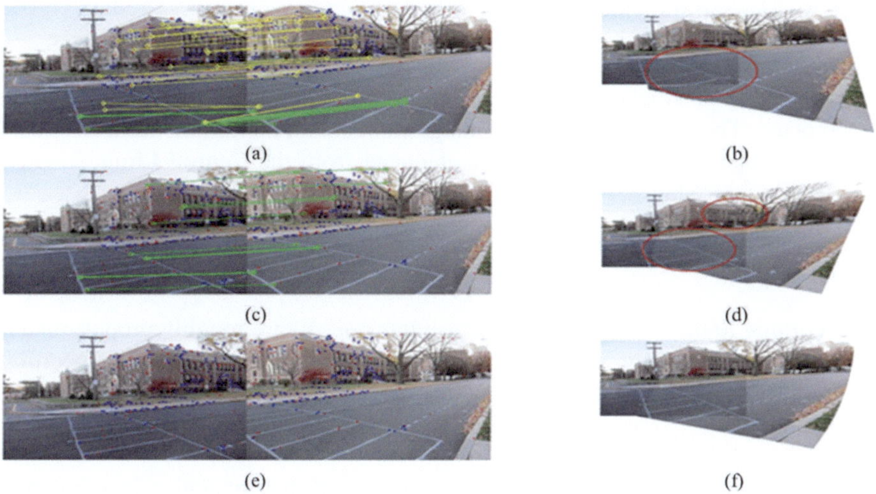

Fig. 6.12 **a** Results using RANSAC, **b** APAP splicing results, **c** (R = 50 and $\gamma = 5$) localized, DLT outlier removal results, **d** demonstrating APAP splicing results combined with local DLT, **e** robust feature refinement results, **f** corresponding splicing result plots

because there are not enough matches around some of them to support judging their correctness. Figure 6.12d shows the splicing results of APAP combined with local DLT.

For structural deformation and TPS, the registration terms are constructed based on a small number of anchor points. Therefore, the deformation is very sensitive to the matching error. In other words, the presence of outliers in the matched data can lead to severe distortion. Outlier removal in image elasticity registration is very important. Due to these shortcomings, the application of deformation functions in natural image stitching is severely limited.

Analyzing the distribution of weights ω_i (x-direction) and υ_i (y-direction) attached to the RBF $\varphi_i(x)$ leads to a probabilistic model for robust feature refinement. As a post follow-up step after the relaxed global RANSAC, it is able to adaptively remove local outliers. Since ω_i and υ_i are computed globally, the model is not affected by problems caused by insufficient matching. The feature refinement results of the model are shown in Fig. 6.12e, and the results are better than those in Fig. 6.12a, c. It is worth noting that the proposed feature refinement model involves the manipulation of the elastic deformation model weights. Therefore, it has to be used together with the elastic local registration model.

To facilitate the quantitative analysis, two idealized assumptions are first made on the matching data.

Assumption 6.1 The changes in projection deviation between two neighboring pixels are independent and identically distributed.

Assumption 6.2 Each anchor point is separated from all other anchor points by a sufficient number of pixels.

In the above assumption, the term "pixel" is not a real unit of measure for a digital image, but is an abstract concept that refers to the projection of a field point onto the image plane. Therefore, an idealized image can contain an infinite number of such pixels. An interesting corollary can be deduced from the Central Limit Theorem based on Assumptions 6.1 and 6.2.

Corollary 6.1 *The difference between the deviations*g_i *and* g_j, $i \neq j$ *obeys a normal distribution.*

According to the nature of TPS, the weights (ω_i, υ_i) are roughly proportional to the difference between g_i and the projection deviation of other anchors. According to Corollary 6.1, the values ω_i and υ_i also roughly obey the normal distribution. Specifically, $\omega_i \sim N(0, \sigma_\omega)$ and $\upsilon_i \sim N(0, \sigma_\upsilon)$. ω_i are equal to 0 on average, i.e.,$\sum_{i=1}^{n} \omega_i = 0$.

Given a pair of matches $\{p_i, q_i\}$, the correctness of their matching is represented by a binary variable $\tau_i \in \{0, 1\}$. We denote by p_1 the proportion of interior points in all matches, i.e., $p(\tau_i = 1) = p_1$ and $p(\tau_i = 0) = 1 - p_1$. Considering the event $A_i = \{|\omega_i| > t\sigma_\omega\}$, we have $p(A_i|\tau_i = 1) = 2(1 - \Phi(t))$, where $\Phi(\cdot)$ is the standard normal distribution function. The probability A_i meets

$$\begin{aligned} p(A_i) &= \sum_{k=\{1,2\}} p(A_i|\tau_i = k)p(\tau_i = k) \\ &\le p(A_i|\tau_i = 1)p(\tau_i = 1) + p(\tau_i = 0) \\ &= 1 - (2\Phi(t) - 1)p_1 \end{aligned} \tag{6.41}$$

Under the condition of $|\omega_i| > t\sigma_\omega$, the posterior probability of $\tau_i = 0$ can be calculated according to Bayes' rule as

$$p(\tau_i = 0|A_i) = \frac{p(\tau_i = 0)}{p(A_i)} \ge \frac{1 - p_1}{1 - (2\Phi(t) - 1)p_1} \tag{6.42}$$

According to the 3σ rule, t is set to 3 such that $\Phi(t) = 0.9987$. If $p_1 \in [0, 0.9973]$, then $p(\tau_i = 0|A_i) = 0.5$.

6.3.4 Global Similarity Constraints

In the next section, methods are presented to reduce distortion so that the panorama looks more natural. The Natural Image Stitching based on Global Similarity Prior (NISwGSP) method solves the problem of exact image registration more generically by smoothing the non-overlapping regions by global similarity and combining it with a local similarity optimization term to minimize the distortion and form a more natural stitched image.

If the global similarity transform can approximate the camera motion between the target image and the reference image, then the estimated similarity transform can be used to compensate for the camera motion. However, using all point matching to find the global similarity transform may lead to a non-optimal solution, especially if the overlapping region contains different image planes. This problem is evident in the SPHP stitching results. SPHP uses the global homography transform to uniquely determine the global similarity transform, which may not approximate the camera motion well.

Given that there are multiple planes in the scene as well as the image projection plane at focal length, each plane in the scene has an angle of intersection with the image projection plane, and each plane in the scene corresponds to a local homography transform. The homography transform corresponding to the plane most parallel (with the smallest angle of intersection) to the image projection plane can be used to derive the optimal similarity transform that represents the camera motion (Fig. 6.13).

A method to compute the optimal similarity transform between the reference image and the target image is described below by the following way. After obtaining feature point matches, outliers are first removed using the RANSAC algorithm with a threshold of ε_g. Then, the RANSAC algorithm with a threshold of ε_l is used to find the homographyness of the plane with the maximum number of interior points, where $\varepsilon_l < \varepsilon_g$, and remove these interior points. This process is repeated until the number of interior points is less than η. Each set of matched interior points is used to compute separate similarity transformations. Then, the rotation angles corresponding to these transformations are checked and the one with the smallest rotation angle is selected.

Figure 6.14 shows an example of the grouping results. The green and yellow circles in the figure belong to two different groups of point correspondences. The red circles do not belong to either group. In this example, the group in which the yellow points are located generates the optimal global similarity transform with the smallest rotation angle.

After the global similarity transform is computed, it is used to adjust the distortion of the target image to reduce the overall perspective distortion of the panorama. If the transform is adjusted only in non-overlapping regions, the stitching result

Fig. 6.13 Results of temple stitching using SPHP

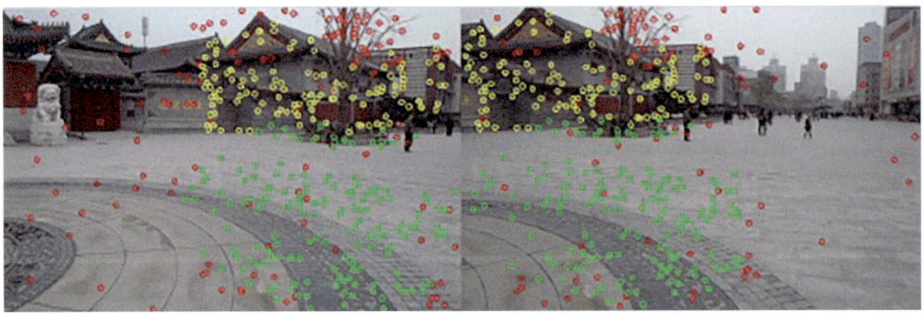

Fig. 6.14 Feature point groupings used to compute the optimal global similarity transformation

may have an unnatural visual effect. Therefore, it is necessary to update the local transformations gradually.

Use the following equations to convert the entire target image to a global similarity transform as

$$\hat{H}_i^{(t)} = \mu_h H_i^{(t)} + \mu_s S \tag{6.43}$$

where, $H_i^{(t)}$ is the first ith local homography matrix and $\hat{H}_i^{(t)}$ is the updated local homography matrix. S is the global similarity transform. μ_h and μ_s are the weighting factors. The superscript (t) denotes the target image and the superscript (r) denotes the reference image. Also constrain $\mu_h + \mu_s = 1$, where μ_h and μ_s are between 0 and 1. They are computed as

$$\mu_h(i) = \frac{\left\langle \overline{\kappa_m p(i)}, \overline{\kappa_m \kappa_{\overline{M}}} \right\rangle}{\left| \overline{\kappa_m \kappa_{\overline{M}}} \right|} \tag{6.44}$$

$$\mu_s(i) = 1 - \mu_h(i) \tag{6.45}$$

where κ is the projection point of the distorted target image on the vector $\overrightarrow{o_r o_t}$. o_r and o_t are the center points of the reference image and the distorted target image. κ_m and κ_M are the points where the value of $\langle \overrightarrow{o_r p(i)}, \overrightarrow{o_r o_t} \rangle$ is the smallest and the largest, respectively. Here, $p(i)$ is the index of the ith position in the final panorama.

Updating the distortion of the target image using the global similarity transform will result in misregistration of the overlapping regions of the reference image and the target image that have been previously aligned. Therefore, we need to compensate for these changes by properly propagating the changes in the target image to the reference image. The local transform of the reference image can now be obtained as

$$\widehat{H}_i^{(r)} = \widehat{H}_i^{(t)} \left(H_i^{(t)} \right)^{-1} \tag{6.46}$$

The robust elastic distortion model for aligning overlapping regions has been introduced earlier. However, directly extrapolating the distortion to non-overlapping regions can lead to overfitting problems. This tendency can be mitigated by smoothly transitioning the local distortions to the estimated global transformations, in particular the global homography matrix H. This can be done by using the local distortions in the non-overlapping regions. This can be easily achieved by gradually reducing the distortion function $g(x, y)$ to zero in the non-overlapping region. The suppressed distortion $g_s(x, y)$ can be expressed as follows (Fig. 6.15)

$$g_s(x, y) = \eta g(x, y) \tag{6.47}$$

The scaling parameter η changes smoothly from 1 to 0 as the point (x, y) moves gradually away from the overlapping region as

$$\eta = 1 - \frac{\max(0, x - x_u, x_l - x, y - y_u, y_l - y)}{d_s} \tag{6.48}$$

where x_l and x_u are the boundaries of the overlapping region in the x and y directions, respectively, and d_s is the smooth transition width (in pixels), whose setting is proportional to the maximum deviation. Specifically

$$d_s = K \times \max\{|g_1|, \ldots, |g_n|, |h_1|, \ldots, |h_n|\} \tag{6.49}$$

where K is a scale parameter. Outside the excess region, the suppressed distortion $g_s(x, y)$ becomes 0, thus maintaining the global projectivity of the non-overlapping region well. Figure 6.15 presents some comparison results of image stitching.

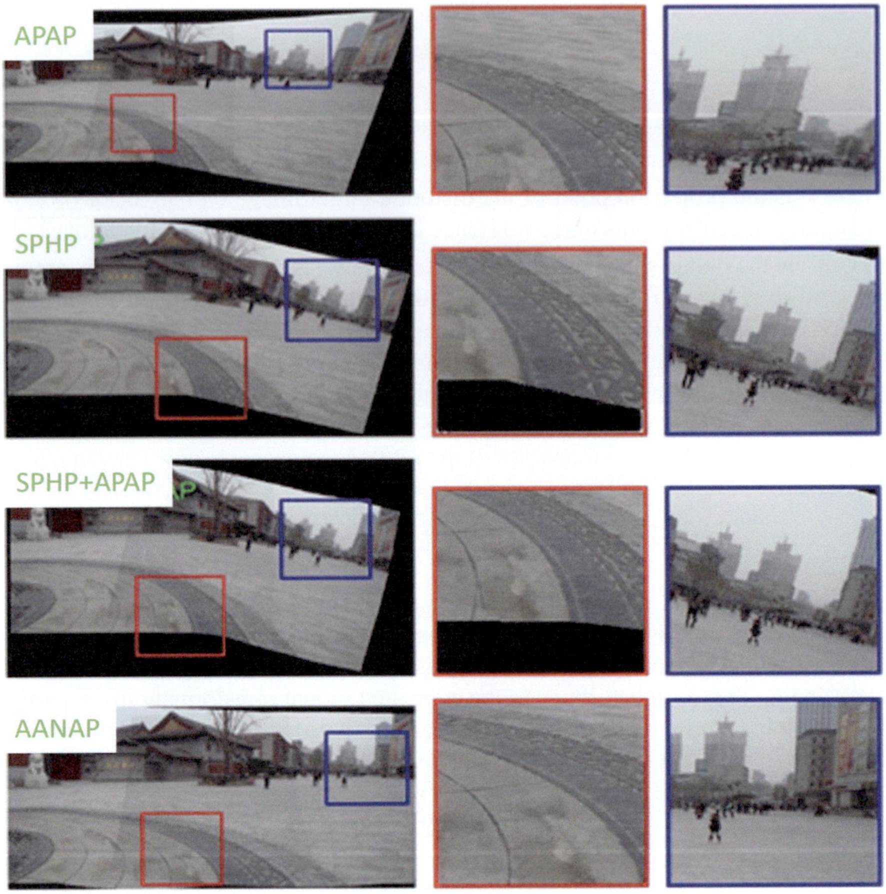

Fig. 6.15 Comparison of several image stitching techniques

Compared with other methods, the AANAP algorithm demonstrates better shape preservation ability in the non-overlapping area.

The above distortions involve only the manipulation of the image content, independent of the type of transform. The global homography transform is used here because it provides a good approximation of the local projectivity of the overlapping regions. However, when stitching wide-field-of-view images, the global homography transform may lead to projection distortions in non-overlapping regions, as shown in the extensively tested temple dataset in Fig. 6.16a. Some state-of-the-art methods use the similarity transform because it performs well in mitigating projection distortion. By combining these two transformations, it is possible to construct an overall spatially varying patchwork field that is both accurately aligned and has natural projectivity, as Fig. 6.16b, which looks more natural than Fig. 6.16a. Unlike the local model, distorting the input image before reprojection removes the misregistration before

Fig. 6.16 Comparison of different global transforms: **a** using homography as global transform, **b** result of combining distortion with similarity transform

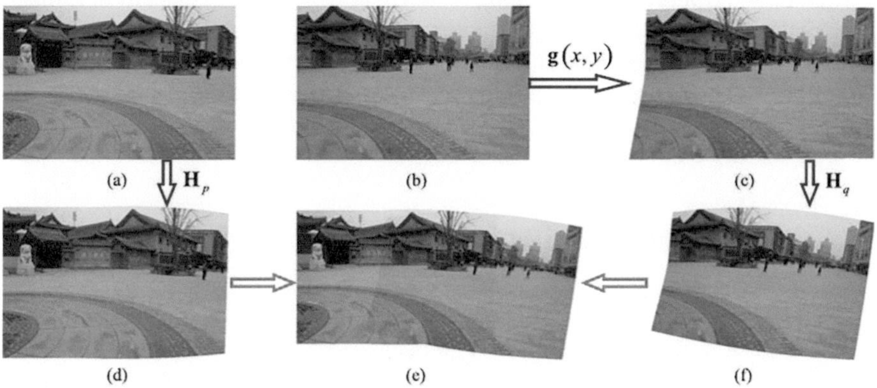

Fig. 6.17 Elastic deformation combined with global similarity transformation: **a**, **b** are the input images, **b** is transformed using the deformation functionc number $g(x, y)$ to deform to get the deformed image (**c**), **c** utilizes H_q to transform to get (**f**), (**d**) and (**f**) fuse to form (**e**)

applying the transform. The subsequent stitching process can then be interpreted as a manipulation of the parallax-free image, which means that any global transform can be applied without affecting the registration accuracy. As a result, the proposed warping model has a high compatibility with various transform types.

After receiving the refined point matching $\{p_i, q_i\}$, $i = 1, \ldots, n_r$, the estimation is performed. Meanwhile, Fig. 6.17a is transformed using H_q to obtain Fig. 6.17d. Finally, the panoramic image Fig. 6.17e is synthesized by Fig. 6.17d, f.

6.4 Image Fusion and Splice Elimination

Image fusion is a key step in the image stitching process, and its core task is to eliminate the visual inconsistencies in the overlapping regions, including brightness differences, color differences, seam traces, and ghosting, to produce a seamless, visually coherent panoramic image after completing the geometric registration. Essentially, it is the use of mathematical methods to achieve a smooth transition of pixel values so that images from different sensors are fused into a single comprehensive image.

Seam removal differs from image fusion in that only brightness and contrast are adjusted, and no additional information is added by the process. Image seams usually appear at the boundary between two images, causing large pixel variance or gradient differences. Critical seam removal is needed to reduce this difference (variance) to an acceptable range and make the transition in the overlapping region smoother.

There are pixel weighting based methods, transform domain methods and seam optimization methods. Pixel-weighting based seam removal operates directly on the overlapping region, but does not achieve the desired results when the external environment or camera settings of the two images differ significantly.

Therefore, the optimal seam method is used to find the optimal seam by optimizing an objective function that aims to minimize the differences near the overlapping regions. Removing seams in the transform domain first decomposes the image by a specific transform, then applies different rules for different frequency bands, and finally reconstructs and outputs the image after removing the seams.

6.4.1 Pixel Level Fusion Method

When using pixel level fusion methods, the images can be further categorized into: pixel level fusion, feature level fusion and decision level fusion based on the representation layer of the image. Pixel level fusion is the commonly used fusion method, which refers to the direct averaging, weighted averaging and other operations on the pixel gray values, which is computationally intensive. Feature-level fusion is to first extract features such as contours, textures, corners, etc. to the image, and then analyze, process and integrate these features to get the fused features, which is faster.

Decision-level fusion is based on feature-level fusion, the feature information is recognized and classified, and then based on cognitive models and discriminative systems to judge the credibility measure of the decision, the fusion effect is better and faster. However, the algorithms of feature-level fusion and decision-level fusion are complex, more original image information is lost in the fusion process, and it may cause distortion of the fused image, which is less applied.

The simplest method in pixel-level fusion is the direct averaging method, which averages the gray values of the overlapping regions of the image directly after registration, avoiding the overflow caused by the direct addition of gray values. Let $f(x, y)$ be the fused image of $f_1(x, y)$ and $f_2(x, y)$, then the direct averaging method can be expressed as follows

$$f(x, y) = \begin{cases} f_1(x, y), & (x, y) \in f_1 \\ [f_1(x, y) + f_2(x, y)]/2, & (x, y) \in (f_1 \cap f_2) \\ f_2(x, y), & (x, y) \in f_2 \end{cases} \tag{6.50}$$

The algorithm is simple and easy to implement, but the fusion quality is poor, and there may still be obvious splicing gaps in the fused image.

Weighted averaging method is to calculate the average value after weighting the grayscale values of the overlapped areas of two images according to certain rules. Let $f(x, y)$ be the fused image of $f_1(x, y)$ and $f_2(x, y)$, then the weighted average method is expressed as follows

$$f(x, y) = \begin{cases} f_1(x, y), & (x, y) \in f_1 \\ w_1 f_1(x, y) + w_2 f_2(x, y), & (x, y) \in (f_1 \cap f_2) \\ f_2(x, y), & (x, y) \in f_2 \end{cases} \tag{6.51}$$

where w_1 and w_2 are the weights of the gray values of the corresponding pixels of the images f_1 and f_2 respectively, and $w_1 + w_2 = 1, 0 \leq w_1$ and $w_2 \leq 1$ are often used to dynamically select the weights w_1 and w_2 using the asymptotic method, and $w_1 = x_2 - x_i/x_2 - x_1$ and $w_2 = 1 - w_1 = x_i - x_1/x_2 - x_1$, where x_1 is the left boundary of the part to be fused, x_2 is the right boundary of the part to be fused, x_i is the horizontal coordinate of the pixel to be fused, and $x_1 \leq x_i \leq x_2$.

The weighted average method is simple, efficient and has good fusion effect, so it is more widely used.

6.4.2 Transform Domain Fusion Method

Fusion methods in transform domain are used to decompose the image by specific transforms and apply different rules for different frequency bands. By fusing in the transform domain, the seams in the stitched image can be reduced. The common transforms are Fourier Transform, Wavelet Transform and Curvelet Transform.

The core process is as follows: firstly, N-level decomposition is realized and different frequency coefficients corresponding to each level are obtained. Then, different rules are applied to the high-frequency and low-frequency bands, respectively, in order to preserve both contour and detail information, since the high-frequency band contains details while the low-frequency band contains contours. Finally, the image is reconstructed by inverse transform.

The curvilinear wavelet transform is an extension of the wavelet transform and is a non-adaptive technique for multi-scale object representation. Compared to wavelets, curved waves are able to represent curves more smoothly. Local energy based fusion rules are better than single pixel based fusion rules. Detail extraction is enhanced by image chunking and fine fusion is achieved by applying ridge wave transform to each sub-chunk. The combination of curve wave transform and genetic algorithm is used for medical image fusion.

6.4.3 Seam Optimization Based on Graph Cutting

In practical applications, optimal seam splicing can usually be regarded as a shortest path problem or an energy minimization problem. This method can effectively eliminate the artifacts that may arise during the splicing process and optimize the image splicing effect. The optimal seam stitching method aims to solve the problems of image misregistration, image blurring, and motion-generated artifacts.

One idea of the method is to try to maintain the image consistency in the region near the seams through precise registration. The seam driven image stitching method is based on modifying the RANSAC algorithm to obtain a series of singular susceptibility matrices with variance and stitching the image based on different geometric transformations, from which the optimal seams are selected. The optimal seams are crucial for improving the visual effect of the spliced images, but there is still room for improving the efficiency of this method.

For image stitching with large parallax, the image stitching method based on seam driven, which combines global transformation with content-keeping warping [5], uses the cost of the seams as a metric of the stitching quality, which dramatically improves the performance of stitching in large-parallax scenes, but the efficiency of the stitching is low.

The seam-guided local registration (SEAGULL) method for large parallax image stitching [6], uses estimated seams to guide the optimization process of local registration, and gradually improves the quality of the seams in each iteration. However, the method may fail when the parallax outside the overlapping region is too large or when the local region contains rich salient structures but few feature matching points.

Another idea of the optimal seam splicing method is to utilize the difference of pixels in the overlapping regions of two images to generate a difference matrix, and select the best path that meets the requirements as a seam through a suitable search strategy. This search strategy can usually be regarded as a shortest path problem or an energy minimization problem. Commonly used optimization methods include Dijkstra's algorithm, Snake model, dynamic programming algorithm, graph cut method and so on.

The Fast and Robust Seam Estimation (FARSE) method [7], utilizes gray-weighted distances and differential gradient domains to estimate the optimal seams. The method is computationally efficient since it searches the seam paths directly.

Dijkstra's algorithm can be applied to image stitching, but the resulting seams are not optimal. In order to improve the efficiency of Dijkstra's algorithm, a subregion with limited maximum cost is found within the overlapping region and Dijkstra's algorithm is applied to search for seams within the subregion. Since Dijkstra's algorithm uses exhaustive computation and needs to traverse a large number of nodes, it has the disadvantages of high computation and low efficiency.

The twins nake model is used to search for seams, but this method is more sensitive to the parameter settings in the energy function, and the final seams obtained may not be the global minimum.

Dynamic Planning Algorithm [8] By solving for the seam with the smallest difference in pixel values in the rectangular overlapping region, seams can be effectively avoided to pass through moving objects, thus reducing artifacts and other phenomena. An optimal seam selection method with enhanced dynamic planning can effectively handle significant structural misregistrations. Based on the region saliency of human attention model, image similarity metrics such as color, edge and texture are used to generate a difference matrix, and then a dynamic programming algorithm is used to search for seams, but the method is computationally intensive. The advantage of the dynamic programming algorithm is that it can determine the end point of the seam line, but the path it obtains may not be globally optimal.

Graph cut is an effective energy optimization algorithm. The problem of searching for the best seam between images can be transformed into an energy optimization problem based on graph cut, where the search for the global best seam is equivalent to the minimization process of the graph cut energy function. The basic idea of graph cut is to construct a weighted graph, where the weight of each edge represents the corresponding cost energy value, and the minimum value of the difference is found in the graph by the max-flow or min-cut algorithm (max-flow/min-cut).

As Fig. 6.18a, I_1 and I_2 are two images with overlapping regions, which are shaded. The search process for the optimal seam is to find a cut (cut) within the overlap region that minimizes the total energy value. The cut passes through the nodes and cuts the weighted edges between the nodes so that the nodes in the overlap region are divided into two parts, one from I_1 and the other from I_2 as Fig. 6.18b.

An optimal seam finding method is described below. The optimal seam is a continuous path through the overlapping region starting from the first row and column to the last row and column. The weights of the pixels in the overlapping regions are optimized by using a more compact and descriptive data structure, the tensor (a symmetric and semi-positive definite matrix T). A Riemannian paradigm is used to measure the distance between two tensors. In an image of height M, the seam cost function Cost is defined as

$$\text{Cost} = \text{Cost}(I_s) = \sum_{i=1}^{M} c(I(s_i)) \tag{6.52}$$

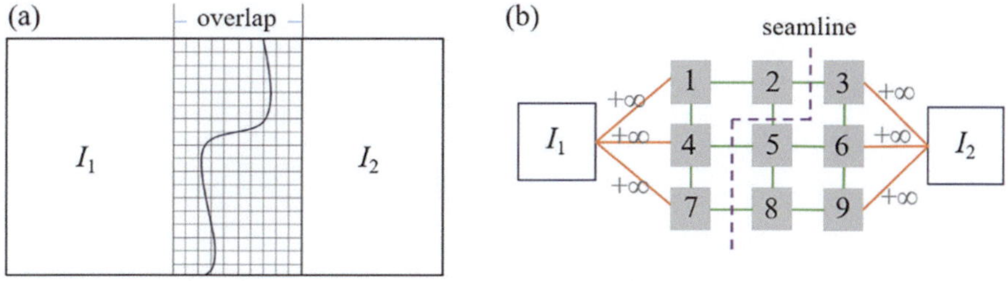

Fig. 6.18 Schematic diagram of solving the optimal seam based on graph cut: **a** the original image and the solvedc the optimal seam line, **b** using graph cut to solve the optimal seam line

where $I(s_i)$ is the ith pixel on the seam. $c(i, j)$ It is defined as

$$c(i,j) = d(T_1(i,j), T_2(i,j)) + \left|I_1''(i,j) + I_2''(i,j)\right| \tag{6.53}$$

where T_1 and T_2 are the tensors of I_1 and I_2, respectively, and I'' is the second-order derivative of the image I. The optimal seam is defined as

$$s^* = \min_s \mathrm{Cos}t(s) = \min_s \sum_{i=1}^{M} c(I(s_i)) \tag{6.54}$$

The demonstration results of this method are shown in Fig. 6.19. There are some other optimal seam methods that have achieved better results. An improved seam finding method approximates the seams by reducing the overlap region. There are also methods to find optimal seams for image stitching by energy aggregation and traversal strategies. The optimal seam line between adjacent input images is detected by graph cut energy minimization framework. The color, gradient magnitude and texture information of the image is fused into the graph cut. The optimal seam pair is selected by comparing the cross-correlation of multiple seams detected by different cost weights [9].

The image splicing method based on fast marching method further improves the image splicing efficiency. Firstly, an image region weight map is constructed based on the image similarity in color, edge, texture and saliency, and the region weight map is used as the cost value of the seams, and the optimal seams are continuously optimized using the fast marching method. This new method can better locate the seams in

Fig. 6.19 Removing seams using only the optimal seam method

reasonable regions, which can more effectively reduce defects such as ghosting, misregistration and color difference, and achieve high-quality image splicing.

However, there is a problem in the above method: there is a parallax between different images, and if the seam passes through an obvious object area, it will lead to defects such as geometric misregistration and color difference on both sides of the seam area, which will affect the visual perception of the spliced image. In order to avoid these problems, it should be preferred to cross the texture-smooth region as the seam path and avoid crossing the obvious object region.

A DLink Net neural network was used to generate a probabilistic map of the overlapping regions so that the seams would be along the wider road regions as much as possible, avoiding going through the region of salient objects, and to find the final seams using the Dijkstra's shortest path algorithm. However, the number of training samples for this method has a large impact on the determination of the final seam.

Image segmentation can be used to achieve image region optimization, the use of mean shift (mean shift) algorithm to achieve image segmentation, and then according to the characteristics of different objects to set the priority to pass through the region, to determine the starting point and end point of the seams, and then use the Dijkstra algorithm to search for the best seams. However, this method is very sensitive to the mean shift segmentation parameter, and needs to manually select the priority region for the image, and the algorithm is less efficient.

To address the above problems, a convolutional neural network can be used to pixel-label the building regions and create their probability maps, and then find the best seams as the shortest paths, which avoids the seams going through the building regions. However, this method cannot avoid the effects caused by shadows.

On this basis, the convolutional neural network is further applied to the best seam detection for semantic segmentation of the scene by independently categorizing the overlapping regions of the left and right images through the trained network, obtaining the probability of each pixel belonging to a specified category, and finally finding the best seam by graph cut method, which avoids the seam crossing the region of obvious objects.

6.5 Comparison of Feature Detection and Matching Algorithms

6.5.1 Corner Detection Algorithm

Corner points, which are usually generated by the intersection of lines, tend to be locations with significant brightness differences or great curvature. Corner points have scale invariance and are less affected by illumination, making them suitable for detection. The commonly used operators are Forstner operator, Moravec operator and Harris operator. Forstner operator was first proposed by Wolfgang Forstner in where q and w are calculated as

$$q = \frac{4\,\det(C)}{[\mathrm{trace}(C)]^2} \tag{6.55}$$

$$w = \frac{\det(C)}{\mathrm{trace}(C)} \tag{6.56}$$

where C is the intensity covariance matrix. Suppose t is a threshold (some value). If $q > t$, then the pixel is a candidate point. Then, the extreme point of q is selected as the feature. The best candidate window is the center window of the point. The exact position of the corner point is calculated using the regular equation, where g_x and g_y are the Robert's gradients in the x and y directions, respectively, as

$$C\begin{bmatrix} x \\ y \end{bmatrix} = \begin{bmatrix} x\sum g_x^2 + y\sum g_x g_y \\ x\sum g_x g_y + y\sum g_y^2 \end{bmatrix} \tag{6.57}$$

The Moravec operator selects the point with the lowest variance of intensity in the four main directions. The Harris operator is proposed based on the Moravec method. In Harris operator, corner point detector D_{h} is used to detect the corner points as

$$D_{\mathrm{h}}(i,j) = \sum_x \sum_y w(x,y)\big[I(x+i, y+i) - I(x,y)\big]^2 \tag{6.58}$$

where $I(x, y)$ is a pixel in the image I. $w(x, y)$ is a window function which can be a rectangular window or a Gaussian window. Assuming that

$$M = \sum_x \sum_y w(x,y)\begin{bmatrix} I_x^2 & I_x I_y \\ I_x I_y & I_y^2 \end{bmatrix} \tag{6.59}$$

where I_x and I_y are the gradient values in the directions of x and y, respectively. The detection operator can be approximated as

$$D_{\mathrm{h}}(i,j) = [i,j]M\begin{bmatrix} i \\ j \end{bmatrix} \tag{6.60}$$

The detected response value R_e is

$$R_e = \det(M) - k[\mathrm{trace}(M)]^2 \tag{6.61}$$

Typically, the value of k is from 0.04 to 0.06. Setting the value of k less than the threshold t to 0 and suppressing non-maxima in the 3×3 or 5×5 region, the local maxima in this region are the corner points.

The Harris operator is invariant to translations and rotations. However, the value of t depends on image properties and cannot be measured when a specific threshold is set. In addition, points with large eigenvalues are usually clustered within a certain range, which can lead to an uneven distribution of detected corner points. Since more corner points can be detected in texture-rich regions and fewer corner points are detected in less textured regions, corner points tend to be distributed in regions with richer texture information. To improve Harris corner point detection, a new response function is introduced. It is redefined as

$$R_e = \frac{\det(M)}{[\mathrm{trace}(M)]^2 + \varepsilon} \tag{6.62}$$

where ε is used to prevent $trace(M)$ from being zero. It is possible to take a very small value of ε, e.g. 10 to the negative 6th power. In addition, there is no need to select the value of k in the redefined R_e. There are some applications of corner detection algorithms in image stitching. Adaptive Harris corner point detection method. The Harris corner point detector is customized to meet the requirements of real-time viewing of panoramic videos. Corner point detection is used to help extract features. Improved Harris algorithm which can be used to stitch images related to traffic accidents.

6.5.2 Scale Invariant Feature Transform

The Scale Invariant Feature Transform (SIFT) method was initially used for feature detection. Later it was developed and used for image registration and image stitching. The SIFT algorithm detects the extremes in the intensity and scale domains and selects the critical points using a Gaussian kernel G. The SIFT algorithm can be used to detect the extremes in the intensity and scale domains. Differential Gaussian kernel (DoG) at different scales σ is used to detect extreme values as

$$G(x, y, \sigma) = \frac{1}{\sqrt{2\pi\sigma}} \exp\left(-\frac{x^2 + y^2}{2\sigma^2}\right) \tag{6.63}$$

$$\begin{aligned} \mathrm{DoG}(x, y, \sigma) &= \left[G(x, y, k\sigma) - G(x, y, \sigma)\right] * I(x, y) \\ &= L(x, y, k\sigma) - L(x, y, \sigma) \end{aligned} \tag{6.64}$$

For each $L(x, y, \sigma)$, the magnitude and direction of the key points are then computed. Then, an 8×8 window is taken centered on the key points and an 8-direction gradient histogram is computed for each 4×4 direction. The key point descriptor is a 128-dimensional vector.

The key point descriptors are matched by a metric such as Euclidean distance. Once the matching is complete, a RANSAC algorithm is typically used to remove mismatches.

The image is then affine transformed, in which one image can be combined with another. Usually, a homography matrix is used in this step to describe the class of 2D planar projective transforms by matrix multiplication. The projective transform can be defined as a matrix with 8° of freedom called H as

$$\begin{bmatrix} x' \\ y' \\ 1 \end{bmatrix} = H \begin{bmatrix} x \\ y \\ 1 \end{bmatrix} = \begin{bmatrix} a_1 & a_2 & a_3 \\ a_4 & a_5 & a_6 \\ a_7 & a_8 & 1 \end{bmatrix} \begin{bmatrix} x \\ y \\ 1 \end{bmatrix} \tag{6.65}$$

where x' and y' can be obtained by

$$\begin{cases} x' = \frac{a_1 x + a_2 y + a_3}{a_7 x + a_8 y + 1} \\ y' = \frac{a_4 x + a_5 y + a_6}{a_7 x + a_8 y + 1} \end{cases} \tag{6.66}$$

By affine transformation and finding the right parameters, two images can be combined so that their main features match exactly. The homographyness matrix is determined by eight parameters and the coordinates of the points should be chosen to estimate these parameters. However, in pixel-based algorithms, image registration is used to select templates from one image and measure their correlation in another image. Therefore, affine transformation cannot be used. This is the reason for the existence of a large number of mismatches.

In many cases, image stitching using SIFT works well. It runs very fast and can handle translation, rotation and scaling.

The PCA-SIFT algorithm is an improved version of SIFT. It accepts the same inputs as the standard SIFT descriptors. PCA (Principal Component Analysis) is a dimensionality reduction technique that has been widely used in the field of computer vision. PCA-SIFT has three main steps: firstly, a feature space is computed to represent the gradient image of a local patch; secondly, the gradient of the local image of the given patch is computed; and lastly, the vector of gradient images is projected using the feature space to obtain a compact feature vector.

The SIFT descriptor is invariant to deformations such as translation, rotation and scaling of images. It is also robust to moderate perspective transformations and illumination changes. However, in some cases, it cannot meet the computational speed requirements. As an improvement, PCASIFT takes less time than SIFT, but at the cost of being less robust to scaling.

6.5.3 Accelerated Robust Feature Algorithm

Accelerated Robust Feature (SURF) is a robust image recognition algorithm for pattern recognition. And there is also 3D reconstruction. Experiments show that SURF is several times faster compared to SIFT. In recent years, the research on image registration and image stitching based on SURF have achieved good results.

SURF utilizes the integral image to convert a second order Gaussian integral stencil filter to a modification of the integral image. The matrix $H(x, \sigma)$ at x when the scale is σ is then applied to detect the feature points as

$$H(x, \sigma) = \begin{bmatrix} L_{xx}(x, \sigma) & L_{xy}(x, \sigma) \\ L_{yx}(x, \sigma) & L_{yy}(x, \sigma) \end{bmatrix} \tag{6.67}$$

The second order derivatives L_{xx}, L_{xy}, L_{yy} in the direction of xx, xy, yy, respectively, can be obtained by convolving the image with a second order Gaussian gradient template. In SURF, this template is approximated by a box function, usually a 9×9 box function equivalent to a Gaussian filter ($\sigma = 1.2$). This approximation is referred to as a speckle. The response plot represented by B_{xx}, B_{yy}, B_{xy}.

Thus, the determinant $\det(H)$ can be approximated as

$$\det(H) = B_{xx}B_{yy} - w(B_{xy})^2 \tag{6.68}$$

Next, Haar wavelets are used to create feature descriptors. The integral image is used to simplify the computation. The descriptor is a 64-dimensional matrix containing coordinates, amplitudes, and rotation angles.

SURF detects fewer feature points than SIFT. The application of Hessian matrix and box function makes SURF faster than SIFT. However, SURF is less robust to large angle rotations compared to SIFT.

Template Convolutional Speeded Up Robust Features (TSURF) is an improvement of SURF and is designed to increase the speed of stitching while maintaining the accuracy. The maximum value of the determinant of the Hessian matrix ($H_{\max}$) minus the secondary value ($H_{\sec}$) is used as

$$H_{\max} - H_{\sec} \geq \theta \tag{6.69}$$

In the algorithm, θ is set to 0.5 when the contrast of the image is 0.14. If $H_{\max} - H_{\sec} \geq \theta$, the point is considered as a candidate point. The higher the contrast, the larger the value of θ.

SURF based registration takes less time than SIFT based registration. However, it is not as robust to image rotation as SIFT. In addition, using SURF produces more misregistrations than using SIFT.

6.5.4 KAZE Features

KAZE feature is a multiscale 2D feature detection and description algorithm in nonlinear scale space, in which 2D features are detected and described in nonlinear scale space by nonlinear diffusion filtering. In KAZE characterization, the nonlinear scale space with maximum evolution time is first constructed. Then, the 2D features whose scale-normalized Hessian response determinant reaches the maximum value in the nonlinear scale space are detected. Finally, the principal directions of the key points are computed to obtain scale- and rotation-invariant descriptors.

In the KAZE characterization, the scale space is discretized logarithmically into O octaves and S sub-levels. The set of octaves and sub-levels are identified by o (discrete octave index) and s (sub-level octave index). They are mapped to the corresponding scale ε_{as} as

$$\varepsilon_i(o, s) = \varepsilon_0 2^{o+\frac{s}{S}}, \quad o \in [0, O-1], \quad s = [0, S-1], \quad i \in [0, N] \tag{6.70}$$

where ε_0 is the base scale level and N is the total number of filtered images.

The scale level in pixels is then converted to a scale level in time as

$$t_i = \frac{\varepsilon_i^2}{2}, \quad i \in [0, N]. \tag{6.71}$$

First, the input image is convolved with a Gaussian kernel with a standard deviation of ε_0 to reduce noise and possible image artifacts. Then, the contrast parameter is obtained by automatically computing its gradient histogram k. Given the contrast parameters and a set of evolution times t, the nonlinear space is constructed as

$$L^{i+1} = \left(I - (t_{i+1} - t_i)\sum_{l=1}^{m} A_l L_l\right)^{-1} L_i \tag{6.72}$$

To detect points of interest, the scale-normalized Hessian matrix determinant response is computed at multiple scale levels as

$$L_{\text{Hessian}} = \varepsilon^2\left(L_{xx}L_{yy} - L_{xy}^2\right) \tag{6.73}$$

where L_{xx} and L_{yy} denote second-order horizontal and vertical derivatives, respectively, and L_{xy} denotes second-order cross derivatives.

Then, the maximum value on the scale and spatial location is found. Extreme points are also found in all filtered images (except top and bottom layers). Each extreme point is searched within a rectangular window of $\varepsilon_i \times \varepsilon_i$ size on the current i layer, the filtered image of the previous $i+1$ layer and the next $i-1$ layer.

Finally descriptors are generated. The main direction is found in a circular area with a radius of $6\varepsilon_i$ in steps of ε_i. For each sample within the circular region, the first order derivatives L_x and L_y are weighted with a Gaussian function centered on the point of interest. The derivative responses are then represented as points in vector space. The principal directions are found by summing within a sliding circular segment covering the angle $\pi/3$. The principal direction is obtained from the longest vector.

For the detected features with ε_i scale, first order derivatives L_x and L_y are first computed on a rectangular grid of $24\varepsilon_i \times 24\varepsilon_i$, which is divided into 4×4 sub-regions of size $9\varepsilon_i \times 9\varepsilon_i$ with $2\varepsilon_i$ overlap between the sub-regions. The derivative response of each subregion is weighted with a Gaussian function ($\varepsilon_l = 2\varepsilon_i$) centered at the center of the subregion) and aggregated into a descriptor vector $dy = \left(\Sigma L_x;\ \Sigma L_y;\ \Sigma L_z;\ \Sigma L_y\right)$. Each subregion vector is then weighted using a Gaussian function $\varepsilon_2 = 1.5\varepsilon_i$, defined on a mask of 4×4, centered on the key point of interest. When considering the dominant direction of the key point, each sample in the rectangular grid is rotated according to the dominant direction. The derivatives are also computed according to the dominant direction. Finally, the 64-dimensional descriptor vectors are normalized to unit vectors for contrast invariance.

The KAZE feature description is robust to noise and ambiguity. However, it employs nonlinear diffusion at low levels and takes longer computation time than SURF. To reduce the computational effort, A-KAZE, or Accelerated KAZE, is proposed. It employs the idea of Fast Explicit Diffusion (FED) in a pyramidal framework, and the binary descriptor accelerates the computation and is suitable for portable devices. Once the KAZE characterization was proposed, many studies focused on its applications. In the literature, KAZE features are used for matching of frame block feature points in videos. In literature, S-AKAZE is proposed as an improved version of KAZE for image matching. A-KAZE is applied for object recognition. KAZE is used for assisted image segmentation. In literature AKAZE is used for image stitching.

6.6 Comparison of Region Based Matching Algorithms

6.6.1 *Pixel Matching Algorithm*

The basic pixel matching is performed on a monochrome image. The algorithm selects one image as a reference, aligns it using overlapping regions, and determines the matching region. Some methods use one image as a reference so that the overlapping region moves within the reference image. The correlation is calculated every time a pixel is moved. Suppose I_1 is the reference image, I_2 is another image, and the overlapping regions are Ω_1 and Ω_2, both of size $M \times N$. The correlation R can be expressed as

$$R = \sum_{x=1}^{M} \sum_{y=1}^{N} |\Omega_1(x, y) - \Omega_2(x, y)| \tag{6.74}$$

Then, the rows (columns) with the largest correlation are selected for registration. Obviously, the computation is very large. Also this type of method is difficult to handle rotation and scaling.

In order to reduce the amount of computation, the search region is modified to a single region and it is applied in. Image stitching method based on correlation is utilized to calculate the displacement between the source images.

Dynamic Time Warping (DTW) for image stitching is a method for aligning two sequences. The time sequences are ordered by dynamically planning the elements of the matching time sequences while minimizing the total distance metric. The method first selects two images with overlapping regions. Then a sub-image of $m \times$ n is cropped from one of the images as a template from which features are extracted and normalized to a one-dimensional sequence. The features are extracted from the other image and compared using the DTW method to determine the location. Finally, the stitched image is created.

Using the two rows with a certain interval in the overlapping region in the image I_1 as a template, the best match is searched in the corresponding overlapping region I_2. The number of matching rows is then adjusted to three to improve the search strategy and make the results more accurate.

Another proposed algorithm for stitching biomedical images achieves alignment by balancing adjacent edges and forcing luminance values at corner points to be consistent. Luminance correction is performed through a specific matrix prior to registration to make the algorithm less sensitive to light variations. Overall, this method is relatively simple to implement but is computationally intensive. Since the images are often distorted when taken and no transformations are performed during the registration process, the results are not as good as expected.

6.6.2 Mutual Information Based Image Registration

Mutual information (MI) is a measure of the degree of dependence between two variables. If two variables are independent of each other, the mutual information is 0. If there is a strong dependency between two variables, the mutual information is larger. In recent years, mutual information has been widely used in the fields of image registration, image retrieval, pattern recognition and image stitching. The MI of images I_1 and I_2 is defined as

$$\mathrm{MI}(I_1, I_2) = \sum_{I_1, I_2} P_{I_1, I_2}(I_1, I_2) \log_{10} \frac{P_{I_1, I_2}(I_1, I_2)}{P_{I_1}(I_1) P_{I_2}(I_2)} \tag{6.75}$$

where $P_{I_1,I_2}(I_1, I_2)$ is the joint probability density and $P_{I_1}(I_1)$ and $P_{I_2}(I_2)$ are the edge probability densities. According to the definition of MI, two images should be considered as variables and if the similarity between the images is large, their mutual information value is also large. Therefore, the image registration problem is now transformed into a problem of searching for the maximum value of mutual information between a template image and a sub-image.

In order to improve the MI algorithm, some concepts such as Normalized Mutual Information (NMI), Local Mutual Information (LMI) have been proposed. An implementation of α-Mutual Information (α-MI) based on self-similarity weighted graphs for non-rigid image registration is applied to the registration of preoperative magnetic resonance (MR) images with intraoperative ultrasound (US) images. α-MI is used to detect the similarity between images. Registration using MI is highly accurate, less intrusive and reliable. However, its computational process is very complex and less robust to rotation.

6.6.3 Laplace Pyramid Based Image Registration

Laplace pyramid-based image registration was first proposed in in which a series of low-pass filtered images can be obtained by repeatedly convolving a small low-pass filter $I_1, \ldots, I_{Ldec}$. The weighting function is determined from a sample image I_0. By this method, the image sample density is also reduced at each iteration, thus gradually reducing the bandwidth by a uniform octave. The reduction of samples also means that the computational cost is kept to a minimum. The pixel values in the image I_0 correspond to the 5×5 weighted average transition values in the I_0 level. Iterating in this way results in a series of low-pass filtered images (Gaussian low-pass filtered image sequence): $I_0, \ldots, I_{Ldec}$. The pixel (x, y) at level l in the image I_{Ldec} is defined as

$$I_l(x, y) = \sum_{i=1}^{5}\sum_{j=1}^{5} w(i, j) I_{l-1}(2x + i, 2y + j) \tag{6.76}$$

where $w(i, j)$ is referred to as the restricted generating kernel function as

$$\sum_{i=-2}^{2}\sum_{j=-2}^{2} w_i(x - 2i, y - 2j) = 1 \tag{6.77}$$

The image stitching localization algorithm based on the classical Laplace method proposes multi-resolution templates, which are able to compute multiple templates in a single scanning process and obtain descriptors at different scales. Experiments show that the use of structural information limits the decomposition and reconstruction process and reduces the amount of computation.

6.6.4 Registration in Frequency Domains

The common methods used for image registration in the transform domain are Fourier transform (FT) and wavelet transform (WT). Here, the Fourier transform based registration methods are introduced first, and then the wavelet transform based registration methods are also introduced. FT is applied to underwater sonar image stitching, and computed tomography (CT) image set stitching. The WT based image registration method uses Dual Tree Complex Wavelet Transform (DT-CWT) to decompose the image to be aligned. Starting from the coarsest level, an initial estimate of the transform vector $v = [\alpha, t_x, t_y]$ is found, where α is the rotation angle and t_x and t_y are the translation parameters in the x and y directions, respectively. Please note that no scaling transformation has been applied here. Mutual correlation is chosen as the matching criterion. Next, the edge maps of the two low-pass images E_{map1} and E_{map2} are extracted. A search is performed to determine the best initial transform vector ν_{init}

$$\nu_{\text{init}} = \arg\max_{v} c(Emap_1, T(Emap_2)) \tag{6.78}$$

where $c(x, y)$ denotes the mutual correlation of x and y, and $T(Emap_2)$ denotes the image that is convolved using vector v. The transformation vector v_l is found by means of

$$v_l = \arg\max_{v} \left[\frac{I(\text{Re}\{D_1, I\}, T(\text{Re}\{D_2, I\}))}{I(\|C_P(D_1, I)\|, T\|C_P(D_2, I)\|)} \right] \tag{6.79}$$

where $\text{Re}\{D_2, I\}$ and $C_P(D_2, I)$ are the real and imaginary parts of x at level l, respectively. D_1 denotes the decomposition of the original image and D_2 denotes the decomposition of the image to be aligned. To further reduce the computational burden, the proposed alternative search method can be applied. Registration in the transform domain is fast and robust and can solve most of the deformation problems in images. However, the results may not be accurate enough when the overlapping regions of the images are small.

6.7 General Conclusions

This chapter presented the core technical processes and key challenges of image stitching. The chapter includes an overview of image stitching, image registration techniques, image transformation and deformation optimization, image fusion and seam optimization, as well as a comparative analysis of deep learning-based stitching methods and various types of feature detection and region matching algorithms. Strategies for solving problems such as parallax, dynamic object ghosting, illumination differences, and non-overlapping region distortion are highlighted. It is convenient for readers to have a more in-depth research on the field of image stitching.

Acknowledgments This work was supported by the National Natural Science Foundation of China (12374271, 62375065, 62405079, 62575091); Fundamental Research Funds for the Central Universities (HIT.OCEF.2024055, AUGN5750300325); the Fund of National Laboratory on Adaptive Optics, China (FNLAO-24-MS-O07); Natural Science Foundation of Heilongjiang Province of China (LH2024A009); Key Laboratory of Image Processing and Intelligent Control (Huazhong University of Science and Technology), Ministry of Education, Wuhan, China (IPIC2025-05); Key Laboratory of Advanced Medical Imaging and Intelligent Computing of Guizhou Province (AMIIC250204); the China Postdoctoral Science Foundation-CCTEG Joint Support Program (2025T011ZGMK).

References

1. Zaragoza TJC, Brown MS, Suter D (2014) As-projective-as-possible image stitching with moving DLT. IEEE Trans Pattern Anal Mach Intell 36(7):1285–1298
2. Liao T, Li N (2020) Single-perspective warps in natural image stitching. IEEE Trans Image Process 29:724–735
3. Chang CH, Sato Y, Chuang YY (2014) IEEE in shape-preserving half-projective warps for image stitching. In: 27th IEEE conference on computer vision and pattern recognition (CVPR), Columbus, OH, Jun 23–28, pp 3254–3261
4. Junhong G, Seon Joo K, Brown MS (2011) Constructing image panoramas using dual-homography warping. In: IEEE conference on computer vision and pattern recognition (CVPR), pp 49–56
5. Zhang F, Liu F (2014) IEEE in parallax-tolerant image stitching. In: 27th IEEE conference on computer vision and pattern recognition (CVPR), Columbus, OH, Jun 23–28, pp 3262–3269
6. Lin KM, Jiang NJ, Cheong LF, Do M, Lu JB (2016) In SEAGULL: seam-guided local alignment for parallax-tolerant image stitching. In: 14th European conference on computer vision (ECCV), Amsterdam, NETHERLANDS, Oct 08–16, pp 370–385
7. Hejazifar H, Khotanlou H (2018) Fast and robust seam estimation to seamless image stitching. Signal Image Video Process 12(5):885–893
8. Hua G, Yue Y, Weidong S (2009) A new optimal seam selection method for airborne image stitching. In: IEEE international workshop on imaging systems and techniques (IST 2009), pp 159–163
9. Li L, Yao J, Shi SZ, Yuan SG, Zhang YX, Li J (2018) Superpixel-based optimal seamline detection in the gradient domain via graph cuts for orthoimage mosaicking. Int J Remote Sens 39(12):3908–3925

Chapter 7
Super-Resolution Imaging

Zhengyu Wu, Zhengjun Liu, and Yutong Li

Abstract This chapter provides an overview of super-resolution imaging techniques. It covers the fundamental concepts of digital resolution, image interpolation algorithms, registration methods, image super-resolution techniques, structured illumination microscopy, Fourier ptychographic microscopy, and stimulated emission depletion (STED) for super-resolution imaging. Studying these topics helps readers develop computational programs for building corresponding imaging systems.

7.1 Imaging Resolution and Image Degradation Issues

Resolution is a core concept in describing image quality. It not only determines the level of detail that can be presented in an image but also affects the practical value of an imaging system in fields such as scientific research, industrial inspection, and medical diagnostics. Optical resolution is fundamentally constrained by the physical structure of the imaging system and governed by the principles of wave optics, such as the diffraction limit and numerical aperture. It defines the minimum ability to distinguish between two adjacent details. With the advancement of optical imaging technologies and image reconstruction algorithms, it is possible to enhance the effective resolution of images through computational methods, even under limited optical conditions, enabling clearer and more detailed visual information representation.

Z. Wu · Z. Liu · Y. Li (✉)
School of Physics, Harbin Institute of Technology, Harbin, Heilongjiang, China
e-mail: yutong_li@hit.edu.cn

Z. Wu
e-mail: 976056734@qq.com

Z. Liu
e-mail: zjliu@hit.edu.cn

Z. Liu and Y. Li (eds.), *Intelligent Analysis of Optical Images*, Scientific Computation,
https://doi.org/10.1007/978-981-95-3720-4_7

7.1.1 Diffraction-Limited Resolution of Optical Systems

Optical resolution refers to the minimum distance at the physical level that an imaging system can distinguish between two adjacent target points. It is a key metric for evaluating the fineness of an imaging system. Due to the wave nature of light, any light passing through a finite aperture undergoes diffraction, which limits the resolving power of the imaging system. As a result of this diffraction, even an ideal point source will no longer appear as a perfect point on the image plane but instead spreads into a bright spot with a specific intensity distribution, known as the Airy disk in Fig. 7.1a.

The intensity distribution of the Airy disk appears as concentric rings, with the brightest point at the center, followed by alternating bright and dark rings. The radius of the Airy disk is given by the following equation [1]

$$r = \frac{0.61\lambda}{\mathrm{NA}} \tag{7.1}$$

where r is the radius of the Airy disk, λ is the imaging wavelength, and NA is the numerical aperture of the system. This radius defines the "blur region" of a point source on the image plane, representing the minimum response to a single point.

When two point sources are too close together, their Airy disks overlap. According to the Rayleigh criterion, two points are just resolvable when the central maximum of one Airy disk coincides with the first minimum of the other as Fig. 7.1b. This criterion is widely used to estimate the theoretical resolution limit of optical systems

$$\Delta r = \frac{0.61\lambda}{\mathrm{NA}} \tag{7.2}$$

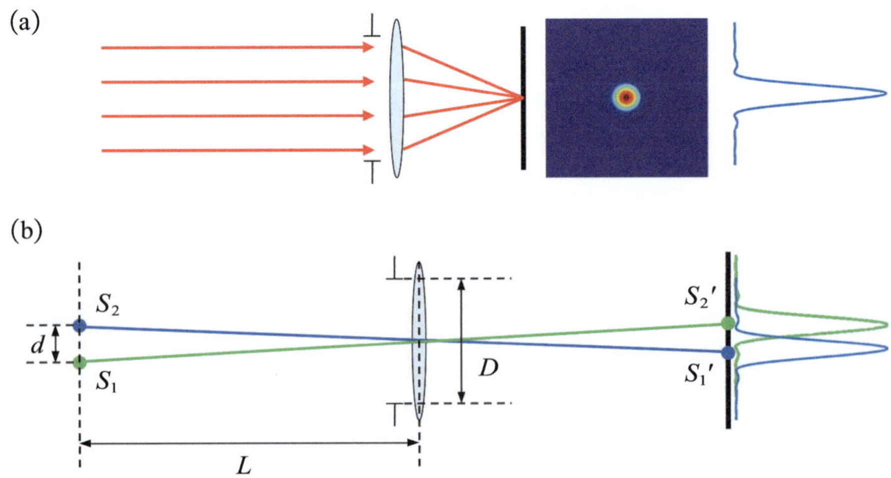

Fig. 7.1 Airy disk and resolution: **a** schematic of the Airy disk, **b** illustration of the diffraction-limited resolution

From this equation, resolution can be improved by either reducing the wavelength, for example by using blue light, ultraviolet light, or even electron beams, or by increasing the numerical aperture, such as employing large-aperture lenses or immersion objectives to expand the incident angle of light.

In practical applications, optical resolution is not only limited by wavelength and numerical aperture, but also affected by factors such as system aberrations, alignment errors, and sensor response characteristics. Moreover, under low illumination or high noise conditions, even a system with high theoretical resolution may fail to distinguish fine details effectively.

7.1.2 Image Degradation Model

Hardware imaging systems exhibit low-pass filtering characteristics, meaning that during image acquisition, frequency components above the system's cutoff frequency are filtered out, resulting in the loss of high-frequency image information. This cutoff frequency is determined by the system's optical parameters, such as lens aperture, sensor pixel size, and imaging wavelength. Low-resolution images can be viewed as captured by low-bandwidth systems with narrow frequency response ranges, preserving limited information; in contrast, high-resolution images come from high-bandwidth systems that retain more spatial details. Therefore, increasing the bandwidth of an imaging system from the hardware perspective is a direct way to obtain high-resolution images. However, this approach faces clear limitations due to physical constraints and cost considerations. In practical digital image acquisition, the imaging system is not only limited by its low-pass nature but also suffers from additional image degradation caused by factors such as aberrations, motion blur, geometric distortions, downsampling, and random noise. Furthermore, various forms of noise, including Gaussian noise and salt-and-pepper noise, may be introduced during image acquisition, transmission, and storage processes.

As shown in Fig. 7.2, assume that the original scene is ideally imaged as a high-resolution image $x \in \mathbb{R}^{r_1 N_1 \times r_2 N_2}$, where r_1 and r_2 are the downsampling factors in the horizontal and vertical directions, respectively. By vectorizing x into a one-dimensional column vector, multiple images of the scene are captured to obtain n frames of low-resolution images $\{y_k\}_{k=1}^{n}$. Each low-resolution frame undergoes a different degradation process, and the entire imaging process can be expressed as [2]:

$$y_k = DB_kM_kx + n_k, \quad k = 1, \ldots, n \tag{7.3}$$

For convenience, let $H_k = DB_kM_k$ represent the overall degradation process. Then, the degradation process of the low-resolution images can be rewritten as:

$$y_k = H_kx + n_k, \quad k = 1, \ldots, n \tag{7.4}$$

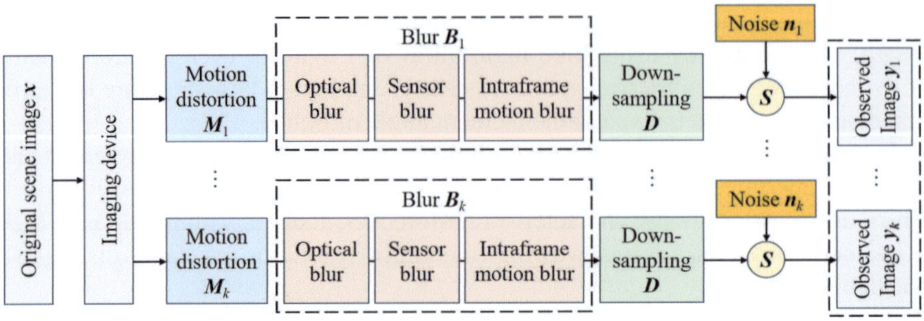

Fig. 7.2 Image acquisition and degradation process

7.1.3 Theoretical Foundations of Super-Resolution Reconstruction

The goal of super-resolution reconstruction technology is to recover the high-frequency information lost during image acquisition due to the physical limitations of the imaging system, that is, the frequency components beyond the system's cutoff frequency. The imaging system can be modeled as a linear space-invariant system, whose imaging process in the spatial domain can be expressed as:

$$g(x) = h(x) * f(x) \tag{7.5}$$

where $g(x)$ is the observed image, $h(x)$ is the system's degradation function, and $f(x)$ represents the original scene image. The expression $*$ denotes the convolution operation. Applying the Fourier transform to the above expression converts it to the frequency domain:

$$G(\omega) = H(\omega) \cdot F(\omega) \tag{7.6}$$

The original image spectrum can then be obtained as:

$$F(\omega) = \frac{G(\omega)}{H(\omega)} \tag{7.7}$$

However, beyond the cutoff frequency, the frequency response of a low-pass imaging system satisfies $H(\omega) = 0$, making it extremely difficult to invert and recover $F(\omega)$, thereby hindering the recovery of high-frequency information.

Due to the low-pass characteristics of imaging systems and the downsampling operations during the sampling process, the observed image typically exhibits spectral aliasing in the frequency domain. Let the spatial information of the image be represented by the function $f(x)$, which, due to the non-negativity and boundedness of images, can be expressed as

$$\begin{cases} f(x) > 0, \ x \in X \\ f(x) = 0, \ x \notin X \end{cases} \tag{7.8}$$

By fusing multiple observed images taken from different downsampling perspectives and incorporating prior knowledge of the image, the high-frequency components can be effectively estimated, achieving the goal of super-resolution reconstruction.

In Eq. (7.8), X denotes the domain of the function $f(x)$, i.e., the spatial extent of the image. If a rectangular function rect is introduced, the function can be formally written as:

$$f(x) = \text{rect}(x/X) \tag{7.9}$$

Applying the Fourier transform to the above expression yields the spectrum function $F(\omega)$. According to the system's cutoff frequency, the spectrum can be divided into two parts: $F_b(\omega)$ represents the low-frequency information below the cutoff, and $F_a(\omega)$ represents the high-frequency information beyond the cutoff. Thus, the spectrum function can be expressed as:

$$F(\omega) = [F_b(\omega) + F_a(\omega)] * \text{sinc}(X\omega) \tag{7.10}$$

Here, sinc is an infinitely extending function, indicating that $F_a(\omega)$ is aliased into $F_b(\omega)$ through convolution. This means that the observed low-frequency data not only contains true low-frequency information but also includes part of the high-frequency components. If the aliased high-frequency components can be separated, it becomes possible to fully recover the image spectrum and achieve super-resolution reconstruction.

Assuming that the function $f(x)$ is bounded, that is, it is identically zero outside its domain, its Fourier spectrum $F(\omega)$ is an analytic function. For an analytic function, if its values are known over a finite interval, the function's values over the entire complex plane can be uniquely determined via analytic continuation. Therefore, if the Fourier spectrum $F_b(\omega)$ of the image below the cutoff frequency is known, it is theoretically possible to recover the high-frequency portion $F_a(\omega)$ beyond the cutoff through mathematical modeling and function extrapolation.

7.2 Image Interpolation Method

As a fundamental technique for image reconstruction, interpolation methods are widely used in tasks such as image scaling, pixel value estimation, and super-resolution imaging. The core idea of interpolation is to infer new pixel values between known pixels to fill in missing information in the image. Due to their simplicity and high computational efficiency, interpolation methods are often employed during the initial reconstruction phase or for initial value estimation. This section introduces common image interpolation methods, including nearest-neighbor interpolation, bilinear interpolation, bicubic interpolation, and Lanczos interpolation. The key differences among these methods lie in the different kernel functions they use, resulting in varying levels of computational efficiency and interpolation performance.

7.2.1 Nearest-Neighbor Interpolation

Nearest-neighbor interpolation is a highly efficient image interpolation method. Its main idea is to directly assign the gray value or color value of the nearest pixel in the original image to the target interpolation position in the output image. Since this method does not require weighted averaging or complex computations, it is very simple to implement and is particularly suitable for scenarios where computational speed is critical and high image precision is not required.

The process of nearest-neighbor interpolation is illustrated in Fig. 7.3. Let the original image be denoted as $I(x, y)$ with dimensions $W \times H$, and the target image have dimensions $W' \times H'$. The scaling factors are given by $s_x = W'/W$ and $s_y = H'/H$. For a pixel (x', y') in the target image, its corresponding coordinate in the original image is:

$$(x, y) = \left(\frac{x'}{s_x}, \frac{y'}{s_y}\right) \tag{7.11}$$

Since the coordinate (x, y) is not necessarily an integer, the nearest integer pixel location is determined as:

$$(\hat{x}, \hat{y}) = \left(\text{round}\left(\frac{x'}{s_x}\right), \text{round}\left(\frac{y'}{s_y}\right)\right) \tag{7.12}$$

where round $(\cdot)$ denotes the rounding operation. Then, the pixel value at this point is assigned to the position (x', y') in the target image:

$$I'(x', y') = I(\hat{x}, \hat{y}) \tag{7.13}$$

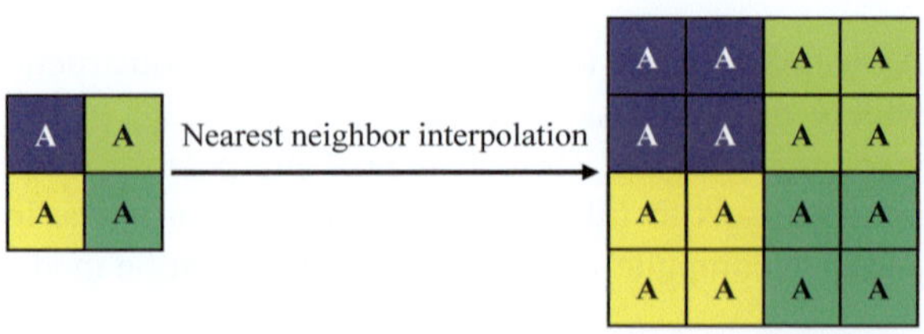

Fig. 7.3 Nearest neighbor interpolation diagram

This method preserves the structure of the original image but produces relatively coarse interpolation results. It is especially prone to jagged edges and blocky artifacts at image boundaries and along diagonal lines. The edges are not smooth, and the visual quality is generally poor, making it unsuitable for applications that require high image fidelity. However, due to its simplicity and extremely fast processing speed, it is still widely used in real-time image processing, quick image preview, or coarse image scaling tasks.

7.2.2 Bilinear Interpolation

Bilinear interpolation is a widely used method in image scaling and geometric transformations. Essentially, it is an extension of linear interpolation to two-dimensional space. This method performs one linear interpolation in the horizontal direction and another in the vertical direction to estimate the pixel value at any non-integer position.

As illustrated in Fig. 7.4a, suppose the target point $P(x, y)$ lies within the rectangle formed by four known pixels in the original image: $Q_{11}(x_1, y_1)$, $Q_{12}(x_1, y_2)$, $Q_{21}(x_2, y_1)$, and $Q_{22}(x_2, y_2)$, with $x_1 \le x \le x_2$ and $y_1 \le y \le y_2$. First, linear interpolation is performed along the x-direction on the two horizontal edges to obtain:

$$\begin{aligned} f(R_1) &\approx \frac{x_2 - x}{x_2 - x_1} f(Q_{11}) + \frac{x - x_1}{x_2 - x_1} f(Q_{21}) \\ f(R_2) &\approx \frac{x_2 - x}{x_2 - x_1} f(Q_{12}) + \frac{x - x_1}{x_2 - x_1} f(Q_{22}) \end{aligned} \tag{7.14}$$

Then, linear interpolation is performed along the y-direction using the two values obtained above, resulting in the final estimated value at the target point as

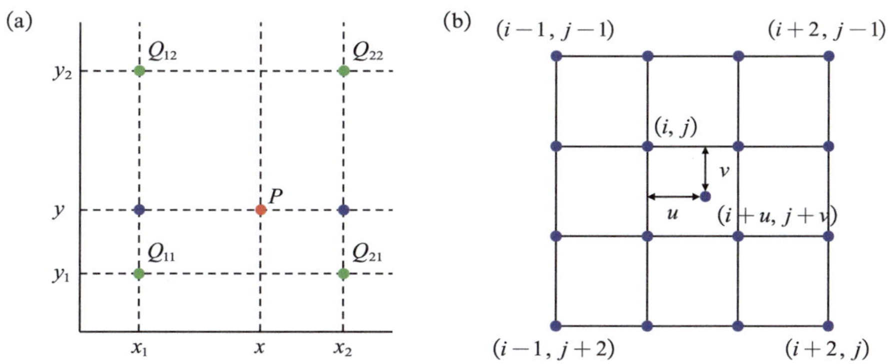

Fig. 7.4 Interpolation: **a** bilinear interpolation, **b** bicubic interpolation

$$
\begin{aligned}
f(P) &\approx \frac{y_2 - y}{y_2 - y_1} f(R_1) + \frac{y - y_1}{y_2 - y_1} f(R_2) \\
&= \frac{y_2 - y}{y_2 - y_1} \left(\frac{x_2 - x}{x_2 - x_1} f(Q_{11}) + \frac{x - x_1}{x_2 - x_1} f(Q_{21}) \right) \\
&+ \frac{y - y_1}{y_2 - y_1} \left(\frac{x_2 - x}{x_2 - x_1} f(Q_{12}) + \frac{x - x_1}{x_2 - x_1} f(Q_{22}) \right) \\
&= \frac{1}{(x_2 - x_1)(y_2 - y_1)} [x_2 - x x - x_1] \begin{bmatrix} f(Q_{11}) f(Q_{12}) \\ f(Q_{21}) f(Q_{22}) \end{bmatrix} \begin{bmatrix} y_2 - y \\ y - y_1 \end{bmatrix}
\end{aligned} \tag{7.15}
$$

The bilinear interpolation method maintains a reasonable balance between interpolation accuracy and computational complexity. It offers good image quality with relatively low computational cost, making it well-suited for real-time image processing, image warping, and resolution enhancement tasks. Compared to nearest-neighbor interpolation, bilinear interpolation produces smoother image edges, though it may introduce slight blurring in regions with sharp transitions.

7.2.3 Bicubic Interpolation

Compared to bilinear interpolation, bicubic interpolation performs significantly better when dealing with complex images, particularly in preserving fine details and edge sharpness. This method takes into account a 4×4 neighborhood around the target pixel, involving the gray values of 16 surrounding pixels. By applying cubic polynomial functions separately in the horizontal and vertical directions for weighted averaging, bicubic interpolation produces smoother and more detail-preserving reconstruction results.

As Fig. 7.4b, let the coordinates of the interpolation point be $(i + u, j + v)$. The gray values of all pixels within the surrounding 4×4 grid are assumed to be known. For each neighboring pixel, its weight is determined by its distance from the target point. Taking the pixel at (i, j) as an example, its distances from the target point in the x and y directions are u and v, respectively, so its weight is expressed as $w(u) \cdot w(v)$, where $w(\cdot)$ is the interpolation kernel function. The weights for the remaining 15 pixels are calculated similarly.

Based on these weights, the gray value of the target point $f(i + u, j + v)$ can be expressed in matrix multiplication form as

$$f(i + u, j + v) = A \cdot B \cdot C \tag{7.16}$$

where

$$A = \begin{bmatrix} w(1 + u) & w(u) & w(1 - u) & w(2 - u) \end{bmatrix} \tag{7.17}$$

B is a 4×4 matrix containing the 16 pixel values centered at (i, j) as

$$B = \begin{bmatrix} f(i-1,j-1) & f(i-1,j) & f(i-1,j+1) & f(i-1,j+2) \\ f(i,j-1) & f(i,j) & f(i,j+1) & f(i,j+2) \\ f(i+1,j-1) & f(i+1,j) & f(i+1,j+1) & f(i+1,j+2) \\ f(i+2,j-1) & f(i+2,j) & f(i+2,j+1) & f(i+2,j+2) \end{bmatrix} \quad (7.18)$$

$$C = \left[w(1+v)\ w(v)\ w(1-v)\ w(2-v) \right]^T \quad (7.19)$$

The interpolation kernel function $w(x)$ is defined as

$$w(x) = \begin{cases} 1 - 2|x|^2 + |x|^3, & 0 \leq |x| < 1 \\ 4 - 8|x| + 5|x|^2 - |x|^3, & 1 \leq |x| < 2 \\ 0, & |x| \geq 2 \end{cases} \quad (7.20)$$

Because this method incorporates a larger range of pixel information and more complex polynomial calculations during interpolation, it demonstrates excellent performance in image enlargement, sharpening, and super-resolution reconstruction. Bicubic interpolation can effectively preserve smooth edges and rich details, providing ideal visual results.

However, the computational cost of bicubic interpolation is significantly higher than that of bilinear interpolation, which may become a limiting factor in scenarios with strict efficiency requirements. Additionally, since it emphasizes the restoration of image edges and textures, it may cause over-sharpening in noisy images, thereby amplifying noise and negatively impacting the final image quality. Therefore, in practical applications, its performance and computational complexity should be balanced according to specific needs.

7.2.4 Lanczos Interpolation

Lanczos interpolation is a high-quality interpolation method based on resampling theory, commonly used for estimating pixel values during image enlargement and reduction. This method is built upon the sinc function, and by introducing a window function within a finite support range, it suppresses high-frequency oscillations, achieving an interpolation effect that combines smoothness with strong edge-preserving capability. Compared to bilinear and bicubic interpolation, Lanczos interpolation typically offers superior detail preservation and anti-aliasing performance. The core idea of Lanczos interpolation is to use a sinc-based spline kernel weighted by a window function to compute a weighted average at the interpolation point. Its general form is as follows

$$f(x) = \sum_{i=[x]-a+1}^{[x]+a} f(i) \cdot L(x-i) \quad (7.21)$$

where $L(x)$ is the Lanczos kernel function, and defined as

$$L(x) = \begin{cases} \mathrm{sinc}(x) \cdot \mathrm{sinc}\left(\frac{x}{a}\right), & -a < x < a \\ 0, & x - a, xa \end{cases} \quad (7.22)$$

The sinc function is defined as

$$\mathrm{sinc}(x) = \begin{cases} \frac{\sin(\pi x)}{\pi x}, & \text{if } x \neq 0 \\ 1, & \text{if } x = 0 \end{cases} \quad (7.23)$$

Here, the parameter a represents the kernel width, typically set to 2 or 3, corresponding to the Lanczos-2 and Lanczos-3 algorithms, respectively. This parameter determines the width of the neighborhood considered during interpolation. A wider kernel results in smoother interpolation but also increases the computational cost.

In two-dimensional image interpolation, Lanczos interpolation is usually performed by applying one-dimensional interpolation separately in the horizontal and vertical directions, and then combining the results as

$$f(x, y) = \sum_{m=-a+1}^{a} \sum_{n=-a+1}^{a} f(i+m, j+n) \cdot L(x-(i+m)) \cdot L(y-(j+n)) \quad (7.24)$$

Through this approach, the gray value of the target pixel is computed as a weighted combination of $(2a)^2$ neighboring pixels, with the weights controlled by the sinc window function. This gives Lanczos interpolation a distinct advantage in preserving image sharpness and reducing mosaic artifacts and jagged edges, making it especially suitable for high-quality applications such as image enlargement, high-definition reconstruction, printing, and video scaling.

Lanczos interpolation also has some limitations. Its interpolation kernel has a relatively large support region, which leads to higher computational complexity. In addition, the sinc function inherently exhibits a ringing effect that may introduce slight artifacts near edges or in high-frequency regions, especially when noise is present. When neighboring pixels vary sharply, the interpolation result may show slight overshoot or undershoot. Even so, Lanczos interpolation performs exceptionally well in preserving image details and edge sharpness, and remains one of the most widely used high-quality interpolation algorithms in image processing, suitable for applications requiring high interpolation accuracy.

7.3 Image Registration Methods

A key element of super-resolution reconstruction technology is the precise estimation of sub-pixel displacements across multiple images. Only through accurate sub-pixel registration can the similar and redundant information contained in multiple low-resolution images be fully utilized, providing sub-pixel-level complementary information for high-resolution image reconstruction.

7.3.1 *Image Registration*

The purpose of image registration is to transform multiple images that cover the same region but are in different coordinate systems into a common coordinate system, thereby achieving pixel-level correspondence between images. Let the reference image and the image to be registered be denoted as $I_1(x, y)$ and $I_2(u, v)$, respectively. Then, the spatial geometric mapping relationship between the two images can be expressed as

$$I_1(x, y) = I_2(T\{(u, v)\}) \tag{7.25}$$

where $T(\cdot)$ represents the spatial transformation between I_1 and I_2. Image registration involves computing the transformation function $T(\cdot)$ between the reference image and the image to be registered and performing image correction accordingly.

Image registration methods are generally classified into two categories: area-based and feature-based methods. Area-based registration methods mainly include those based on intensity information and those based on the transform domain. These methods are primarily applied in scenarios where the image pairs have small intensity differences and involve only spatial transformations. Feature-based methods are more commonly used for registering multi-source remote sensing images with significant geometric differences.

7.3.2 *Spatial Transformation Models for Image Registration*

The spatial transformation model in image registration is used to describe and implement the geometric correspondence between two or more images. It enables the processing of spatial relationships among images according to different application needs, allowing them to be compared or fused within a common coordinate system. Depending on the relationship between images and the nature of the changes, spatial transformation models can be either global or local. Global geometric transformation models applies to overall image transformations. It assumes that the changes between images are consistent across the entire image and can be described by a single function representing the spatial correspondence. The key is to identify homologous points through feature matching and use these points to solve for the transformation model. Local geometric transformation models, on the other hand, describes changes in specific regions of the image. It is suitable for large-scale images or cases involving local deformations between images, such as local elastic or linear deformations. These models may involve multiple forms or functions to describe the localized variations.

(1) Rigid transformation

Rigid transformation, also known as Euclidean transformation, involves only translation and rotation, without scaling or shearing. It is used to handle changes in image position and orientation without altering the image's shape or size. The expression is as follows

$$\begin{bmatrix} x' \\ y' \end{bmatrix} = \begin{bmatrix} \cos\theta & -\sin\theta \\ \sin\theta & \cos\theta \end{bmatrix} \begin{bmatrix} x \\ y \end{bmatrix} + \begin{bmatrix} \Delta x \\ \Delta y \end{bmatrix} \tag{7.26}$$

Here, (x', y') represents the transformed coordinates, (x, y) denotes the original coordinates, $(\Delta x, \Delta y)$ is the translation vector, and θ is the rotation angle. After the transformation, the distances and angles between points remain unchanged.

Similarity transformation is an extension of rigid transformation, with the addition of a scaling operation. It changes the size of the image while preserving its shape and proportions. The expression is as follows

$$\begin{bmatrix} x' \\ y' \end{bmatrix} = s \cdot \begin{bmatrix} \cos\theta & -\sin\theta \\ \sin\theta & \cos\theta \end{bmatrix} \begin{bmatrix} x \\ y \end{bmatrix} + \begin{bmatrix} \Delta x \\ \Delta y \end{bmatrix} \tag{7.27}$$

Here, s is the scaling factor.

(2) Affine transformation

Affine transformation allows for rotation, translation, scaling, and shearing of an image while preserving properties such as collinearity, straight lines, and parallelism. It is a combination of a linear transformation and a translation, and the image coordinates can be transformed using a 3×3 matrix of the following form

$$\begin{bmatrix} x' \\ y' \\ 1 \end{bmatrix} = \begin{bmatrix} a & b & \Delta x \\ c & d & \Delta y \\ 0 & 0 & 1 \end{bmatrix} \begin{bmatrix} x \\ y \\ 1 \end{bmatrix} \tag{7.28}$$

Here, (x, y) are the pixel coordinates in the original image, and (x', y') are the transformed coordinates. The coefficients a, b, c, and d in the matrix control rotation, scaling, and shearing of the image, while Δx and Δy represent the translation components.

(3) Projective transformation

Projective transformation, also known as perspective transformation, can simulate the visual effects of a camera lens and handle perspective distortions in images. It preserves straight lines after transformation but does not maintain the parallelism or orthogonality of lines, reflecting how the real 3D world is projected onto a 2D imaging plane. The projection process from 3D to a 2D plane is expressed as follows

$$\begin{bmatrix} x' \\ y' \\ 1 \end{bmatrix} = \begin{bmatrix} m_0 & m_1 & m_2 \\ m_3 & m_4 & m_5 \\ m_6 & m_7 & m_8 \end{bmatrix} \begin{bmatrix} x \\ y \\ 1 \end{bmatrix} \tag{7.29}$$

According to this formula, the 2D projective transformation maps a pixel coordinate (x, y) to the target pixel coordinate $\left(x', y'\right)$ as

$$\begin{cases} x' = \dfrac{m_0 x + m_1 y + m_2}{m_6 x + m_7 y + m_8} \\ y' = \dfrac{m_3 x + m_4 y + m_5}{m_6 x + m_7 y + m_8} \end{cases} \tag{7.30}$$

The transformation parameters m_i $(i = 0, 1, \ldots, 8)$ are constants determined by the specific scene and image.

(4) Nonlinear transformation

After undergoing a nonlinear transformation, a straight line in one image may be mapped to a curve in another image. In two-dimensional coordinates, a nonlinear transformation can be expressed as

$$\left(x', y'\right) = F(x, y) \tag{7.31}$$

In this expression, $F(x, y)$ represents the transformation function that maps one image to another. Polynomial transformation is a typical form of nonlinear transformation. There are various types of polynomial functions, such as quadratic and cubic functions. A general polynomial function can be expressed as

$$f(x) = a_0 + a_1 x + a_2 x^2 + \cdots + a_{n-1} x^{n-1} + a_n x^n \tag{7.32}$$

where a_i are the polynomial coefficients. Once the transformation parameters between two images are determined, the input image is transformed accordingly so that it aligns with the reference image in the same coordinate system, enabling further image matching. Common types of geometric image transformations are illustrated in Fig. 7.5.

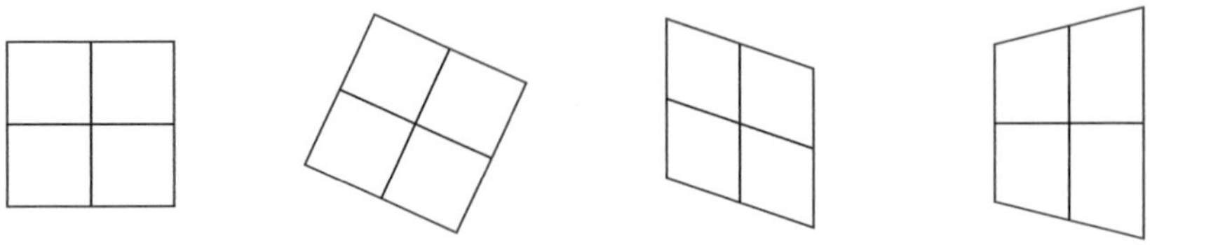

Fig. 7.5 Diagram of geometric transform: **a** input, **b** rigid, **c** affine, **d** projective

7.3.3 Intensity-Based Image Registration Methods

The intensity-based method is a registration approach based on the similarity of image grayscale values. This method directly utilizes the grayscale similarity between two images, relying on the internal grayscale information of the images combined with a search strategy to find corresponding points with maximum or minimum similarity. From these correspondences, the geometric transformation parameters between the image to be registered and the reference image can be calculated. Such methods are relatively simple to implement, requiring no complex preprocessing of images, and can perform registration directly using the original grayscale information. Common intensity-based image registration methods mainly include Mutual Information (MI) and Sequential Similarity Detection Algorithm (SSDA) [3, 4].

(1) Mutual information method

The mutual information method is a typical intensity-based image registration approach with good registration accuracy and strong applicability. The amount of information in an image is usually represented by entropy, which measures the uncertainty of a system and can effectively reflect the total information contained within the system. Its mathematical expression is

$$H(X) = -\sum_{i=1}^{L} p(a_i) \log p(a_i) \tag{7.33}$$

For an image, the entropy can be expressed as

$$H(Y) = -\sum_{i=0}^{N-1} p_i \log p_i, \quad p_i = h_i \Big/ \sum_{i=0}^{N-1} h_i \tag{7.34}$$

where h_i represents the number of pixels in image Y with grayscale value i, N is the number of grayscale levels in the image, and p_i is the probability of grayscale value i occurring.

Mutual information (MI) is primarily used to measure the correlation between two systems or to evaluate how much information one system contains about another. In image registration, mutual information can be represented using the entropies and joint entropy of images. For two images R and F, mutual information is defined as

$$MI(R, F) = H(R) + H(F) - H(R, F) \tag{7.35}$$

where $H(R)$ and $H(F)$ are the entropies of images R and F, respectively, and $H(R, F)$ is their joint entropy, which can be computed from the joint histogram. The higher the mutual information value, the more similar the images are or the larger their overlapping region, indicating stronger correlation. The registration is achieved by finding the template and sub-image pair that maximizes mutual information.

(2) Sequential similarity detection algorithm

Sequential similarity detection algorithm (SSDA) is a fast image registration method based on image residuals. Its basic idea is: when comparing the sum of residuals between two images, set an upper threshold T_k. During the calculation of the residual sum at a candidate position, if the accumulated residual exceeds this threshold, that position is immediately discarded as it cannot be the best match. The process then moves on to the next candidate position. Ultimately, the position with the smallest residual sum and slowest growth rate is chosen as the matching point.

Let the reference image S have size $N \times N$, and the template image T have size $M \times M$. Denote S_{ij} as the sub-image of S covered by the template positioned at (i, j) $(1 < i, j < N - M + 1)$. $S_{ij}(m, n)$ and $T(m, n)$ represent the pixel intensities at position (m, n) within the sub-image and template, respectively.

In SSDA, the key to matching is evaluating the grayscale differences between the template and the sub-region of the reference image. Define the absolute error ε to measure the difference as

$$\varepsilon(i, j, m_k, n_k) = \left|S_{ij}(m_k, n_k) - \overline{S}(i, j) - T(m_k, n_k) + \overline{T}\right| \tag{7.36}$$

where $\overline{S}(i, j)$ and $\overline{T}$ are the average grayscale values of the search sub-image S_{ij} and template T, calculated as

$$\overline{S}(i, j) = \frac{1}{M^2}\sum_{m=1}^{M}\sum_{n=1}^{M} S_{ij}(m, n), \quad \overline{T} = \frac{1}{M^2}\sum_{m=1}^{M}\sum_{n=1}^{M} T(m, n) \tag{7.37}$$

During the algorithm execution, a fixed error threshold T_k is set. In each comparison, a number of non-repeated pixels are randomly selected from the template. Their absolute errors ε with the corresponding pixels in the current search sub-image are computed and accumulated. Once the accumulated error exceeds the threshold T_k, the current position is rejected, and the number of pixels r required to reach this threshold is recorded.

Based on this, the detection surface of SSDA can be defined as follows

$$I(i,j) = \left\{ r \middle| \min\left[\sum_{k=1}^{r} \varepsilon(i,j,m_k,n_k) \right] \geq T_k \right\}, \quad 1 \leq r \leq M^2 \tag{7.38}$$

During the search process over the entire image, points on the detection surface $I(i,j)$ with larger values, i.e., positions where the error accumulation grows slowest, are considered the best matching points. This is because these positions require more pixels to be accumulated before the total error exceeds the preset threshold, indicating a higher similarity to the template image.

Intensity-based image registration methods are relatively simple to implement and can directly utilize grayscale information for matching calculations. However, they also have some notable drawbacks. First, these methods are very sensitive to grayscale variations, especially under adverse conditions such as illumination changes or noise interference, leading to poor robustness and significantly reduced computational efficiency. Second, due to the large number of pixels in an image, intensity-based registration often faces a very high computational burden in practical applications. Additionally, these methods have weak adaptability to physical changes in the target image such as scaling, rotation, and occlusion deformation, making it difficult to ensure accuracy and stability in registration.

7.3.4 Fourier Transform Domain-Based Image Registration Methods

Image registration based on the Fourier transform generally leverages the properties of translation, rotation, and scaling inherent in the Fourier domain. When dealing with translation, suppose the image $f_2(x, y)$ is generated by shifting the image $f_1(x, y)$, then the relationship between the two images is

$$f_2(x, y) = f_1(x - x_0, y - y_0) \tag{7.39}$$

The corresponding Fourier spectra $F_1(\xi, \eta)$ and $F_2(\xi, \eta)$ satisfy:

$$F_2(\xi, \eta) = \exp\left[-\mathrm{i}2\pi(\xi x_0 + \eta y_0)\right] F_1(\xi, \eta) \tag{7.40}$$

The cross-power spectrum of the two images in the frequency domain is

$$\frac{F_1(\xi, \eta) F_2^*(\xi, \eta)}{\left|F_1(\xi, \eta) F_2^*(\xi, \eta)\right|} = \exp\left[\mathrm{i}2\pi(\xi x_0 + \eta y_0)\right] \tag{7.41}$$

where ξ and η are frequency variables, and $F_2^*(\xi, \eta)$ is the complex conjugate of $F_2(\xi, \eta)$. According to the translation property, the phase of the cross-power spectrum equals the phase difference between the two images. By applying the inverse Fourier transform to the cross-power spectrum, one obtains an impulse function that features a sharp peak at the offset position (x_0, y_0), while values elsewhere tend to zero, thus determining the displacement between the two images. When there is no significant scale change, suppose $f_2(x, y)$ is obtained by translating by (x_0, y_0) and rotating by θ_0 the image $f_1(x, y)$, namely

$$f_2(x, y) = f_1(x \cos\theta_0 + y \sin\theta_0 - x_0, -x \sin\theta_0 + y \cos\theta_0 - y_0) \tag{7.42}$$

Using the Fourier transform properties of translation and rotation, the relationship between the transformed images in the frequency domain is

$$F_2(\xi, \eta) = e^{-i2\pi(\xi x_0 + \eta y_0)} F_1(\xi \cos\theta_0 + \eta \sin\theta_0, -\xi \sin\theta_0 + \eta \cos\theta_0) \tag{7.43}$$

Let $M_1(\xi, \eta)$ and $M_2(\xi, \eta)$ denote the power spectra of $F_1(\xi, \eta)$ and $F_2(\xi, \eta)$ respectively. Their relationship can be expressed as

$$M_2(\xi, \eta) = M_1(\xi \cos\theta_0 + \eta \sin\theta_0, -\xi \sin\theta_0 + \eta \cos\theta_0) \tag{7.44}$$

Since $F_1(\xi, \eta)$ and $F_2(\xi, \eta)$ have identical power, converting the above formula from Cartesian to polar coordinates yields

$$M_1(\rho, \theta) = M_2(\rho, \theta - \theta_0) \tag{7.45}$$

Using phase correlation theory, θ_0 can be successfully derived and applied.

7.3.5 Feature-Based Image Registration Methods

Feature-based image registration methods [5] mainly follow the process in Fig. 7.6. This approach does not require consideration of the vast number of pixels in the image; instead, it extracts prominent feature information such as points, lines, and edges from the image to be registered, and completes registration without any additional auxiliary information. While significantly reducing computational load and simplifying the process, this method demonstrates certain robustness to changes in grayscale values of the images to be registered. It also exhibits resilience to adverse conditions like illumination changes and shadows, resulting in relatively good registration performance. Unlike the enormous number of pixels in an image, the number of feature points is relatively small. This allows feature-based registration methods to achieve high precision and accuracy in feature extraction and matching, as well as strong resistance to noise and other interferences. Since these algorithms typically extract and use only a small portion of the features in the image, the availability and quality of feature points are critical for improving accuracy.

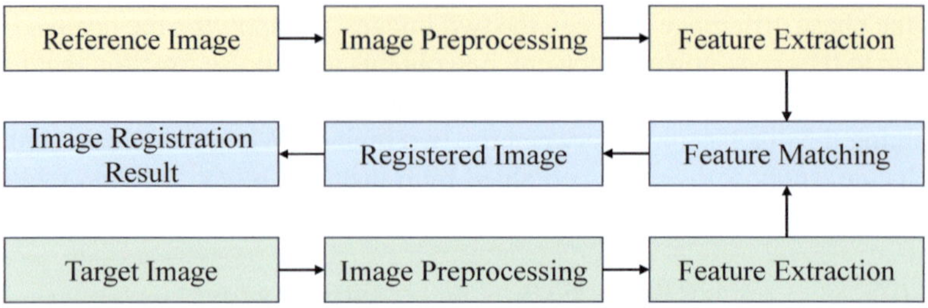

Fig. 7.6 Workflow diagram of feature-based registration method

(1) Point feature-based image registration

Point features generally include corners, endpoints, and extrema. Point feature registration algorithms typically detect feature points that are robust to noise and disturbances and establish correspondences between the reference and target images based on these points, utilizing geometric transformations to achieve matching.

When selecting feature points, priority is given to pixels with distinctive characteristics relative to their local neighborhood. Because point features contain limited information, usually only representing approximate spatial locations, a sufficient number of effective feature points must be extracted for image registration. Common detectors include Harris, Susan, Hessian, and FAST. Among these, Harris corners consider gradient differences in multiple directions, providing good rotation invariance and partial affine transform stability.

Point-based feature registration is suitable for images with distinct corners, high curvature points, and edge endpoints. However, for images with weak grayscale variations and smooth edge contours, it is often difficult to extract enough valid feature points. In particular, when images contain significant noise, this method may misclassify noise points as features, causing registration failure.

(2) Line feature-based image registration

Line feature-based registration extracts line features via edge detection algorithms. Line features typically refer to image edges and the pairs of edges that form lines. Edges can be understood as boundary lines separating regions with differing local characteristics, while lines refer to edge pairs with minimal width and similar features in the enclosed region. The task of edge detection is to find pixel sets exhibiting step or roof-like changes. As shown in Fig. 7.7, the basic method for edge detection is differential analysis: for step changes, peaks in the first derivative correspond to edges, and zero crossings in the second derivative also indicate edges; for roof changes, zero crossings of the first derivative and peaks in the second derivative correspond to edges.

After line features are extracted, descriptors are constructed to facilitate feature matching, ultimately identifying matching line pairs.

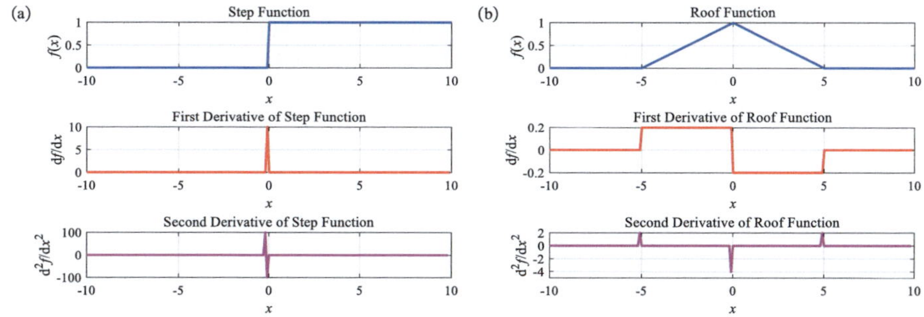

Fig. 7.7 Shape functions with their derivatives: **a** step function, **b** roof function

Based on the matched lines' corresponding homologous points, the registration transformation is determined: if a line is closed, its centroid is taken as the homologous point; if not closed, the start point and midpoint serve as homologous points. Afterwards, error correction is applied to the image. Common edge detection operators used in line feature matching include: Robert, Prewitt, Sobel, Laplacian, Canny, and the Hough transform.

The Sobel operator combines difference operations with weighted averaging in the neighboring region. Considering the varying influence of adjacent pixels, it uses a weighted average approach. Its convolution kernels are

$$G_x = \begin{bmatrix} -1 & 0 & 1 \\ -2 & 0 & 2 \\ -1 & 0 & 1 \end{bmatrix}, \quad G_y = \begin{bmatrix} -1 & -2 & -1 \\ 0 & 0 & 0 \\ 1 & 2 & 1 \end{bmatrix} \tag{7.46}$$

The Roberts operator is a gradient operator that uses the difference between diagonal neighboring pixel values. Its computation is defined as

$$G_x = f(i, j) - f(i - 1, j - 1) \tag{7.47}$$

$$G_y = f(i - 1, j) - f(i, j - 1) \tag{7.48}$$

$$|G(x, y)| = \sqrt{G_x^2 + G_y^2} \tag{7.49}$$

Its convolution kernels are

$$G_x = \begin{bmatrix} 0 & -1 \\ 1 & 0 \end{bmatrix}, \quad G_y = \begin{bmatrix} -1 & 0 \\ 0 & 1 \end{bmatrix} \tag{7.50}$$

The Canny operator is an edge detection method that applies smoothing before differentiation. It first smooths the image using a Gaussian filter to reduce noise, then computes the edge direction and magnitude via first-order differences. Its convolution kernels are

$$H_1 = \begin{bmatrix} -1 & -1 \\ 1 & 1 \end{bmatrix}, \quad H_2 = \begin{bmatrix} 1 & -1 \\ 1 & -1 \end{bmatrix} \tag{7.51}$$

Knowing the global gradient direction alone is not sufficient to determine edge locations, so it is necessary to suppress non-maximum gradient values to highlight true edges. The gradient direction obtained previously spans a full circle 2π, which is used to divide the image into four directional regions. A dual-threshold strategy is then applied using two thresholds, Th_1 and Th_2. If $Th_1 = 0.4Th_2$ pixels with gradient values below the thresholds are set to zero, resulting in two new images A and B. Image B, with the higher threshold, effectively removes most noise, while image A retains more edge information and complements image B. Repeating this process allows for the extraction of complete edge contours, and all such contours are then extracted to complete the detection. Figure 7.8 illustrates the visual results of several typical edge detection operators.

Feature registration methods based on line features are well-suited for images with prominent linear structures, such as roadways, coastlines, or building outlines.

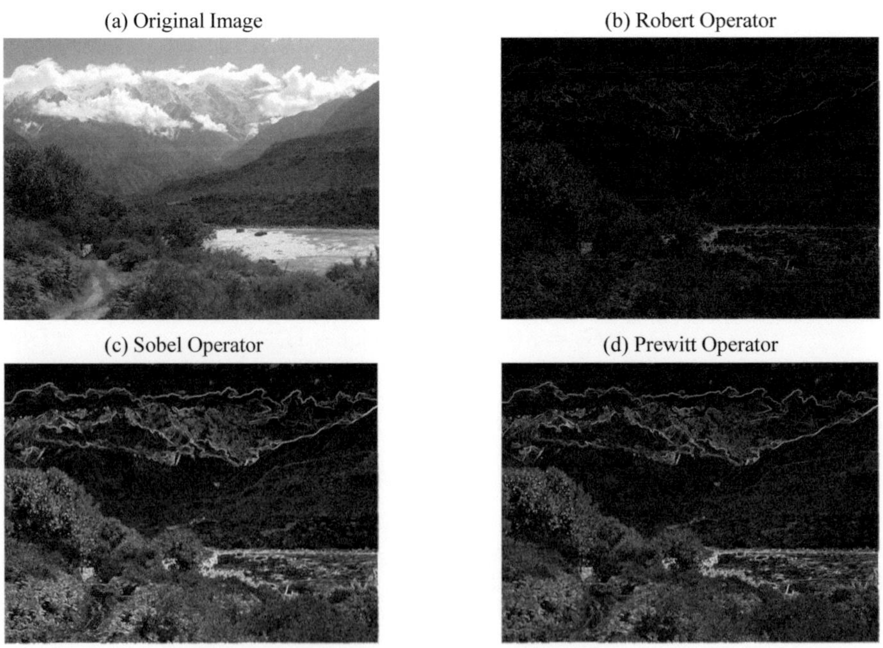

Fig. 7.8 Effect diagrams of several typical operators

However, they also have limitations: when there are many repeated and similar straight lines in the images to be registered, both the algorithm's efficiency and registration accuracy can be significantly affected.

7.4 Digital Image Super-Resolution Techniques

In practical imaging processes, due to limitations imposed by sensor resolution, imaging system bandwidth, and noise, the acquired images often suffer from insufficient resolution. To enhance the spatial resolution of images without altering the hardware setup, digital image super-resolution techniques have emerged. These techniques process and reconstruct multiple low-resolution images within the computational domain to recover image details with higher spatial resolution. In recent years, super-resolution reconstruction technology has developed rapidly. Among them, traditional methods based on prior knowledge and image degradation models still play a significant role in practical applications. This section focuses on three classical image super-resolution approaches: the Projection Onto Convex Sets [6] (POCS) method, the Iterative Back Projection [7] (IBP) algorithm, and the Maximum a Posteriori [8] (MAP) estimation method. Their fundamental principles and characteristics will be analyzed respectively.

7.4.1 Projections onto Convex Sets

The goal of super-resolution reconstruction algorithms is to recover a high-resolution image from a sequence of low-resolution images. Super-resolution reconstruction is an ill-posed problem, characterized by its invertibility and instability. To address this ill-posed nature, prior knowledge about the image must be incorporated during the reconstruction process.

Each image carries its own distinct prior knowledge, such as data reliability, energy boundedness, smoothness, and positivity. These priors are also referred to as prior constraints, and they play various constraining roles in the super-resolution reconstruction process. In the Projection Onto Convex Sets (POCS) theory, each prior constraint corresponds to a convex set, and the intersection of all convex sets associated with different constraints forms the solution space for the high-resolution image.

Assuming that the estimated high-resolution image is an element in a Hilbert space, then all convex sets corresponding to the prior constraints are also contained within this Hilbert space. According to the degradation model described in previous sections, the degradation process of the sequence of low-resolution images can be expressed as

$$y_k = Hx + n_k, \quad k = 0, 1, \ldots, p \tag{7.52}$$

Here, H represents the overall degradation process, and in the POCS algorithm, it specifically takes the form of a Toeplitz matrix.

Assuming there are m pieces of prior knowledge in the Hilbert space, we can define m convex sets C_i accordingly, where $i = 1, 2, \ldots, m$, and each convex set C_i is associated with a projection operator P_i. Thus, the ideal high-resolution image x is considered to be a point within the intersection of all these convex sets

$$x \in \bigcap_{i=1}^{m} C_i \tag{7.53}$$

The fundamental idea of the POCS algorithm is to iteratively project the estimated high-resolution image onto each convex set C_i using the corresponding projection operators P_i, thereby approaching the optimal solution. The iterative process can be expressed as

$$x^{(n+1)} = P_m \cdots P_2 P_1 x^{(n)} \tag{7.54}$$

By continuously projecting the current estimate onto each convex set, the solution gradually converges to a reconstructed image that satisfies all prior constraints. This algorithm does not rely on an explicit analytical form of the image but instead incorporates various prior constraints through a geometric approach during image estimation. It has strong adaptability and theoretical convergence guarantees.

Common prior constraints include data consistency, amplitude constraint, and similarity constraint. Below, we introduce data consistency constraint and its corresponding convex set and projection operator.

(1) Data consistency constraint

The pixel values of the ideal high-resolution image x at a given location should agree with the observed values from all low-resolution images y_k at the corresponding positions. Based on this, the convex set C_D corresponding to the data consistency constraint is defined as

$$C_D = \{\hat{x}(s,t) | R(i,j) \le \delta_0\} \tag{7.55}$$

Here, δ_0 is a residual threshold determined by the statistical properties of noise, representing the confidence interval for the estimated values. $R(i, j)$ denotes the residual at position (i, j) in the low-resolution image, i.e., the difference between the observed value and the simulated value. Suppose that the pixel (i, j) in the low-resolution image y_k corresponds to the position $(s\prime, t\prime)$ in the high-resolution image $\hat{x}$, then the residual $R(i, j)$ is defined as

$$R(i,j) = y_k(i,j) - \sum_{s=s'-(w-1)/2}^{s'+(w-1)/2} \sum_{t=t'-(w-1)/2}^{t'+(w-1)/2} \hat{x}(s,t) \cdot H\left(s, t; s', t'\right) \tag{7.56}$$

Here, $H(s, t; s', t')$ is the normalized point spread function (PSF), and w is the size of the convolution kernel. The first term on the right is the observed value, while the second term is the simulated value derived from the estimated high-resolution image after applying degradation.

During the iterative projection, if the residual $R(i, j)$ exceeds the threshold δ_0, it indicates that the estimated pixel value $\hat{x}(s, t)$ deviates significantly and needs adjustment. If $R(i, j) > \delta_0$, the estimate is too low and should be increased; if $R(i, j) < -\delta_0$, the estimate is too high and should be decreased. By repeatedly correcting through projection, the residual between the estimated and observed values is confined within a credible range, thereby enforcing data consistency. Based on the relationship between $R(i, j)$ and δ_0, the projection operator P_D for data consistency is defined as

$$P_D[\hat{x}(s,t)] = \begin{cases} \hat{x}(s,t) + \left[R(i,j) + \delta_0\right] \cdot H\left(s,t;s',t'\right), & R(i,j) < -\delta_0 \\ \hat{x}(s,t), & -\delta_0 \le R(i,j) \le \delta_0 \\ \hat{x}(s,t) + \left[R(i,j) - \delta_0\right] \cdot H\left(s,t;s',t'\right), & R(i,j) > \delta_0 \end{cases} \quad (7.57)$$

This operator slightly adjusts the estimated value $\hat{x}(s, t)$ in the direction of the residual to gradually reduce the difference between the observed and simulated values, effectively enforcing the data consistency constraint within the POCS framework.

(2) Amplitude constraint

The amplitude constraint refers to utilizing the known grayscale value range of the image as prior knowledge to construct a corresponding constraint convex set. The goal of this constraint is to restrict the pixel intensity values of the reconstructed image within a reasonable interval, thereby avoiding grayscale distortion caused by noise or abnormal perturbations. For example, in a typical 8-bit grayscale image, pixel values should lie within the interval [0, 255]. Based on this, the convex set C_A for the amplitude constraint can be defined as

$$C_A = \{\hat{x}(s,t) | 0 \le \hat{x}(s,t) \le 255\} \quad (7.58)$$

The corresponding projection operator P_A is

$$P_A[\hat{x}(s,t)] = \begin{cases} 0, & \hat{x}(s,t) < 0 \\ \hat{x}(s,t), & 0 \le \hat{x}(s,t) \le 255 \\ 255, & \hat{x}(s,t) > 255 \end{cases} \quad (7.59)$$

This operator performs a clipping operation on pixel values that exceed the grayscale range after each iteration, effectively suppressing numerical drift caused by overfitting or noise.

7.4.2 Iterative Back Projection Algorithm

Iterative Back Projection (IBP) is an iterative reconstruction algorithm whose core concept is as follows: ideally, the reconstructed high-resolution (HR) image should be almost indistinguishable from the original HR image, or even completely identical. Accordingly, a simulated low-resolution (LR) image generated by applying a degradation process to the reconstructed HR image should also match the real observed LR image precisely.

However, in practice, due to imperfections in the degradation model and the presence of noise, discrepancies between the simulated LR image and the actual observed image are inevitable. The IBP algorithm addresses this by "back-projecting" the reconstruction error into the HR image estimate. Through repeated iterations, the algorithm incrementally corrects the estimated HR image so that it gradually approaches the true original image, thereby achieving super-resolution reconstruction.

The overall reconstruction process of the IBP algorithm is illustrated in Fig. 7.9. First, consider the imaging degradation model: assume that under this model, there are p low-resolution image frames observed from the same scene. Suppose the k-th low-resolution image y_k is obtained by applying motion transformation M_k, blur operation B_k, and downsampling operation D to a high-resolution image x, with added noise n_k. The imaging model is expressed as

$$y_k = DB_kM_kx + n_k, \quad k = 0, 1, \ldots, p \tag{7.60}$$

For simplicity, let $H = DB_kM_k$ represent the overall degradation system matrix, and n_k be the total noise term. The model can then be simplified as

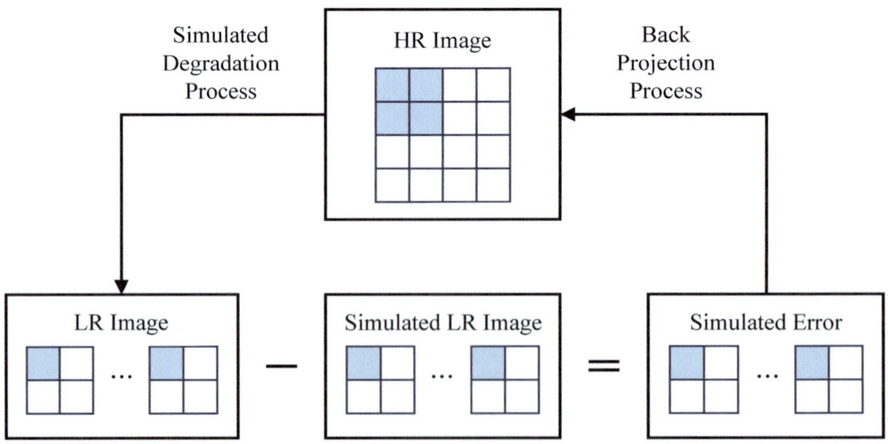

Fig. 7.9 Reconstruction process of the IBP algorithm

$$y_k = Hx + n_k, \quad k = 0, 1, \ldots, p \tag{7.61}$$

In each iteration, the simulated low-resolution image $\hat{y}_k^{(n)}$ generated from the current high-resolution estimate $\hat{x}^{(n)}$ is compared with the corresponding observed image y_k to compute an error image. This error image is then projected back to the high-resolution image space through an inverse degradation process to correct the estimate, gradually converging to an optimal reconstruction result.

Let $H^{-1} = M_k^{-1} B_k^{-1} D^{-1}$ denote the overall inverse degradation operation. Then the steps of the IBP algorithm are as follows

$$\hat{y}_k^{(n)} = H\hat{x}^{(n)} + n_k \tag{7.62}$$

$$E_I^{(n)} = \frac{1}{p} \sum_{k=1}^{p} H^{-1}\left(y_k - \hat{y}_k^{(n)}\right) \tag{7.63}$$

$$\hat{x}^{(n+1)} = \hat{x}^{(n)} + \lambda E_I^{(n)} \tag{7.64}$$

Here, $\hat{x}^{(n)}$ and $\hat{x}^{(n+1)}$ denote the high-resolution image estimates at the n-th and $(n+1)$-th iterations, respectively. λ is a step size that controls the convergence speed, and $E_I^{(n)}$ is the correction term generated by back-projecting the error in the n-th iteration. Through this iterative update process, the IBP algorithm gradually drives the estimate $\hat{x}$ closer to the true high-resolution image, thereby accomplishing super-resolution reconstruction.

7.4.3 Maximum A Posteriori Estimation

Maximum a Posteriori (MAP) Super-Resolution Reconstruction aims to estimate the high-resolution image x that maximizes the posterior probability $\Pr(x|y_k)$, given a sequence of low-resolution observed images y_k. According to Bayes' theorem, the reconstructed image $\hat{x}$ under MAP estimation is expressed as

$$\hat{x} = \arg\max_x \left[\Pr(x|y_k)\right] = \arg\max_x \left[\frac{\Pr(y_k|x)\Pr(x)}{\Pr(y_k)}\right] \tag{7.65}$$

In this expression, x denotes the original high-resolution image, $\hat{x}$ is the reconstructed high-resolution image, y_k is the sequence of low-resolution images, $\Pr(x)$ is the prior probability of x, $\Pr(y_k)$ is the prior probability of y_k, $\Pr(y_k|x)$ is the conditional probability of observing y_k given x, and $\Pr(x|y_k)$ is the posterior probability. Therefore, Eq. (7.65) aims to obtain the MAP estimate $\hat{x}$ by maximizing $\Pr(x|y_k)$ given the observed low-resolution image sequence y_k.

For convenience, take the natural logarithm of both sides of Eq. (7.65). Since the logarithm is a monotonic increasing function, it does not affect the location of the maximum with respect to x, yielding

$$\begin{aligned}\hat{x} &= \arg\max_{x} \left[\log \Pr(x|y_k)\right] \\ &= \arg\max_{x} \left[\log \Pr(y_k|x) + \log \Pr(x) - \log \Pr(y_k)\right]\end{aligned} \tag{7.66}$$

Since $\Pr(y_k)$ is independent of x, $\log \Pr(y_k)$ can be omitted, simplifying the computation as

$$\begin{aligned}\hat{x} &= \arg\max_{x} \left[\log \Pr(y_k|x) + \log \Pr(x)\right] \\ &= \arg\max_{x} \left[-\log \Pr(y_k|x) - \log \Pr(x)\right]\end{aligned} \tag{7.67}$$

As shown above, in order to obtain the optimal solution $\hat{x}$ that maximizes the posterior probability $\Pr(x|y_k)$, we must first determine the conditional probability $\Pr(y_k|x)$ and the prior probability $\Pr(x)$. In Bayesian analysis, an image is treated as a random field, and its properties are described using probability distributions. Thus, we must construct a conditional random field model for $\Pr(y_k|x)$ and a prior random field model for $\Pr(x)$.

Common random field models include the Gaussian model, Markov model, Gibbs model, and Gaussian-Markov model, each offering different constraint capabilities. In this section, we use a Gaussian random field to model the conditional probability $\Pr(y_k|x)$ and a Gibbs random field to model the prior probability $\Pr(x)$.

According to the MAP formulation, the probability density of random noise in the observed low-resolution images determines the conditional probability of those images. Assume the noise in the low-resolution sequence y_k is zero-mean Gaussian noise with variance σ_k^2. Then $\Pr(y_k - Hx) = \Pr(n)$. Considering the noise between image frames is independent and identically distributed, the observed image y_k can be viewed as a random field with mean $\Pr(y_k - Hx) = \Pr(n)$. Let $\hat{y}_k$ denote the simulated observed image after the degradation process. Then, the conditional probability distribution of the observed low-resolution images given the high-resolution image can be written as

$$\begin{aligned}\Pr(y_k|x) &= \Pr(y_k - Hx) = \Pr(n) \\ &= \prod_{k} \frac{1}{\sqrt{2\pi}\sigma_k} \exp\left(-\frac{1}{2\sigma_k^2}\|\hat{y}_k - y_k\|^2\right)\end{aligned} \tag{7.68}$$

For the prior probability distribution $\Pr(x)$, we use a Gibbs random field. Each pixel in the image is considered a point in the random field. Although each pixel value is random, it exhibits consistent statistical properties across multiple images. This is known as the Markov property, which states that nearby pixels tend to be more similar. Since Markov random fields are equivalent to Gibbs random fields, and the latter is computationally more convenient, we use the Gibbs random field to model the prior probability $\Pr(x)$ as

$$\Pr(x) = \frac{1}{Z} \exp\left(-\frac{U(x)}{T}\right), \quad Z = \sum_x \exp\left(-\frac{U(x)}{T}\right) \tag{7.69}$$

where, Z is the normalization factor, T is a constant, and $U(x)$ is the prior constraint function representing the Markov property. In this section, we use the Laplacian operator L to construct $U(x)$ as

$$U(x) = \|Lx\|_2^2, \quad L = \begin{bmatrix} -1 & -1 & -1 \\ -1 & 8 & -1 \\ -1 & -1 & -1 \end{bmatrix} \tag{7.70}$$

Substituting Eqs. (7.68), (7.69) and (7.70) into (7.67), we obtain the objective function for MAP-based super-resolution reconstruction

$$\hat{x} = \arg\max_x \left[\sum\nolimits_{k=1}^{p} \|y_k - DB_k M_k x\|^2 + \alpha_M \|Lx\|^2\right] \tag{7.71}$$

In this objective function, the first term is called the data term, which describes the fidelity between the observed image and the simulated observation after degradation, i.e., the data fidelity. The second term is the regularization term, which incorporates prior knowledge of the solution and suppresses high-frequency noise to make the reconstruction smoother. The parameter α_M is the regularization weight that balances the data term and the regularization term. If α_M is too large, the regularization dominates and the reconstruction will be overly smoothed. If α_M is too small, noise suppression will be ineffective, resulting in severe distortion.

7.5 Optical Super-Resolution Imaging Methods

Conventional optical imaging systems are constrained by the diffraction limit, where spatial resolution is jointly restricted by the wavelength and the numerical aperture, making it difficult to capture sub-wavelength scale information. To overcome this limitation, researchers have proposed various optical super-resolution imaging techniques. By introducing external modulation mechanisms or leveraging nonlinear optical effects, these methods achieve imaging capabilities far beyond the diffraction limit. Optical super-resolution technologies play a crucial role not only in fundamental scientific research but also show great potential in practical fields such as life sciences and nanofabrication. This section introduces several representative optical super-resolution imaging methods, including Fourier Ptychographic Microscopy (FPM), Structured Illumination Microscopy (SIM), and Stimulated Emission Depletion Microscopy (STED), with a focus on their operating principles and imaging characteristics.

7.5.1 Fourier Ptychographic Microscopy

Fourier Ptychographic Microscopy (FPM) aims to achieve high-resolution and high-throughput imaging using visible light. A typical Fourier ptychographic microscopy system consists of a programmable LED array and a standard optical microscope equipped with a low numerical aperture (NA) objective lens [9, 10]. As shown in Fig. 7.10, the system includes three key planes: the object plane where the sample is located, an aperture at the Fourier plane, and a detector at the image plane. The objective lens first performs a Fourier transform on the light waves from the object plane, mapping them to the Fourier plane. Then, the tube lens applies another Fourier transform to project the light onto the image plane. In this imaging model, the aperture at the Fourier plane effectively serves as a low-pass filter for measurements at the image plane.

During image acquisition, the LED array illuminates the sample from various incident angles, and the system captures corresponding low-resolution intensity images. Each image represents information from a circular sub-aperture region in the Fourier domain. The size of this sub-aperture is determined by the NA of the objective lens, while its position relative to the origin is determined by the incident wave vector of the LED.

The FPM model assumes an ideal coherent imaging system and treats the sample as a thin object with a distribution denoted by $o(r)$, where $r = (x, y)$ represents the 2D coordinates in the object plane. The modulation of the incident light intensity by the sample reflects its absorption or reflectance properties, while the phase modulation represents information such as thickness or refractive index. The wave emitted after modulation by the sample can be expressed as $\exp(-\mathrm{i}k_m r)$, where k_m is the incident wave vector of the m-th LED. Since the LED array is located far from the sample, the emitted light can be approximated as a plane wave. After low-pass filtering by the objective lens, the resulting Fourier-transformed spectrum is

$$G_m(k) = O(k - k_m)P(k) \tag{7.72}$$

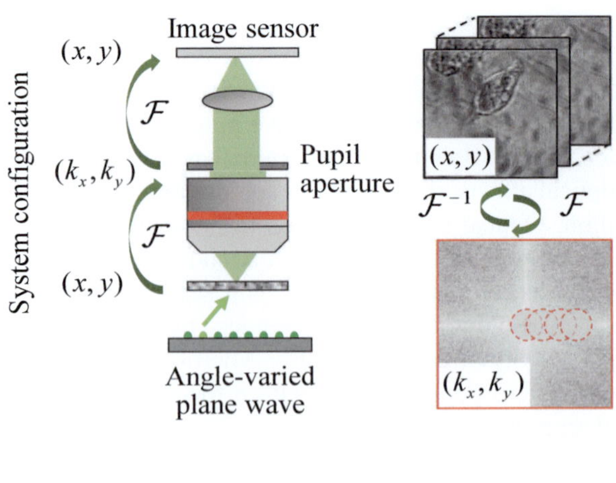

Fig. 7.10 Schematic diagram of Fourier ptychographic microscopy principle

Here, $O(k)$ is the Fourier transform of the sample function, and $P(k)$ is the coherent transfer function of the optical system, representing the aperture in the frequency domain. Its ideal form is

$$P(k) = \begin{cases} 1, & |k| \le k_{obj} \\ 0, & |k| > k_{obj} \end{cases} \tag{7.73}$$

The image captured by the detector can be expressed via inverse Fourier transform as

$$\begin{aligned} I_m(r) &= \left|\mathcal{F}^{-1}\{\mathcal{F}\{o(r)\exp(-ik_m r)\}P(k)\}\right|^2 \\ &= \left|\mathcal{F}^{-1}\{O(k-k_m)P(k)\}\right|^2 \end{aligned} \tag{7.74}$$

The incident wave vector $k_m = \left(k_{xm}, k_{ym}\right)$ can be calculated based on the geometric position of the LED as

$$\begin{aligned} k_{xm} &= -\frac{2\pi}{\lambda}\frac{x_0 - x_m}{(x_0 - x_m)^2 + (y_0 - y_m)^2 + d^2} \\ k_{ym} &= -\frac{2\pi}{\lambda}\frac{y_0 - y_m}{(x_0 - x_m)^2 + (y_0 - y_m)^2 + d^2} \end{aligned} \tag{7.75}$$

Through an iterative phase retrieval process, one can synthesize measurement data in the Fourier domain and reconstruct a high-resolution image of the object. The resolution of the reconstructed image depends on the maximum incident angle of the LED array, rather than the numerical aperture of the objective lens, enabling a balance between high resolution and wide field of view. Fourier ptychographic microscopy has evolved from a microscopy tool into a versatile technique widely applied in 3D imaging, remote sensing, X-ray nano-imaging, and other fields.

7.5.2 Structured Illumination Microscopy

In the spatial domain, the imaging process of an incoherent optical system can be represented as a convolution between the sample and the system's Point Spread Function (PSF). In the frequency domain, this corresponds to a multiplication between the sample's Fourier spectrum and the system's Optical Transfer Function (OTF). The PSF and OTF are related through Fourier and inverse Fourier transforms, and this relationship holds in both spatial and frequency domains. In the visible wavelength range, the resolution limit of conventional optical systems is typically around 200 nm. When two periodic patterns with different spatial frequencies are multiplied, a lower-frequency Moiré pattern is generated. If one of the patterns is known, it is possible to algebraically extract information about the other pattern from the Moiré pattern [11, 12]. A schematic diagram of a structured illumination imaging system is

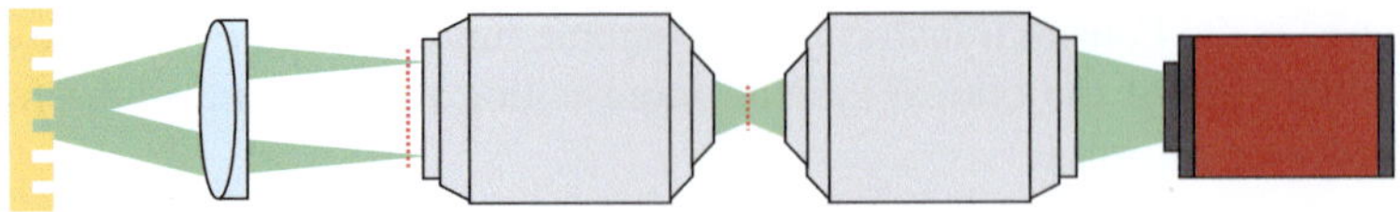

Fig. 7.11 Diagram of 2D structured illumination imaging system

shown in Fig. 7.11. By using modulated illumination, high spatial frequency information from the sample, originally beyond the detection range of the objective lens, can be encoded into detectable low-frequency images. If the intensity distribution of the modulated illumination field and the resulting low-frequency encoded fringes containing the sample's high-frequency information are known, then the original high-frequency content of the sample can be computationally retrieved.

During the imaging process, an optical microscope functions as a low-pass filter, making it unable to directly resolve high-frequency structural information within the sample. However, this detailed information can be indirectly recovered from the low-frequency Moiré patterns that are formed. The resolution limit of the microscope corresponds to the maximum frequency value in the image's frequency domain, known as the cutoff frequency. This frequency band defines the range that the system's Optical Transfer Function (OTF) can transmit; information above this frequency is filtered out. Due to diffraction effects, the imaging process from an object point to an image point can be described as the convolution of the sample's fluorescence distribution $S(r)$ with the point spread function $h(r)$, which in the spatial domain is expressed as

$$D(r) = [S(r) \cdot I(r)] * h(r) \tag{7.76}$$

The corresponding frequency domain expression is

$$D(k) = [S(k) * I(k)] \cdot H(k) \tag{7.77}$$

Under conventional wide-field illumination, the light intensity distribution is constant as

$$I(r) = 1 \tag{7.78}$$

Its frequency domain representation is

$$I(k) = \delta(k) \tag{7.79}$$

Substituting into Eq. (7.77) yields

$$D(k) = S(k) \cdot H(k) \tag{7.80}$$

With sinusoidal structured illumination introduced, the normalized intensity distribution becomes

$$I(r) = 1 + m\cos(2\pi k_0 r + \phi) \tag{7.81}$$

Its frequency domain expression is

$$I(k) = \delta(k) + \frac{m}{2}\exp(-\mathrm{i}\phi)\delta(k - k_0) + \frac{m}{2}\exp(\mathrm{i}\phi)\delta(k + k_0) \tag{7.82}$$

Here, k_0 is the spatial frequency of the fringe pattern, ϕ is the phase, and m is the modulation contrast. Substituting Eq. (7.82) into (7.77) gives

$$D(k) = \left[S(k) + \frac{m}{2}\exp(\mathrm{i}\phi)S(k - k_0) + \frac{m}{2}\exp(-\mathrm{i}\phi)S(k - k_0)\right] \cdot H(k) \tag{7.83}$$

Structured illumination introduces two additional high-frequency components whose directions depend on the fringe orientation, and whose amplitudes are related to the modulation contrast m and phase ϕ. As shown in Fig. 7.12, to cover the entire 2D frequency spectrum, three sets of fringe patterns with different orientations are typically used, each separated by an angle of $2\pi/3$. The phases within each set are defined as

$$\phi_3 - \phi_2 = \phi_2 - \phi_1 = \Delta\phi = \frac{2\pi}{3} \tag{7.84}$$

The corresponding frequency spectrum representation can be written in matrix form

$$\begin{bmatrix} D_1(k) \\ D_2(k) \\ D_3(k) \end{bmatrix} = \mathbf{M} \cdot \begin{bmatrix} S(k) \cdot H(k) \\ S(k - k_0) \cdot H(k) \\ S(k + k_0) \cdot H(k) \end{bmatrix} \tag{7.85}$$

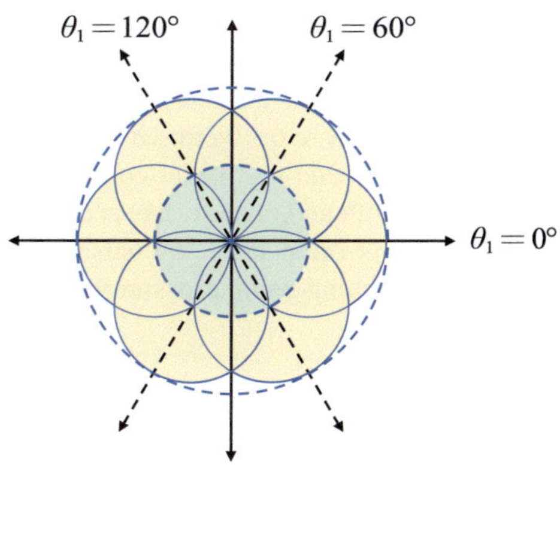

Fig. 7.12 Structured illumination spectrum expansion diagram

where the coefficient matrix **M** is

$$\mathbf{M} = \begin{bmatrix} 1 & \frac{m}{2}e^{i\phi_1} & \frac{m}{2}e^{-i\phi_1} \\ 1 & \frac{m}{2}e^{i\phi_2} & \frac{m}{2}e^{-i\phi_2} \\ 1 & \frac{m}{2}e^{i\phi_3} & \frac{m}{2}e^{-i\phi_3} \end{bmatrix} \tag{7.86}$$

By inverting the matrix, the three spectral components can be decoupled as

$$\begin{bmatrix} S(k) \cdot H(k) \\ S(k - k_0) \cdot H(k) \\ S(k + k_0) \cdot H(k) \end{bmatrix} = \mathbf{M}^{-1} \cdot \begin{bmatrix} D_1(k) \\ D_2(k) \\ D_3(k) \end{bmatrix} \tag{7.87}$$

According to the Wiener deconvolution principle based on the minimum mean square error criterion, the separated spectral components are multiplied by the OTF and shifted in frequency space according to their illumination frequency, after which they are fused to form the super-resolved spectrum as

$$S_\Sigma(k) = \frac{\sum_{\theta,n}^{\theta=1,2,3;n=-1,0,+1} H_n^*(k + nk_\theta) S_{\theta,n}(k + nk_\theta)}{\sum_{\theta,n}^{\theta=1,2,3;n=-1,0,+1} |H_n(k + nk_\theta)|^2 + w^2} A(k) \tag{7.88}$$

where, w is the Wiener constant, and $A(k)$ is a window function used to suppress ringing artifacts. Finally, the super-resolved image is obtained via an inverse Fourier transform.

During the reconstruction process, it is first necessary to estimate the stripe-related measurement parameters, including modulation contrast m, fringe spatial frequency k_θ, and initial phase ϕ_θ. Although modern optoelectronic devices can control illumination frequencies with high precision, obtaining an accurate initial phase without prior knowledge is difficult. Additionally, factors such as sample movement, optical aberrations, and photobleaching can cause fringe displacement in the image, complicating parameter estimation. To address this, various post-processing methods have been proposed for estimating structured illumination pattern parameters. Among them, correlation-based approaches are widely adopted. These methods extract frequency vectors by maximizing the correlation between overlapping regions of the zero-order and first-order components in the spectrum, and then calculate the modulation depth and phase offset accordingly. To accelerate the estimation process, a notch filter is often used to coarsely locate the $+1$ spectral peak, followed by fine-tuned matching using cross-correlation optimization.

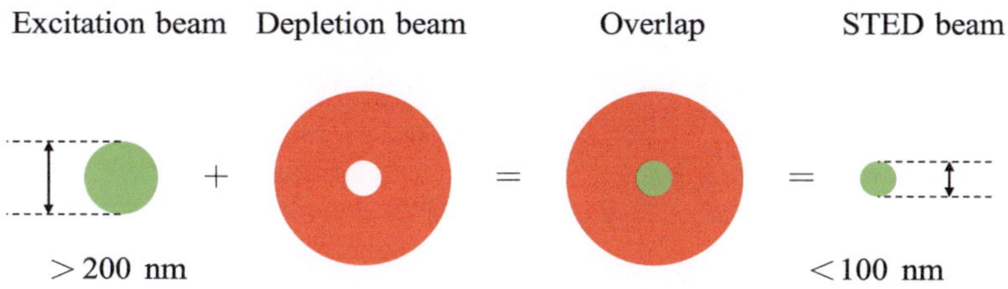

Fig. 7.13 Principle of STED technology

7.5.3 Stimulated Emission Depletion Microscopy

The core principle of Stimulated Emission Depletion (STED) microscopy lies in the coordinated use of two laser beams. First, an excitation beam is used to excite fluorescent molecules in the sample, bringing them to an excited state. Then, a depletion beam with a doughnut-shaped intensity profile induces stimulated emission, selectively forcing part of the excited fluorescent molecules back to the ground state, thereby suppressing their fluorescence emission. This mechanism enables precise control of the excitation region and significantly enhances imaging resolution [13, 14]. As shown in Fig. 7.13, the depletion beam has a longer wavelength than the excitation beam. By depleting the excited states of fluorescent molecules in regions outside the central null intensity zone of the doughnut-shaped beam, only a very small central region is allowed to fluoresce. This effectively compresses the Point Spread Function (PSF), achieving super-resolution imaging.

The resolution enhancement of STED microscopy relies on the degree to which the depletion beam compresses the effective PSF. In conventional microscopes, the resolution is limited by the diffraction limit, with the PSF described by

$$\mathrm{PSF}_{\mathrm{confocal}}(r) = \frac{\lambda}{2\mathrm{NA}} \tag{7.89}$$

where λ is the excitation wavelength and NA is the numerical aperture of the objective lens, typically limiting resolution to about 200–300 nm. In STED microscopy, the effective PSF is further compressed and can be expressed as

$$\mathrm{PSF}_{\mathrm{STED}} = \frac{\mathrm{PSF}_{\mathrm{confocal}}}{\sqrt{1 + I_{\mathrm{STED}}/I_{\mathrm{sat}}}} \tag{7.90}$$

where I_{STED} is the intensity of the depletion beam, and I_{sat} is the saturation intensity, the threshold at which most fluorescent molecules undergo stimulated emission. Increasing the ratio $I_{\mathrm{STED}}/I_{\mathrm{sat}}$ significantly reduces the PSF size, theoretically allowing nanometer-scale resolution. Currently, resolutions of approximately 20–50 nm have been achieved.

Stimulated emission is the physical mechanism behind PSF compression in STED. When fluorescent molecules are in the excited state and are exposed to light of a suitable wavelength (the depletion beam), they are induced to emit photons in the same direction and phase as the depletion light, thereby suppressing spontaneous fluorescence. The rate of this process is given by

$$W_{\text{stim}} = \frac{\sigma I_{\text{STED}}}{h\nu} \tag{7.91}$$

where σ is the stimulated emission cross-section of the fluorescent molecule, h is Planck's constant, and ν is the frequency of the depletion light. This equation indicates that increasing the depletion beam intensity enhances the probability of stimulated emission, thereby enabling control over the system's resolution.

7.6 General Conclusions

This chapter presents commonly used super-resolution imaging methods. The content includes fundamental concepts of imaging resolution, image interpolation techniques, image registration methods, digital image super-resolution techniques, and optical super-resolution imaging approaches, enabling readers to quickly grasp the relevant knowledge.

Acknowledgements This work was supported by the National Natural Science Foundation of China (12374271, 62375065, 62405079, 62575091); Fundamental Research Funds for the Central Universities (HIT.OCEF.2024055, AUGN5750300325); the Fund of National Laboratory on Adaptive Optics, China (FNLAO-24-MS-O07); Natural Science Foundation of Heilongjiang Province of China (LH2024A009); Key Laboratory of Image Processing and Intelligent Control (Huazhong University of Science and Technology), Ministry of Education, Wuhan, China (IPIC2025-05); Key Laboratory of Advanced Medical Imaging and Intelligent Computing of Guizhou Province (AMIIC250204); the China Postdoctoral Science Foundation-CCTEG Joint Support Program (2025T011ZGMK).

References

1. Aflalo K, Gao P, Trivedi V, Sanjeev A, Zalevsky Z (2024) Optical super-resolution imaging: a review and perspective. Opt Lasers Eng 183:108536
2. Yang J, Wright J, Huang TS, Ma Y (2010) Image super-resolution via sparse representation. IEEE Trans Image Process 19(11):2861–2873
3. Maes F, Collignon A, Vandermeulen D, Marchal G, Suetens P (2002) Multimodality image registration by maximization of mutual information. IEEE Trans Med Imaging 16(2):187–198
4. Ma G, Wang C, Liu P, Zhu S (2014) Sequential similarity detection algorithm based on image edge feature. J Shanghai Jiaotong Univ Sci 19:79–83
5. Li Q, Wang G, Liu J, Chen S (2009) Robust scale-invariant feature matching for remote sensing image registration. IEEE Geosci Remote Sens Lett 6(2):287–291
6. Stark H, Oskoui P (1989) High-resolution image recovery from image-plane arrays, using convex projections. J Opt Soc Am A 6(11):1715–1726

7. Irani M, Peleg S (1991) Improving resolution by image registration. Graph Models Image Process 53(3):231–239
8. Nguyen N, Milanfar P, Golub G (2001) A computationally efficient superresolution image reconstruction algorithm. IEEE Trans Image Process 10(4):573–583
9. Zheng G, Horstmeyer R, Yang C (2013) Wide-field, high-resolution Fourier ptychographic microscopy. Nat Photon 7:739–745
10. Zhang S, Wang A, Xu J, Feng T, Zhou J, Pan A (2024) FPM-WSI: Fourier ptychographic whole slide imaging via feature-domain back diffraction. Optica 11(5):634–646
11. Gustafsson MGL (2000) Surpassing the lateral resolution limit by a factor of two using structured illumination microscopy. J Microsc 198:82–87
12. Zhao T, Zhang J, Wang M, Shu M, Fu X, Yan J, Liang Y, Wang S, Lei M (2025) Real-time super-resolution structured illumination microscopy: current progress in joint space and frequency reconstruction. Rep Prog Phys
13. Hell SW, Wichmann J (1994) Breaking the diffraction resolution limit by stimulated emission: stimulated-emission-depletion fluorescence microscopy. Opt Lett 19:780–782
14. Blom H, Widengren J (2017) Stimulated emission depletion microscopy. Chem Rev 117(11):7377–7427

Chapter 8
Shape Analysis

Yulin Xue, Zhengjun Liu, and Yutong Li

Abstract This chapter introduces shape analysis methods, focusing on techniques for extracting object shape features from images. Shape analysis is performed when the object boundaries are known, extracting geometric and topological features through various algorithms. Mastering these concepts enables readers to handle more complex and variable object shapes, providing strong support for image understanding and automation.

8.1 Introduction

Shape analysis is an image processing technique used to extract shape features of objects from images, typically after determining their contours and boundaries. These features, including geometric and topological characteristics, are stored in data structures like contour descriptors or feature points. Shape analysis aids in object recognition, classification, and image retrieval by helping to compare and differentiate objects based on their shapes. Advancements in computer vision and AI have led to more complex algorithms, allowing for the handling of intricate object forms and supporting image understanding and automation [1]. Shape is crucial for visual recognition, yet describing it is challenging since it is often conveyed only by adjectives. Although we recognize shapes easily in daily life, providing a precise description is difficult, especially for complex objects. Defining a shape mathematically is complex, as there is no clear definition that aligns with human perception, which involves both retinal responses and our knowledge of the world.

Y. Xue · Z. Liu · Y. Li (✉)
School of Physics, Harbin Institute of Technology, Harbin, China
e-mail: yutong_li@hit.edu.cn

Y. Xue
e-mail: Kylinnn9@163.com

Z. Liu
e-mail: zjliu@hit.edu.cn

Z. Liu and Y. Li (eds.), *Intelligent Analysis of Optical Images*, Scientific Computation,
https://doi.org/10.1007/978-981-95-3720-4_8

8.2 Shape Preprocessing

Image degradation, caused by factors like optical distortion, noise, and sensor quality, can distort an image and affect subsequent processing tasks. To mitigate these effects, image preprocessing is necessary to remove noise, enhance clarity, and improve overall quality. Techniques such as noise removal, sharpening, edge enhancement, and contrast adjustments help clear up visuals, highlight target regions, and improve object-background distinction. Geometric correction can also address distortions. Effective preprocessing is crucial for accurate shape segmentation and analysis, improving the efficiency and accuracy of shape extraction and feature matching.

8.2.1 Grayscale Conversion

In image processing, the original image is typically stored in the RGB (Red, Green, Blue) color model, which contains a large amount of information. To simplify subsequent processing and reduce the computational load, the image can be converted to grayscale. Grayscale conversion significantly reduces the spatial complexity of the image and decreases the amount of information, making it easier for further analysis and processing. The most common methods for grayscale conversion are linear grayscale conversion and nonlinear grayscale conversion.

(A) Linear Grayscale Conversion

In linear grayscale conversion, the RGB color space of the image is transformed into grayscale values through weighted averaging. Let the original image's color components be R', G', and B', and the grayscale values R, G, and B after the grayscale transformation are calculated as follows:

$$R = G = B = W_R R' + W_G G' + W_B B' \tag{8.1}$$

where W_R, W_G, and W_B are the weights for R, G, and B, respectively, and satisfy the following condition:

$$W_R + W_G + W_B = 1 \tag{8.2}$$

(B) Nonlinear Grayscale Conversion

In nonlinear grayscale conversion, the color components of the image are transformed using a nonlinear function. This method can provide better grayscale results in some applications. Let the original image's color components be R', G', and B', and the transformed grayscale components R, G, and B are calculated using the following formula:

$$R = f(R'),\ G = f(G'),\ B = f(B') \tag{8.3}$$

where $f(\cdot)$ is a nonlinear function that can adjust the image's brightness and contrast based on actual requirements. Common nonlinear functions include logarithmic transformations and exponential transformations, which can enhance image details or improve the contrast of specific regions.

8.2.2 Smoothing and Denoising

During the image acquisition process, various sources of noise can affect the quality of the captured image, impacting all stages of image processing. The images obtained in experiments are often subject to noise interference. Directly performing image segmentation under these conditions may introduce significant errors, affecting the extraction and understanding of shape features. Therefore, before image analysis, it is essential to reduce the noise in the original image and improve the signal-to-noise ratio. The following discusses several commonly used noise reduction methods.

(A) Spatial Domain Filtering

Spatial domain filtering is a commonly used method for noise reduction in images, including techniques such as mean filtering and median filtering. These methods process each pixel in the image and its neighborhood to remove noise. Mean filtering replaces the pixel value with the average of its neighborhood pixels, making it suitable for reducing random noise, although it can blur edges. Median filtering replaces the pixel value with the median of its neighborhood pixels, effectively removing salt-and-pepper noise while preserving edges. For more detailed information on these techniques, please refer to Sect. 1.2.2 of Chap. 1.

(B) Frequency Domain Filtering

Frequency domain filtering involves converting the image from the spatial domain to the frequency domain and then processing it to suppress noise. The details of frequency domain filtering are also discussed in Sect. 1.2.2 of Chap. 1. In addition to common methods such as low-pass filtering and high-pass filtering, Wiener filtering is a classic frequency domain filtering technique that can effectively recover signals contaminated by noise.

Wiener filtering treats both the signal and noise as random processes and aims to minimize the mean square error (MSE) to find the optimal filter. Specifically, the goal of Wiener filtering is to find an estimate that minimizes the statistical error function e as

$$e^2 = E\{(f - \hat{f})^2\} \tag{8.4}$$

where f is the original image, and $\hat{f}$ is the frequency domain estimate. The frequency domain estimate can be represented as

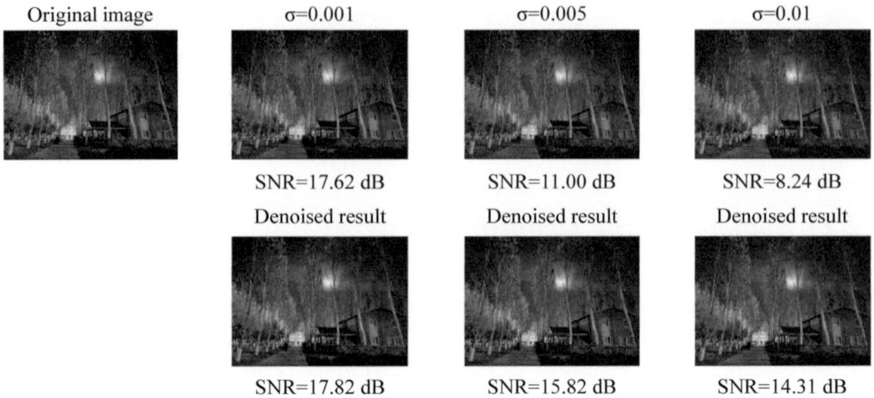

Fig. 8.1 Wiener filtering is used to filter the noise image

$$\hat{F}(u, v) = \frac{H^*(u, v)G(u, v)}{|H(u, v)|^2 + P_n(u, v)/P_f(u, v)} \tag{8.5}$$

where $P_f(u, v)$ is the power spectrum of $f(u, v)$, and $P_n(u, v)$ is the power spectrum of the noise. K represents the noise-to-signal power spectral density ratio. When the statistical properties of the noise are unknown, the following approximation can be used as

$$\hat{F}(u, v) \approx \frac{H^*(u, v)}{|H(u, v)|^2 + K} G(u, v) \tag{8.6}$$

This method analyzes the frequency domain information of the image to suppress the influence of noise while preserving important structural details, resulting in higher-quality image restoration. Figure 8.1 shows the noise images with mean value of 0 and variances of 0.001, 0.005, 0.01 and the results after noise reduction.

8.3 Shape Segmentation

Shape segmentation is a crucial step in shape analysis, directly influencing the accuracy and efficiency of subsequent tasks like feature extraction and matching. Its main goal is to separate the target region from the background and identify objects or shapes of interest for further analysis. Common segmentation methods include thresholding, edge detection, region growing, and watershed algorithms, each suited to specific scenarios. As a foundational element, shape segmentation plays a key role in object recognition and complex analysis tasks, enabling efficient image understanding and automated processing [2].

8.3.1 Region Growing Segmentation Algorithm

The region growing segmentation algorithm is a commonly used method in image segmentation. Its basic idea is to group together pixels with similar properties, gradually forming a region. This process uses predefined criteria to iteratively merge pixels or sub-regions into larger regions. The core of region growing lies in the similarity criteria, which evaluate the similarity between pixels to decide whether adjacent pixels or regions should be merged.

In practical implementation, the region growing algorithm typically starts from one or more seed pixels, which act as the starting point for growth. Then, based on preset growth criteria or similarity standards, the algorithm selects pixels from the neighborhood of the seed pixels that are similar to them, gradually merging these pixels into the target region, until the stopping condition is met. The specific steps are as follows:

(1) Choose suitable seed points

Based on the characteristics of the image and segmentation requirements, select one or more representative seed pixels as the starting point for region growth.

(2) Determine similarity criteria

Set a standard, usually based on pixel color, gray level, or other features, to measure the similarity between pixels.

(3) Set growth stopping conditions

The growth stops when the boundary of the region meets a specific condition, typically when the similarity between adjacent pixels falls below a threshold or when the number of pixels in the region reaches a predetermined range.

For example, in the region growing method shown in Fig. 8.2, the similarity criterion is that the difference between the gray level of neighboring points and the average gray level of the object is less than 2.

Traditional region growing methods require manual seed point selection for each target region, limiting their use in automated segmentation tasks. The region splitting and merging method overcomes this limitation by automatically splitting and merging

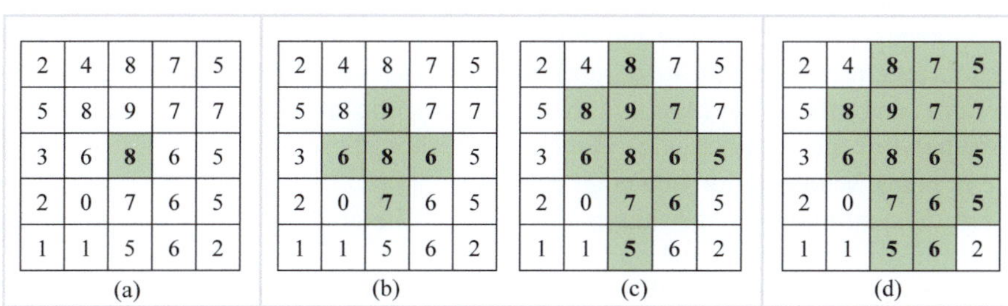

Fig. 8.2 Example of region growing method

regions based on consistency criteria, without manual intervention. This method's core lies in designing effective splitting and merging rules: regions that no longer meet consistency criteria are subdivided, while adjacent regions with similar properties merge. This approach provides a more precise and flexible segmentation solution, allowing for accurate extraction of object boundaries and improving segmentation by eliminating false boundaries. The algorithm steps for this method include:

(1) Set the entire image as the starting region and divide it into four sub-regions.
(2) For each sub-region, if the features of the sub-region do not meet the consistency criteria, further split it into four sub-regions; if the criteria are met, stop splitting.
(3) When splitting stops, for any two adjacent regions, if their properties are similar, they will be merged into one region, continuing until no adjacent regions can be merged, at which point the algorithm ends.

Region growing and the splitting-merging methods provide more flexible and accurate solutions for image segmentation, especially when handling complex and irregularly shaped objects, offering significant advantages [3].

8.3.2 Edge Detection Segmentation Algorithm

Edge detection is a key technique in image segmentation, used to identify object boundaries by capturing significant changes in brightness or color. It provides essential local information for further analysis. Contour detection refines edge detection by accurately delineating object boundaries, directly impacting segmentation precision. Edge detection-based segmentation identifies edge points and connects them into contours to partition the image. This method is fast and simple, ideal for rapid structural analysis. Classical edge detection algorithms include first-order, second-order, and optimal operators, each with specific characteristics and applications. Selecting the right operator is crucial for effective segmentation [4].

(A) First-order edge detection operators

First-order edge detection operators are one of the most basic and widely applied methods in edge detection. The fundamental principle is to identify the edges in an image by calculating the rate of change in pixel intensity. Specifically, the gradient of an image represents the rate and direction of intensity change, and edges typically occur in regions where there are abrupt changes in intensity. First-order edge detection operators can effectively highlight areas with significant intensity changes, which correspond to the boundaries of objects. Therefore, these operators play a crucial role in image segmentation and edge detection.

Let the continuous image function be $f(x, y)$. The gradient at point (x, y) can be represented as a vector. Assuming G_x and G_y represent the gradients in the x-direction and y-direction, respectively, the gradient vector at point (x, y) can be expressed as

$$\nabla f(x, y) = \begin{bmatrix} G_x \\ G_y \end{bmatrix} = \begin{bmatrix} \frac{\partial f(x,y)}{\partial x} \\ \frac{\partial f(x,y)}{\partial y} \end{bmatrix} \tag{8.7}$$

The gradient direction θ_g and magnitude $g(x, y)$ can be calculated as follows

$$\theta_g = \tan^{-1}(f_y/f_x) \tag{8.8}$$

$$g(x, y) = |\nabla f(x, y)| = \sqrt{G_x^2 + G_y^2} = \sqrt{\left(\frac{\partial f(x, y)}{\partial x}\right)^2 + \left(\frac{\partial f(x, y)}{\partial y}\right)^2} \tag{8.9}$$

where g is the gradient operator.

In digital images, the gradient is often approximated using difference operations instead of differentiation, especially in discrete image processing. The common difference operations are

$$G_x = f(x, y) - f(x + 1, y) \tag{8.10}$$

$$G_y = f(x, y) - f(x, y + 1) \tag{8.11}$$

Thus, the approximate calculation of gradient magnitude $g(x, y)$ is

$$\begin{aligned} g(x, y) \approx |G_x| + |G_y| &= \left|\frac{\partial f(x, y)}{\partial x}\right| + \left|\frac{\partial f(x, y)}{\partial y}\right| \\ &= |f(x, y) - f(x + 1, y)| + |f(x, y) - f(x, y + 1)| \end{aligned} \tag{8.12}$$

First-order edge operators typically use convolution with region templates to approximate the gradient for each pixel. This method calculates the gradient values for each pixel in the image, thereby extracting edge information. Specifically, by combining two templates representing the horizontal gradient component G_x and the vertical gradient component G_y, edge features in the image can be effectively captured.

Common first-order edge detection operators include the Robert operator, Prewitt operator, and Sobel operator, which are briefly mentioned in Sect. 1.7.4. The main differences among these operators lie in the template combinations used during computation, and each template is designed with a different emphasis, affecting the performance and accuracy of edge detection. Specifically:

(1) Robert Operator: Typically uses smaller templates, suitable for fast computation but less effective in noisy images.
(2) Prewitt Operator: Uses larger templates, which better smooth noise in the image but may miss fine details.

Fig. 8.3 Several classical first-order edge detection operator templates

Horizontal convolution kernel:

$$\begin{bmatrix}-1 & 0\\ 0 & 1\end{bmatrix} \quad \begin{bmatrix}-1 & -1 & -1\\ 0 & 0 & 0\\ 1 & 1 & 1\end{bmatrix} \quad \begin{bmatrix}-1 & -2 & -1\\ 0 & 0 & 0\\ 1 & 2 & 1\end{bmatrix}$$

Vertical convolution kernel:

$$\begin{bmatrix}0 & -1\\ 1 & 0\end{bmatrix} \quad \begin{bmatrix}-1 & 0 & 1\\ -1 & 0 & 1\\ -1 & 0 & 1\end{bmatrix} \quad \begin{bmatrix}-1 & 0 & 1\\ -2 & 0 & 2\\ -1 & 0 & 1\end{bmatrix}$$

Roberts Prewitt Sobel

(3) Sobel Operator: Improves upon the Prewitt operator by adding weights, providing better noise suppression and gradient computation accuracy in edge detection.

Figure 8.3 shows the commonly used templates for several classic first-order edge detection operators. By performing convolution, these templates operate pixel by pixel with the image to calculate the gradient values at each location. By analyzing the horizontal and vertical gradient components, the edge structure of the image can be revealed. Edges typically appear in regions with large gradient values, so these templates can effectively extract significant edges from the image, providing important information for subsequent image processing tasks such as edge detection and image segmentation.

(B) Second-order edge detection operators

First-order edge detection operators often produce wide responses in edge regions, leading to blurred edges and affecting precision. To improve localization, refinement processes are typically required, but in complex or noisy images, these processes may interfere with accurate edge detection. To address this, second-order differential methods were developed. Second-order derivatives capture the acceleration of grayscale changes, with the maximum rate of change occurring at edges, where the second-order derivative crosses zero. This characteristic allows second-order differential operators to more accurately detect edge locations.

The Laplace operator is one such second-order derivative-based edge detection method. Compared to first-order differential operators, second-order derivatives offer more accurate edge localization due to their sensitivity to the acceleration of grayscale changes. Additionally, second-order derivatives are more effective at suppressing the wide response caused by noise, making edge detection more stable and reliable. The Laplace operator calculates the second-order derivatives at each pixel of the image, effectively capturing the edge information and reducing interference during the refinement process, resulting in more accurate edge detection.

Fig. 8.4 Laplacian operator templates of 4-neighborhood and 8-neighborhood

$$\begin{bmatrix} 0 & -1 & 0 \\ -1 & 4 & -1 \\ 0 & -1 & 0 \end{bmatrix} \qquad \begin{bmatrix} -1 & -1 & -1 \\ -1 & 8 & -1 \\ -1 & -1 & -1 \end{bmatrix}$$

4 adjoining points | 8 adjoining points

Let the continuous image function be $f(x, y)$. The gradient at the pixel point (x, y) can be expressed as

$$\nabla^2 f = \frac{\partial^2 f}{\partial x^2} + \frac{\partial^2 f}{\partial y^2} \tag{8.13}$$

For second-order edge detection, the Laplace operator detects edges by calculating the second-order derivatives of the image. During edge detection, a template is typically used to approximate the second-order derivatives. The Laplace operator's template requires the center coefficient to be positive, while the surrounding neighboring coefficients should be negative, and the sum of all coefficients should be zero. The commonly used Laplace operator templates are the 4-neighborhood template and the 8-neighborhood template. The design of these two templates is based on the contribution of neighborhood pixels to image grayscale changes, as Fig. 8.4.

(C) Optimal edge detection operators

Similar to edges, noise in an image also manifests as points of abrupt intensity changes, which are high-frequency components of the image. Since differential operators are sensitive to changes in image intensity, edge detection is inevitably affected by noise, which compromises the accuracy and stability of edge extraction. To address this issue, researchers have proposed filters based on differential operators that optimize edge extraction according to the signal-to-noise ratio (SNR), leading to the concept of optimal edge operators. The most typical optimal edge detection operators are the Laplacian of Gaussian (LOG) operator and the Canny operator.

The LOG operator combines Gaussian smoothing and the Laplacian operator. It first smooths the image using a Gaussian filter to suppress noise and then performs edge detection using the Laplacian operator. This approach not only reduces the influence of noise on edge extraction but also enhances the visibility of edges in the image. The Canny operator uses a multi-stage process, including smoothing, gradient calculation, non-maximum suppression, and double thresholding, to precisely extract edges from an image. Due to its high edge detection accuracy and robustness to noise, the Canny operator has become a classic method in edge detection.

The LOG operator is essentially an improvement on the Laplacian operator. It first smooths the image using a Gaussian filter to remove noise, then calculates the second-order directional derivatives and finds the extrema or zero-crossings of the second-order derivatives, which correspond to the image edges. The basic steps of the LOG operator are as follows:

(1) Apply Gaussian filtering to smooth the image and reduce noise.
(2) Calculate the second-order derivatives on the smoothed image.
(3) Detect edges by finding the zero-crossings of the second-order derivative, which correspond to maxima of the first-order derivative.
(4) Use linear interpolation at the subpixel level to improve edge detection accuracy.

The Canny operator is based on two-dimensional convolution and computes both the edge strength and direction information. Key steps in the Canny operator include: Gaussian smoothing to reduce noise, calculating the gradient of the image to obtain edge strength, non-maximum suppression to remove non-edge pixels, and double thresholding to further refine edge detection. Through these stages, the Canny operator ensures high precision in edge extraction and strong noise suppression.

In the Canny operator, the Gaussian function is commonly used for image smoothing and is defined as

$$G(x, y) = \frac{1}{2\pi\sigma^2} e^{-\frac{x^2+y^2}{2\sigma^2}} \tag{8.14}$$

where σ controls the standard deviation of the Gaussian filter. This two-dimensional convolution template can be understood as a combination of one-dimensional filters

$$\begin{aligned} \frac{\partial G(x, y)}{\partial x} &= kxe^{-\frac{x^2}{2\sigma^2}} e^{-\frac{y^2}{2\sigma^2}} = h_1(x)h_2(y) \\ \frac{\partial G(x, y)}{\partial y} &= kye^{-\frac{x^2}{2\sigma^2}} e^{-\frac{y^2}{2\sigma^2}} = h_2(y)h_1(x) \end{aligned} \tag{8.15}$$

where

$$\begin{aligned} h_1(x) &= \sqrt{k}xe^{-\frac{x^2}{2\sigma^2}}, \quad h_2(y) = \sqrt{k}e^{-\frac{y^2}{2\sigma^2}} \\ h_1(y) &= \sqrt{k}ye^{-\frac{y^2}{2\sigma^2}}, \quad h_2(x) = \sqrt{k}e^{-\frac{x^2}{2\sigma^2}} \end{aligned} \tag{8.16}$$

These two templates are then convolved with the image to obtain the gradient information:

$$E_x = \frac{\partial G(x, y)}{\partial x} * f, \quad E_y = \frac{\partial G(x, y)}{\partial y} * f \tag{8.17}$$

If $A(i, j) = \sqrt{E_x^2 + E_y^2}$, $a(i, j) = \arctan(E_y(i, j)/E_x(i, j))$, then $A(i, j)$ represents the edge strength information, and $a(i, j)$ indicates the direction perpendicular to the edge. Ultimately, through these calculations, the Canny operator can accurately extract the edges in the image and further refine the edge detection effect based on the image's gradient strength and direction information.

8.3.3 Morphological Segmentation Methods

Mathematical Morphology is a method used to study the structure and topological relationships of objects in digital images. As a nonlinear filtering method, the basic operations of mathematical morphology are Morphological Erosion and Morphological Dilation, along with the derived combined operations: Opening and Closing. The basic principle of mathematical morphology is to scan the image using structural elements with specific shapes and analyze the image information based on the scanning results, applying various transformations to the image structure. With the continuous development and improvement of mathematical morphology theories, morphology-based edge detection algorithms have received widespread attention from researchers. This section briefly introduces the theory of morphology from the perspectives of binary and grayscale morphology, as well as the principles of morphological edge detection algorithms [5].

(A) Binary morphology

In binary image morphological processing, structural elements are used to extract shapes from the image by interacting with each pixel through logical operations. The result of these operations forms the output value at the origin of the structural element, creating a new image. The size and shape of the structural element can be customized based on application needs. Similar to convolution in image processing, binary mathematical morphology replaces arithmetic operations with logical ones, and the results are stored at corresponding pixel positions in the output image. The effectiveness of this processing depends on the structural element's size, shape, and the logical operations used.

(1) Erosion operation

In binary morphology, the set A is eroded by the set S, represented as $A \ominus S$, and defined as:

$$A \ominus S = \{x|(S)_x \subseteq A\} \tag{8.18}$$

where A is the input image, and S is the structural element. When the origin of the structural element S moves to a position in the image, if S is completely contained within A, the output image at that point is 1; otherwise, the point is 0.

(2) Dilation operation

The dilation operation is the inverse of erosion. The set A is dilated by the set S, represented as $A \oplus S$, and defined as:

$$A \oplus S = \{x|(S)_x \cap A \neq \emptyset\} \tag{8.19}$$

When the origin of S moves to a position in the image, if S overlaps with any point in the target image, the output image at that point is 1; otherwise, the point is 0.

(5) Opening and closing operations

The dilation operation expands the image, while the erosion operation shrinks the image. These basic operations give rise to two important morphological operations: Opening and Closing.

Opening: Opening is used to remove small objects, separate objects at small connecting points, and smooth the edges of the target image without changing its shape. It performs erosion followed by dilation, commonly used to remove small noise points from the image. The mathematical definition for opening is as follows.

Opening operation with structural element S on set A, represented as $A \circ S$, defined as

$$A \circ S = (A \ominus S) \oplus S \tag{8.20}$$

Closing: Closing fills small holes within objects, connects adjacent objects, and smooths their edges without significantly changing their area. It performs dilation followed by erosion. The mathematical definition for closing is as follows:

Closing operation with structural element S on set A, represented as $A \bullet S$, defined as

$$A \cdot S = (A \oplus S) \ominus S \tag{8.21}$$

(B) Grayscale morphology

Due to the limitations of binary morphology, it cannot be directly applied to grayscale images. Therefore, extending binary mathematical morphology to grayscale images has great significance. This extension allows morphological operations to be applied to grayscale images, broadening their application to a wider range of image processing tasks.

In grayscale morphology, operations are applied directly to the grayscale values of an image. Grayscale erosion and grayscale dilation are defined by computing the extreme values of the grayscale values within the area covered by the structural element.

Grayscale Erosion: Using structural element b for grayscale erosion on grayscale image f, represented as $f \ominus b$, defined as

$$(f \ominus b)(x, y) = \min_{(u,v)\in b} (f(x + u, y + v)) \tag{8.22}$$

Grayscale Dilation: Using structural element b for grayscale dilation on grayscale image f, represented as $f \oplus b$, defined as

$$(f \oplus b)(x, y) = \min_{(u,v)\in b} (f(x + u, y + v)) \tag{8.23}$$

The opening and closing operations for grayscale images have the same form as the binary counterparts.

Grayscale Opening: Using structural element b for grayscale opening on grayscale image f, represented as $f \circ b$, defined as

$$f \circ b = (f \ominus b) \oplus b \tag{8.24}$$

Grayscale Closing: Using structural element b for grayscale closing on grayscale image f, represented as $f \bullet b$, defined as

$$f \bullet b = (f \oplus b) \ominus b \tag{8.25}$$

Figure 8.5 shows the experimental image of the basic operator of gray-scale morphology. Grayscale morphology has the following characteristics.

(1) If the selected structural element is positive, dilation enhances the brightness of the target image, while erosion reduces the brightness.
(2) By adjusting the size and shape of the structural element, dilation can remove or reduce dark details, while erosion can remove or reduce bright details.
(3) Opening effectively removes small bright spots, while closing removes small dark spots.

(C) Morphological edge detection

Morphological edge detection algorithms involve selecting a suitable structural element and using operations like dilation, erosion, opening, and closing to extract image edges. The choice of edge detector and structural element impacts the final

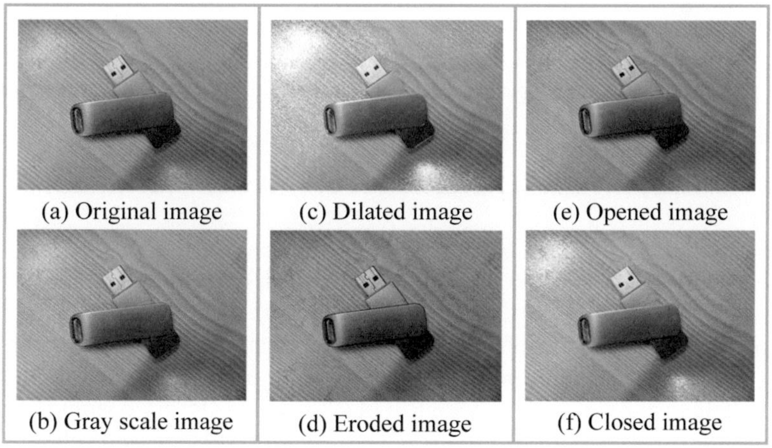

Fig. 8.5 Experimental image of morphological basic operator

processing results. Dilation and closing expand the image, while erosion and opening shrink it. These properties allow for the construction of various morphological edge detectors, each with different functionalities, as represented by specific formulas.

By subtracting the result of eroding the original image from the original image itself, we can obtain the erosion residual edge detection operator. The formula is

$$E_{erosion} = A - (A \ominus B) \tag{8.26}$$

By subtracting the result of dilating the original image from the original image, we can obtain the dilation residual edge detection operator. The formula is

$$E_{\text{dilation}} = (A \oplus B) - A \tag{8.27}$$

By subtracting the result of erosion from the result of dilation of the original image, we can obtain the dilation-erosion residual edge detection operator. The formula is

$$E_{\text{dilation-erosion}} = (A \oplus B) - (A \ominus B) \tag{8.28}$$

By subtracting the result of opening the original image from the original image itself, we can obtain the morphological top-hat edge detection operator. The formula is

$$E_{\text{top-hat}} = A - (A \circ B) \tag{8.29}$$

By subtracting the result of closing the original image from the original image itself, we can obtain the morphological black-hat edge detection operator. The formula is

$$E_{\text{black-hat}} = (A \bullet B) - A \tag{8.30}$$

By subtracting the result of opening the original image from the result of closing the original image, we can obtain the closing-opening edge detection operator. The formula is

$$E_{\text{closing-opening}} = (A \bullet B) - (A \circ B) \tag{8.31}$$

These morphological edge detection operators can be used to extract edge information from the image. Figure 8.6 shows the results of applying different edge detection operators. Based on the results, the erosion residual operator, dilation residual operator, and dilation-erosion residual operator have better edge detection performance, as they are able to capture the edges of the image more effectively. In contrast, the morphological top-hat, morphological black-hat, and closing-opening operators perform poorly, detecting only some of the edge information and missing more details.

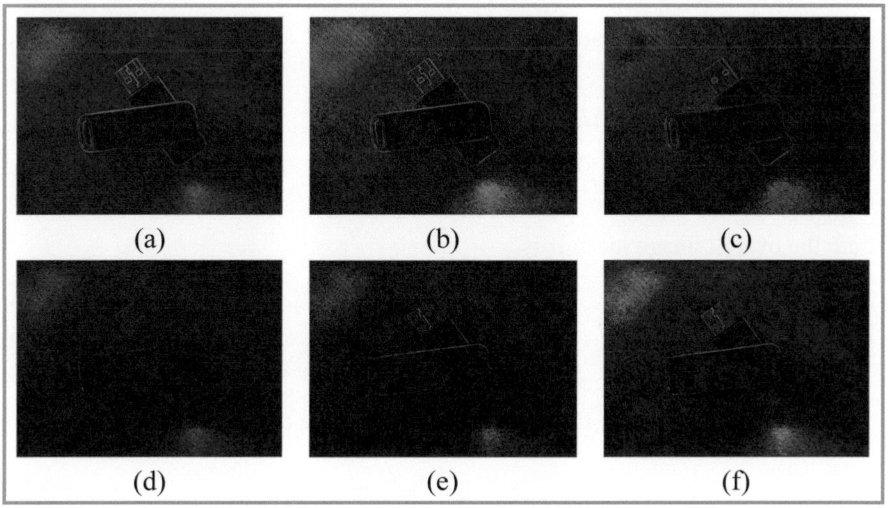

Fig. 8.6 Experimental images of morphological basic operators

8.3.4 Thresholding Segmentation Methods

Thresholding segmentation is a simple and effective method in image segmentation. It involves determining a threshold within the image's grayscale range and classifying pixels into two categories based on whether their grayscale values are above or below the threshold, creating two distinct regions. Thresholding methods are divided into global and local categories. Global thresholding computes a single optimal threshold for the entire image, ideal for images with uniform grayscale distributions, while local thresholding applies global thresholding to smaller sub-images. Common global thresholding methods include OTSU and maximum entropy.

(A) OTSU Method

OTSU Method is also callled as maximum inter-class variance. The OTSU method is based on the concept of separating the grayscale values into two classes using a threshold T. The threshold T is determined by the ratio of the inter-class variance and the intra-class variance of the two classes. Let L be the number of grayscale levels in the image, and let the total number of pixels be N. The number of pixels with grayscale value i is n_i, and the probability of each grayscale value is $p_i = N \big/ n_i$. The histogram of the grayscale image is divided into two classes: background Ω_0 and foreground Ω_1, with their respective probabilities

$$p_0(t) = \sum_{i=0}^{t} p_i, \quad p_1(t) = \sum_{i=t+1}^{L-1} p_i \tag{8.32}$$

The mean values of background and foreground are

$$\mu_0(t) = \frac{\sum_{i=0}^{t} ip_i}{p_0(t)}, \quad \mu_1(t) = \frac{\sum_{i=t+1}^{L-1} ip_i}{p_1(t)} \tag{8.33}$$

The inter-class variance is

$$\sigma_b^2(t) = p_0(t)p_1(t)(\mu_0(t) - \mu_1(t))^2 \tag{8.34}$$

where the overall image mean μ is

$$\mu = \sum_{i=0}^{L-1} ip_i \tag{8.35}$$

The optimal threshold T is the value that maximizes the inter-class variance

$$T = \arg\max_t \sigma_b^2(t) \tag{8.36}$$

Since variance is a measure of the uniformity of the grayscale distribution, a larger variance indicates a greater difference between the background and foreground. Therefore, the threshold that maximizes the inter-class variance leads to the least misclassification. The OTSU method does not require assumptions about the probability density functions of the object and background, and it is based solely on first-order statistical properties of the grayscale histogram. The computation is fast and provides good consistency [6]. Figure 8.7 shows the segmentation result of the OTSU method.

(B) Maximum entropy thresholding method

Entropy is a measure of the average information content. In digital signals, the entropy of an image reflects its overall complexity, and it is generally believed that entropy is maximized at the boundary between the object and the background. Based on

Fig. 8.7 OTSU algorithm segmentation result

this principle, the maximum entropy thresholding method selects the threshold that maximizes the image's entropy for segmentation [7].

Let the image have L grayscale levels, N total pixels, and n_i pixels with grayscale value i, with the probability of each grayscale value being p_i. The image histogram is divided into two classes: background Ω_0 and foreground Ω_1, and their respective entropies are

$$H_0(t) = -\sum_{i=0}^{t} p_i \log p_i, \quad H_1(t) = -\sum_{i=t+1}^{L-1} p_i \log p_i \tag{8.37}$$

The total information entropy of the image is

$$H(t) = H_0(t) + H_1(t) \tag{8.38}$$

The optimal threshold T is the value that maximizes the total entropy

$$T = \arg\max_t H(t) \tag{8.39}$$

This method defines the entropy related to the probability distribution of the object and background and uses the maximum prior estimate of classification validity to select the best threshold. The maximum entropy thresholding method provides good segmentation results for images with high grayscale contrast between the object and background. Compared to the OTSU method, the maximum entropy method involves logarithmic operations, making it slower and less suitable for real-time applications. Additionally, the method only utilizes the original histogram's grayscale information, so its performance degrades when the image's signal-to-noise ratio decreases.

(C) Local Thresholding Methods

When an image is complex or has significant grayscale differences, local thresholding can improve segmentation quality. Local thresholding divides the image into smaller sub-images and computes the optimal threshold for each sub-image separately. This method provides better segmentation results and is more adaptable, but it comes at the cost of higher computational time, especially for larger images.

8.4 Shape Compactness Descriptors

Compactness is an important shape property, closely related to the shape's elongation. Various descriptors have been proposed to measure the compactness of regions, some of which can describe additional properties of the region. These descriptors are generally related to the geometric parameters of the target shape, and therefore are scale-dependent.

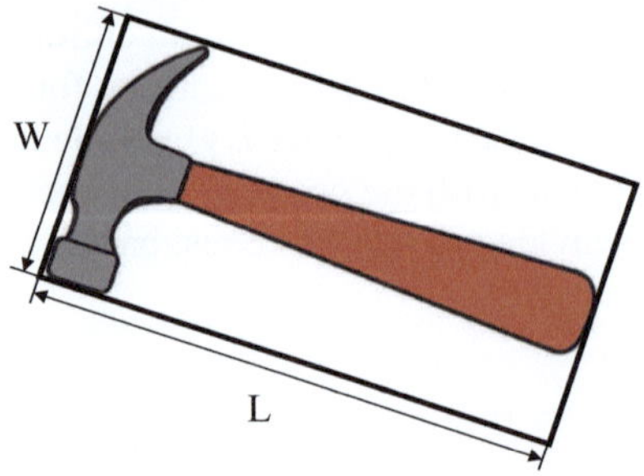

Fig. 8.8 Shape and its bounding box diagram

The compactness of a region can be described directly by calculating its geometric properties or indirectly by comparing the region with typical/ideal shapes, such as circles and rectangles. Below are several commonly used shape compactness descriptors [8].

8.4.1 *Aspect Ratio*

Aspect Ratio is commonly used to describe the shape of an object after plastic deformation, particularly its elongation. As Fig. 8.8, it is defined by the bounding box of the object, as

$$R = \frac{L}{W} \tag{8.40}$$

where L and W are the length and width of the bounding box, respectively. For square or circular objects, the aspect ratio R reaches its minimum value 1, while for more elongated objects, R increases as the elongation of the object increases.

8.4.2 *Form Factor*

The Form Factor F is calculated using the perimeter B and area A of the region:

$$F = \frac{B^2}{4\pi A} \tag{8.41}$$

As Eq. (8.41), when the region is circular, $F = 1$; for other shapes, $F > 1$, indicating that F reaches its minimum value for circular shapes. The form factor describes the compactness of a region to some extent and is dimensionless, making it insensitive to scale variations. However, in some cases, regions with the same form factor value may appear significantly different in shape.

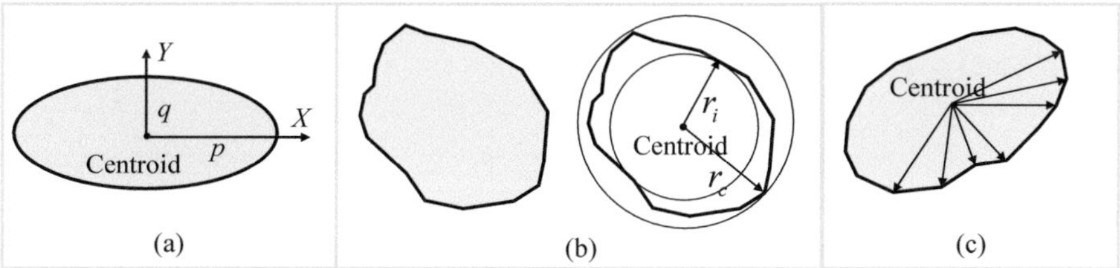

Fig. 8.9 Calculation diagram of eccentricity, sphericity and circularity

8.4.3 Eccentricity

Eccentricity E, also known as elongation, describes the compactness of a region to some extent. Eccentricity can be derived from the region's moments of inertia. Let I $A = \sum m_i y_i^2$ and $B = \sum m_i x_i^2$ be the moments of inertia of the region about the X and Y axes, and $C = \sum m_i x_i y_i$ be the inertia product, as shown in Fig. 8.9a. The two semi-axes p and q of the inertia ellipse corresponding to the region are calculated as

$$p = \frac{\sqrt{2}}{\left[(A+B) - \sqrt{(A-B)^2 + 4C^2}\right]} \tag{8.42}$$

$$q = \frac{\sqrt{2}}{\left[(A+B) + \sqrt{(A-B)^2 + 4C^2}\right]} \tag{8.43}$$

The eccentricity E is then defined as

$$E = \frac{p}{q} \tag{8.44}$$

where the eccentricity E reaches its minimum value 1 when the region is circular.

8.4.4 Sphericity

Sphericity S originally referred to the ratio of surface area to volume for 3D objects, as Fig. 8.9b. For 2D targets, it is defined as

$$S = \frac{C_{\text{circ}}}{C_{\max}} \tag{8.45}$$

where C_{circ} is the perimeter of the circumscribed circle of the target, and $C_{\max}$ is the maximum perimeter of the target. The sphericity S reaches its maximum value 1 for circular targets, and for other shapes, $S < 1$. Sphericity is invariant to translation, rotation, and scale changes.

8.4.5 Circularity

Circularity C is a feature quantity defined by the boundary points of the target region R, as shown in Fig. 8.9c. The calculation formula of circularity C is as follows

$$C = \frac{\mu_R}{\sigma_R} \tag{8.46}$$

where circularity is calculated based on the average distance μ_R and the standard deviation σ_R from the region's centroid to the boundary points as

$$\mu_R = \frac{1}{K}\sum_{k=0}^{K-1} \|(x_k, y_k) - (\overline{x}, \overline{y})\| \tag{8.47}$$

$$\sigma_R = \frac{1}{K}\sum_{k=0}^{K-1} \left[\|(x_k, y_k) - (\overline{x}, \overline{y})\| - \mu_R\right]^2 \tag{8.48}$$

As the region approaches a perfect circle, the circularity CCC increases and approaches infinity. Circularity is invariant to translation, rotation, and scale changes, making it a highly stable shape descriptor.

8.5 Contour-Based Shape Descriptors

Shape descriptors are essential in shape analysis, forming the basis for applications like recognition, classification, and retrieval. The MPEG-7 standard specifies that an ideal shape descriptor should have strong descriptive power, compact feature representation, and robust retrieval performance. It should also be invariant to transformations like rotation, scaling, and translation (RST) to ensure robustness, while considering factors like noise, shape distortion, and local defects. Objects in an image are represented by boundary contours and the enclosed region pixels. Shape description methods are classified into contour-based and region-based approaches, depending on whether features are extracted from the boundary or the entire region [9], as Fig. 8.10.

Contour-based description methods are typically suitable for describing the shape of a single target with a closed contour. These methods are not ideal for describing shapes composed of multiple regions or shapes that contain holes or voids. This

Fig. 8.10 Regions and contours of shapes

section will introduce some typical contour-based shape analysis and description methods in detail.

8.5.1 Shape Signature Descriptor

A shape signature refers to the use of a one-dimensional contour function to describe the boundary (contour) of a two-dimensional shape. Shape signatures are typically used to capture the key visual features of a shape. They play a critical role in shape recognition, classification, retrieval, and similarity measurement. Common shape signatures include centroid distance, complex coordinates, boundary tangent angles, boundary curvature, triangle area, and chord-length functions, among others. These signatures convert the boundary information of a shape into a one-dimensional function, providing an effective description for subsequent shape analysis.

Figure 8.11 illustrates several commonly used shape signatures, where different signature methods describe the shape's contour in different ways. For example, the centroid distance signature describes the shape by calculating the distance of each boundary point from the centroid, while the complex coordinate signature represents the boundary points by converting their coordinates into complex numbers. The boundary curvature describes the shape by measuring the bending of the boundary curve [10].

Before shape matching, shape signatures usually require normalization. This normalization ensures that the shape signature is invariant to translation, scaling, and rotation, meaning that even if the shape undergoes such transformations, the representation remains consistent. Therefore, normalized shape signatures are useful for shape matching and similarity comparison across different transformations.

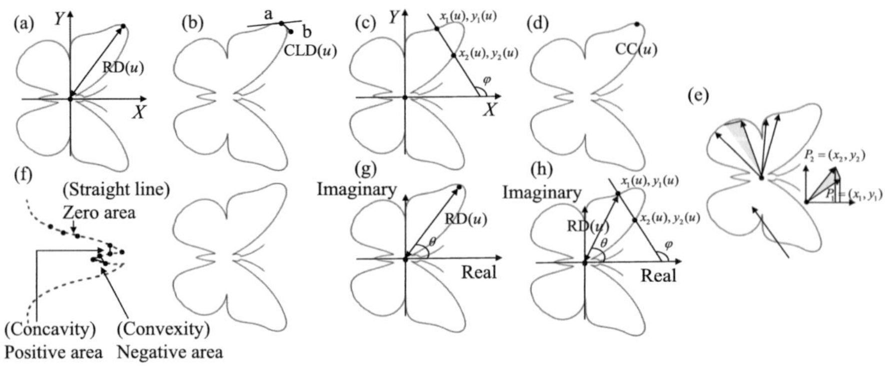

Fig. 8.11 Several commonly used shape signature schematics

8.5.2 *Fourier Descriptors*

Fourier descriptors are widely used methods for describing shapes based on closed contours. They describe the shape's boundary features using a set of Fourier coefficients derived from the Fourier transform. The main advantage of this method is that it simplifies the problem of describing a two-dimensional shape into a one-dimensional problem, significantly reducing the computational complexity. Below, Fourier descriptors are introduced based on the centroid distance shape signature.

Assume that the closed boundary curve of a shape consists of N points, and the coordinates of these points are (x_i, y_i), where $i = 1, 2, \ldots, N$ denotes each point on the boundary. The centroid distance function r_i is defined as the distance from the centroid to each point, given by

$$r_i = \sqrt{(x_i - x_c)^2 + (y_i - y_c)^2} \tag{8.49}$$

where (x_c, y_c) are the centroid coordinates of the shape, and the centroid coordinates (x_c, y_c) are defined as

$$x_c = \frac{1}{N}\sum_{i=1}^{N} x_i, \quad y_c = \frac{1}{N}\sum_{i=1}^{N} y_i \tag{8.50}$$

By applying the Fourier transform to the centroid distance function, we obtain the Fourier coefficients a_n and b_n. These Fourier coefficients represent the boundary features of the shape in the Fourier domain, with the formulas as follows

$$a_n = \frac{1}{N}\sum_{i=1}^{N} r_i \cos\left(\frac{2\pi ni}{N}\right) \tag{8.51}$$

$$b_n = \frac{1}{N}\sum_{i=1}^{N} r_i \sin\left(\frac{2\pi ni}{N}\right) \tag{8.52}$$

By normalizing these Fourier coefficients, we obtain the one-dimensional Fourier Descriptor that describes the shape's boundary. The Fourier descriptor is based on the theory of the Fourier transform and is invariant to translation, rotation, and scaling [11].

8.5.3 *Curvature Scale Space Descriptors*

The Curvature Scale Space (CSS) algorithm is a shape descriptor method that analyzes the curvature of a shape's boundary to extract features, identifying zero-crossings of curvature at different scales. It starts by selecting a pixel on the shape's closed boundary, sampling the boundary at equal intervals, and applying a Gaussian function to smooth the points. The curvature at each point is computed, and zero-crossings are identified. As the Gaussian function's standard deviation increases, the curve smooths, reducing zero-crossings until none remain, and the shape becomes fully convex. A CSS binary image is then generated as a feature descriptor, which is invariant to translation, rotation, and scaling. The process can be described as follows.

Let Γ denote the closed contour curve of a two-dimensional target shape in the plane, and μ represent the normalized arc length parameter. The curve can thus be expressed as a function of the arc length

$$\Gamma(u) = \{(x(u), y(u)), u \in [0, 1]\} \tag{8.53}$$

Evolving the curve via convolution with a 1D Gaussian kernel $g(u, \sigma)$ yields

$$X(u, \sigma) = x(u) * g(u, \sigma) \tag{8.54}$$

$$Y(u, \sigma) = y(u) * g(u, \sigma) \tag{8.55}$$

$$g(u, \sigma) = \frac{1}{\sigma\sqrt{2\pi}} \exp\left(-\frac{u^2}{2\sigma^2}\right) \tag{8.56}$$

where σ is the standard deviation and $*$ denotes the convolution operator. Accordingly, the curvature of the curve can be calculated as

$$k(u, \sigma) = \frac{X_u(u, \sigma)Y_{uu}(u, \sigma) - X_{uu}(u, \sigma)Y_u(u, \sigma)}{\left(X_u(u, \sigma)^2 - Y_u(u, \sigma)^2\right)^{3/2}} \tag{8.57}$$

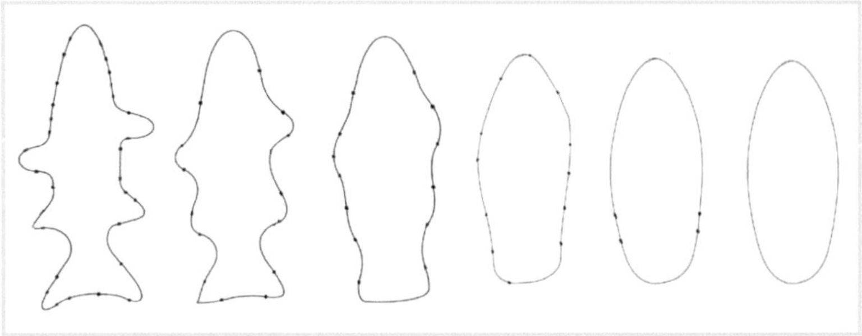

Fig. 8.12 The evolution of CSS images

$$X_u(u,\sigma) = \frac{\partial}{\partial u}[x(u) * g(u,\sigma)] = x(u) * g_u(u,\sigma) \tag{8.58}$$

$$X_{uu}(u,\sigma) = \frac{\partial^2}{\partial u^2}[x(u) * g(u,\sigma)] = x(u) * g_{uu}(u,\sigma) \tag{8.59}$$

$$Y_u(u,\sigma) = \frac{\partial}{\partial u}[y(u) * g(u,\sigma)] = y(u) * g_u(u,\sigma) \tag{8.60}$$

$$Y_{uu}(u,\sigma) = \frac{\partial^2}{\partial u^2}[y(u) * g(u,\sigma)] = y(u) * g_{uu}(u,\sigma) \tag{8.61}$$

As the scale parameter σ increases, the curve becomes smoother, and the number of curvature zero-crossings decreases. Eventually, the curvature zero-crossings disappear, and the curve becomes fully convex [12]. Figure 8.12 illustrates the evolution of the curve in the CSS method and the resulting CSS binary image generated from zero-crossings.

During the evolution process of CSS, the calculation of curvature zero-crossings is not affected by translation or scaling, making the CSS binary image invariant to these transformations. As a result, the CSS descriptor is robust and effective in shape matching and recognition tasks.

8.5.4 Shape Context Method

The Shape Context Method is one of the most effective contour-based shape analysis methods proposed in recent years. This method performs shape matching and recognition by analyzing the corresponding relationships between contour sampling points. The core idea of the method is to establish the context information of the shape by the relative relationships between each contour sampling point and other points, enabling shape matching and recognition.

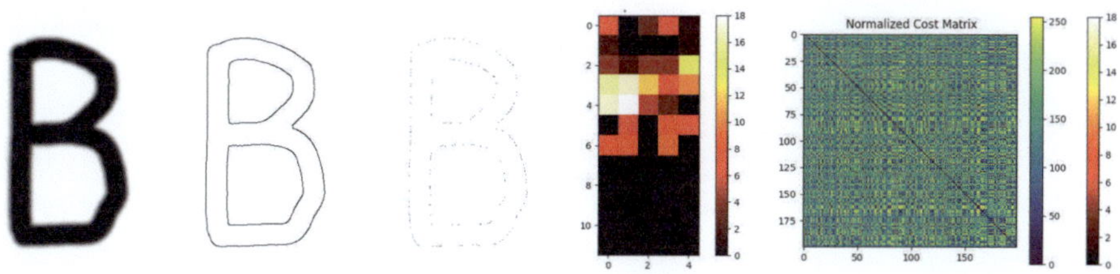

(a) (b) (c) (d) (e)

Fig. 8.13 Shape context diagram

Assume that the shape's contour consists of n sampling points, denoted as the point set $P = \{p_1, p_2, \ldots, p_n\}$. For any point p_i, it can form $n - 1$ vectors with the other points. The distribution of these vectors in the log-polar coordinate system forms a histogram, called the shape context of point p_i, expressed as

$$h_i(k) = \#\{q \neq p_i : (q - p_i) \in \text{bin}(k)\} \tag{8.62}$$

where, $k \in \{1, 2, \ldots, K\}$, K represents the product of the angular and radial parameters in the log-polar coordinate system, while $(q - p_i) \in \text{bin}(k)$ signifies that, with respect to point p_i, the point q falls into the k-th bin of the shape context histogram.

For each point in the point set P, its shape context histogram can be calculated based on Eq. (8.62), resulting in n histograms. These histograms form a matrix of size $n \times K$, which serves as a feature descriptor for the shape [13]. Figure 8.13 shows the shape context of shape A.

The shape distance metric is based on the difference between the shape context histograms of corresponding points. The distance between two corresponding points can be measured as:

$$C_{ij} = C(p_i, q_j) = \frac{1}{2} \sum_{k=1}^{K} \frac{\left[h_i(k) - h_j(k)\right]^2}{h_i(k) + h_j(k)} \tag{8.63}$$

where, $h_i(k)$ and $h_j(k)$ represent the shape context histograms of the corresponding points p_i and q_j, respectively. The cost matrix C encodes the matching costs between all possible point pairs. A bipartite graph matching algorithm is then applied to C to determine the overall distance between the two shape images, which serves as a measure of their similarity.

8.5.5 *Triangle Area Descriptor*

Triangle Area Descriptor (TAR) is a shape descriptor that characterizes the geometry of a contour by representing it in terms of signed triangle areas. TAR uses the signed

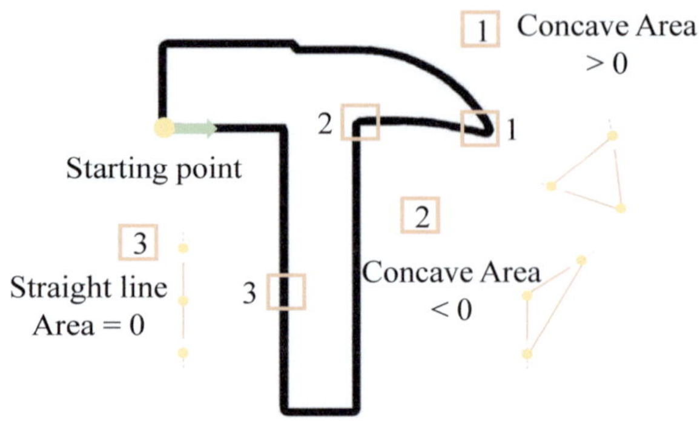

Fig. 8.14 Three types of TAR examples

area of triangles constructed from points on the shape's contour to measure the convexity and concavity of the contour. The TAR method includes both local and global features of the shape, and it is invariant to translation, rotation, and scaling. It also has strong noise resistance and distortion resistance [14].

Assume the shape's contour consists of N sampling points, denoted as the point set $\{(x_1, y_1), (x_2, y_2), \ldots, (x_n, y_n)\}$. The curvature at point (x_n, y_n) on the contour can be represented by TAR. For any three points on the contour, (x_{n-t_s}, y_{n-t_s}), (x_n, y_n), (x_{n+t_s}, y_{n+t_s}), $n \in [1, N]$, $t_s \in [1, N/2 - 1]$, the signed area of the triangle they form can be calculated using the following equation as

$$TAR(n, t_s) = \frac{1}{2} \begin{vmatrix} x_{n-t_s} & y_{n-t_s} & 1 \\ x_n & y_n & 1 \\ x_{n+t_s} & y_{n+t_s} & 1 \end{vmatrix} \tag{8.64}$$

When traversing the contour sampling points in a counterclockwise direction, the computed TAR values of positive, negative, and zero represent the convex, concave, and linear nature of the contour, respectively. Figure 8.14 shows examples of these three types of TAR. By gradually increasing the edge length of the triangle, i.e., increasing the parameter t_s, TAR values at different scales can be obtained. when t_s takes on all values, the multi-scale expression of TAR is obtained. This multi-scale TAR description can capture the changes in shape at different scales and effectively reflect both the local and global features of the shape.

8.5.6 Inertial Axis Descriptor

The Inertial Axis Descriptor (ALI-SF) is a method that describes shapes using symbolic feature vectors based on the minimum inertial axis as a reference. The minimum inertial axis is a line passing through the shape's centroid that minimizes the sum of the squared distances from all points on the shape's boundary to the centroid. As the only reference line, the minimum inertial axis preserves the main

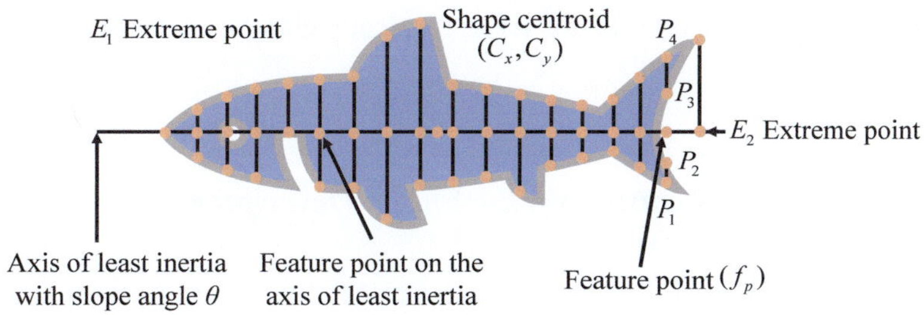

Fig. 8.15 The diagram of ALI-SF method

direction of the shape. Once the minimum inertial axis is determined, the points on the contour can be projected onto it.

The two points which are the farthest apart after projection, are called extreme points, and the Euclidean distance between them is taken as the length of the minimum inertial axis. Next, using the minimum inertial axis as a reference, symbolic features of the shape are extracted. Figure 8.15 shows a schematic diagram of the ALI-SF method. The specific steps are as follows

(1) n feature points are sampled equidistantly along the minimum inertial axis.
(2) For each feature point, a perpendicular line is drawn to the inertial axis and intersects the shape's contour.
(3) For shapes without holes and fully convex shapes, each perpendicular line intersects the contour at two points. The normalized distance (the ratio of the distance between these two intersection points and the length of the minimum inertial axis) is calculated and saved.
(4) For shapes with holes or concave parts, a perpendicular line may intersect the contour at multiple points. In such cases, only the normalized distance between the intersection points through the shape's enclosed regions is calculated and saved.

The feature vector generated by the ALI-SF method has a dimension of n, where each component of the vector represents the distance between the feature point and the contour's intersection points. This forms the ALI-SF feature vector F, where each component f_i represents the length of the segment corresponding to feature point i. It is important to note that in shapes with holes or concavities, some components of the feature vector F may be zero [15].

8.6 Region-Based Shape Descriptors

Region-based shape description methods not only describe the points on the object's contour but also the points within the entire shape area. Therefore, these methods are more suitable for complex shape analysis, such as shapes containing multiple regions or holes. This section will introduce several typical region-based shape analysis and description methods in detail.

8.6.1 Multi-level Feature Vector Shape Descriptor

Multi-level Feature Vector Shape Descriptor (MLEV) is invariant to rotation, translation, and scaling, and its extraction process consists of the following four steps:

Step 1: Obtain Feature Vectors

Given a shape, each pixel within the shape region can be regarded as a random vector $l_i = [x_i, y_i]^{\mathrm{T}}$ in a two-dimensional space, with N representing the number of pixels in the shape region. First, compute the mean position m_L and the covariance matrix C_L of all pixels within the shape region, then calculate the feature vectors e_j and eigenvalues λ_j using the following formulas

$$m_L = \frac{1}{N}\sum_{i=1}^{N} l_i \tag{8.65}$$

$$C_L = \frac{1}{N}\sum_{i=1}^{N} l_i l_i^T - m_L m_L^T \tag{8.66}$$

$$C_L e_j = \lambda_j e_j \tag{8.67}$$

Step 2: Divide the Shape into Four Regions

Assuming the principal axes of the feature vectors are e_1 and e_2, as shown in Fig. 8.16a, the principal axes divide the shape into four regions: $R1$, $R2$, $R3$, $R4$.

Step 3: Secondary Division

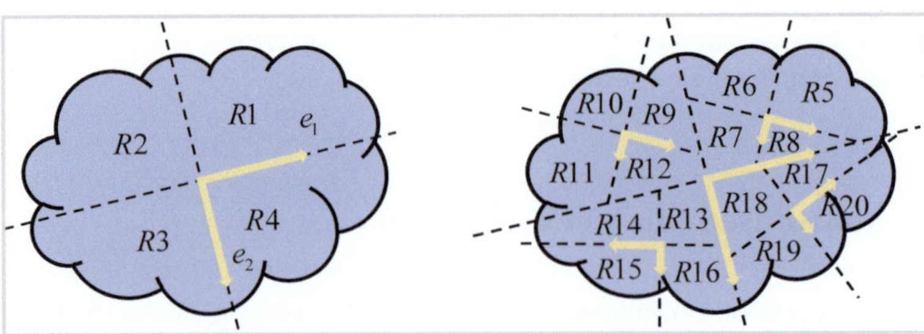

Fig. 8.16 Diagram of shape region division

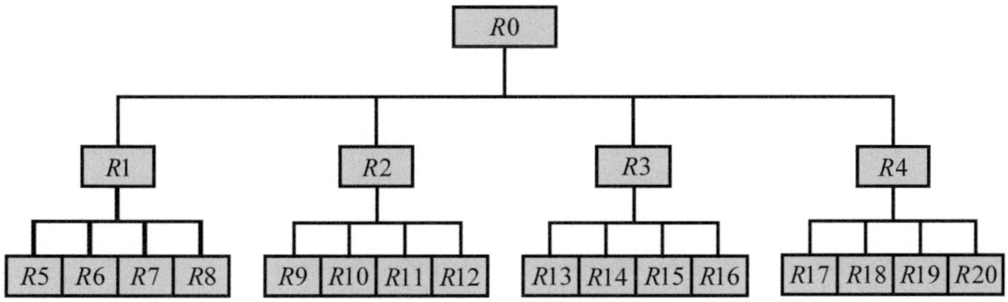

Fig. 8.17 Multi-level feature vector diagram

For each sub-region $R1$, $R2$, $R3$, $R4$, calculate their feature vectors and principal axes, and further divide each sub-region into four sub-regions. Figure 8.16b shows the secondary division results. After these two steps, a multi-level feature vector is formed, as Fig. 8.17.

Step 4: Extract Four Parameters with Invariance to Rotation, Scaling, and Translation

The multi-level feature vector shape descriptor is invariant to rotation, scaling, and translation and uses the following four parameters to support these invariances

(1) Eigen-ratio: The ratio of the eigenvalue of the second feature vector to the eigenvalue of the first feature vector, defined as

$$\text{eigen-ration}(i) = \frac{\lambda_{s,i}}{\lambda_{b,i}} \tag{8.68}$$

(2) Normal-angle: The angle between the principal feature vector of node i and the principal feature vector of the root node. There are two possible angles between these two vectors, and the smaller angle is taken as the normal-angle, as Fig. 8.18.
(3) Center: The ratio of the distance between the center of node i and the center of the root node to the eigenvalue of the principal feature vector of that node, are defined as

$$\text{center}(i) = \frac{|c_i - c_r|}{\lambda_{b,r}} \tag{8.69}$$

Fig. 8.18 Example of normal-angle

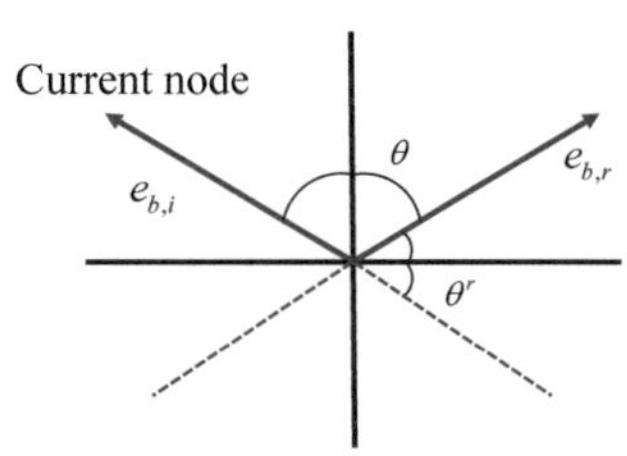

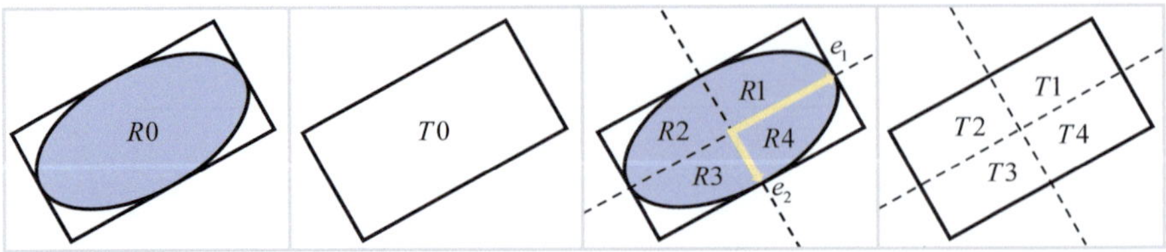

Fig. 8.19 Diagram of regional area and the corresponding border area

(4) Compactness: The ratio of the area of node *i*'s region to the area of the bounding box of that region. Figure 8.19 shows the relationship between region area and bounding box area, which is defined as

$$\text{compactness}(i) = \frac{R_i}{T_i} \tag{8.70}$$

After the division, a shape can be described by 21 regions, with each region containing four parameters. Therefore, a shape can be represented by 84 parameters. Due to the large number of parameters, the computation of the MLEV descriptor is relatively complex [16].

8.6.2 Angle Radial Transform

The Angle Radial Transform (ART) proposed by MPEG-7 is one of the region-based descriptors and belongs to the class of moment invariants for shape description. It is particularly suitable for describing shape regions. The MPEG-7 ART descriptor uses an angular transformation on the unit circle in polar coordinates to obtain moment invariants, and it employs the coefficients of quantized ART basis functions as descriptive features for shape matching.

The ART descriptor is built on a polar coordinate system, where sine basis functions are defined on the unit circle. Given an image function $f(\rho, \theta)$ in polar coordinates, the ART coefficients are defined as

$$F_{mm} = \int_0^{2\pi} \int_0^1 V_{mm}(\rho, \theta) f(\rho, \theta) \rho \, \mathrm{d}\rho \, \mathrm{d}\theta \tag{8.71}$$

where n is the radial order, m is the angular order, and $V_{nm}(\rho, \theta)$ is the ART basis function, which is separable in radial and angular directions, and is defined as

$$V_{nm}(\rho, \theta) = A_m(\theta) R_n(\rho) \tag{8.72}$$

where R_n is the radial basis function, and A_m is the angular basis function. The angular basis function is used to obtain direction invariants, while the radial basis function captures the scale information of the shape. The specific definitions for the radial and angular basis functions are

$$A_m(\theta) = \frac{1}{2\pi} e^{im\theta} \tag{8.73}$$

$$R_n(\rho) = \begin{cases} 1, & \text{if } n = 0 \\ 2\cos(\pi n \rho), & \text{if } n \neq 0 \end{cases} \tag{8.74}$$

According to MPEG-7, the radial order n is limited to $n < 3$, and the angular order mmm is limited to $m < 12$. The ART descriptor is a very compact and effective way of describing shapes. It is not only efficient in shape encoding but also has the ability to segment noise, providing robust performance even in noisy environments [17].

8.6.3 General Fourier Transform

Let $f(x, y)$ be the binary function representing the target shape. The method first calculates the shape's centroid (x_c, y_c) and the radius R of the smallest enclosing circle. The shape is then placed in a polar coordinate system with the centroid as the origin. The image $f(x, y)$ is then transformed into polar coordinates, as shown below

$$r = \sqrt{(x - x_c)^2 + (y - y_c)^2} \tag{8.75}$$

$$\theta = \arctan\left(\frac{y - y_c}{x - x_c}\right) \tag{8.76}$$

$$R = \max\{r(\theta)\} \tag{8.77}$$

By performing equidistant raster sampling along the radial distance, as Fig. 8.20, the shape is unfolded into a corresponding orthogonal Cartesian coordinate system. The function $f(r, \theta)$ is then subjected to a two-dimensional Fourier transform as follows

$$PF(\rho, \varphi) = \sum_i f(\rho, \theta) \exp\left[-i2\pi\left(\frac{r}{R}\rho + \frac{2\pi i}{T}\varphi\right)\right] \tag{8.78}$$

where, $0 \leq r \leq R$, $\theta_i = i(2\pi/T)(0 \leq i \leq T)$, $0 \leq \rho R$, $0 \leq \varphi \leq T$. ρ and φ are the frequencies sampled along the radial and angular directions. The Fourier transform captures the global features of the shape by transforming its representation in polar coordinates into the frequency domain.

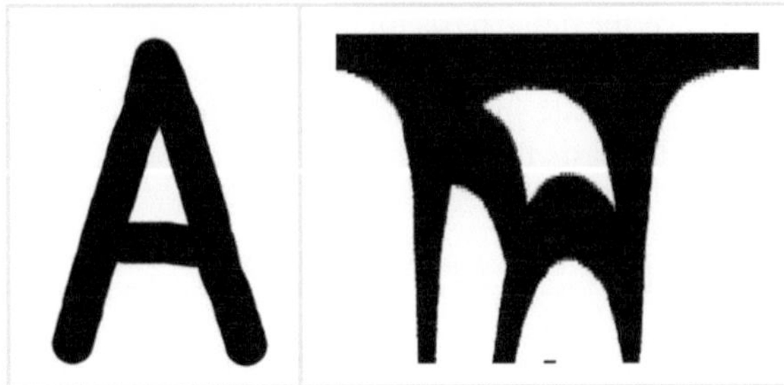

Fig. 8.20 The original image and its polar grating sampling image

After taking the magnitude and normalizing the Fourier coefficients, the desired General Fourier Descriptor (GFD) is obtained, as follows

$$\text{GFD} = \frac{|PF(\rho, \varphi)|}{\text{area}} \tag{8.79}$$

where area is same as the target region [18].

Shape recognition can be performed by directly computing the distance between GFDs, typically using Manhattan distance or other distance metrics. Compared to other methods like orthogonal moments, the General Fourier Descriptor is computationally simpler, and its features are entirely derived from the frequency spectrum, providing multi-resolution analysis in both the radial and angular directions. As a result, it maintains excellent descriptive performance and effectively captures the global characteristics of the shape.

8.6.4 Rotational Projection Descriptor

The Rotational Projection Descriptor (RPD) combines the advantages of both contour-based and region-based shape analysis methods, proposing a unified approach for handling various shapes. The main process of the RPD method involves scanning the shape with a ray passing through the shape's centroid, rotating counterclockwise through 360°, as shown in Fig. 8.21a. During the scanning process, three feature descriptors, $\delta(\theta)$, $\zeta(\theta)$, and $\xi(\theta)$, are extracted from the feature function at different angles, capturing internal structural information of the shape [19].

Let $f(x, y)$ be the binary image function of the target shape, (C_x, C_y) be the coordinates of the shape centroid, and λ_c be the distance from the centroid to the farthest point on the shape's contour. For convenience, the image is first converted to polar coordinates, as shown below

$$g(\rho, \theta) : [0, \lambda_c] \times [0, 2\pi] \to \{0, 1\} \tag{8.80}$$

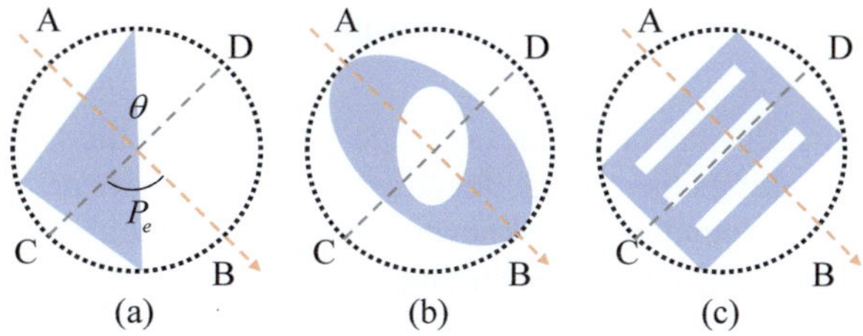

Fig. 8.21 Diagram of the RPD descriptor extraction

The three feature descriptors $\delta(\theta)$, $\zeta(\theta)$, and $\xi(\theta)$ extracted during the scanning process are defined as

$$\delta(\theta) = \frac{L_{\max}(\theta)}{\lambda_c} \tag{8.81}$$

where $L_{\max}(\theta)$ represents the distance between the two most distant intersection points along a scanline within the shape region, θ is the angle between the scanline and the horizontal axis, and λ_c serves as a scale normalization factor. The resulting descriptor $\delta(\theta)$ reflects the radial or wheel-like structural characteristics of the shape as

$$\zeta(\theta) = \begin{cases} \frac{\int_0^{\lambda_c} g(\rho,\theta)\mathrm{d}\rho + \int_0^{\lambda_c} g(\rho,\theta+\pi)\mathrm{d}\rho}{L_{\max}(\theta)}, & \text{if } L_{\max}(\theta) \neq 0 \\ 0, & \text{otherwise} \end{cases} \tag{8.82}$$

In Eq. (8.82), the denominator corresponds to the integral of the shape along the scanline at an angle of 0°, serving as a normalization factor. The descriptor reflects the saturation degree of the shape at a given angle, with larger values indicating a greater number of image pixels along that direction. As Fig. 8.21b, the saturation of the shape along scanline AB is higher than that along scanline CD.

$$\xi(\theta) = \begin{cases} \frac{e^{\left\{-\left[\mathrm{count}\left(\frac{\partial g(\rho,\theta)}{\partial \rho}=0\right) + \mathrm{count}\left(\frac{\partial g(\rho,\theta+\pi)}{\partial \rho}=0\right)\right]\right\}}}{L_{\max}(\theta)}, & \text{if } L_{\max}(\theta) \neq 0 \\ 0, & \text{otherwise} \end{cases} \tag{8.83}$$

The function count$(P(\alpha))$ serves as a counting operator that quantifies the number of intersections between a scanline and the shape region. For example, as Fig. 8.21c, scanline AB intersects the shape at 8 points, whereas scanline CD intersects at 4 points. Descriptor $\xi(\theta)$ measures the fluctuation of the shape along the scanline, whereas descriptors $\xi(\theta)$ and $\zeta(\theta)$ encode internal structural information of the shape region.

8.7 General Conclusions

This chapter introduces shape analysis methods, focusing on techniques for extracting object shape features from images. Shape analysis is performed when the object boundaries are known, extracting geometric and topological features through various algorithms. Mastering these concepts enables readers to handle more complex and variable object shapes, providing strong support for image understanding and automation.

Acknowledgements This work was supported by the National Natural Science Foundation of China (12374271, 62375065, 62405079, 62575091); Fundamental Research Funds for the Central Universities (HIT.OCEF.2024055, AUGN5750300325); the Fund of National Laboratory on Adaptive Optics, China (FNLAO-24-MS-O07); Natural Science Foundation of Heilongjiang Province of China (LH2024A009); Key Laboratory of Image Processing and Intelligent Control (Huazhong University of Science and Technology), Ministry of Education, Wuhan, China (IPIC2025-05); Key Laboratory of Advanced Medical Imaging and Intelligent Computing of Guizhou Province (AMIIC250204); the China Postdoctoral Science Foundation-CCTEG Joint Support Program (2025T011ZGMK).

References

1. Da Fontoura Costa L, Cesar Jr RM (2009) Shape analysis and classification: theory and practice, 2nd edn (Image processing series). CRC Press
2. Benchamardimath B, Hegadi RS (2014) A survey on traditional and graph theoretical techniques for image segmentation. Int J Comput Appl Technol 38–46
3. Wan SY, Higgins W (2003) Symmetric region growing. IEEE Trans Image Process 12(9):1007–1015
4. Tariq N, Hamzah RA, Ng TF, Wang SL, Ibrahim H (2021) Quality assessment methods to evaluate the performance of edge detection algorithms for digital image: a systematic literature review. IEEE Access 9:87763–87776
5. Meyer F, Beucher S (1990) Morphological segmentation. J Visual Commun Image Represent 1(1):21–46
6. Otsu N (1979) A threshold selection method from gray-level histograms. IEEE Trans Syst Man Cybern 9(1):62–66
7. Singh S, Mittal N, Singh H, Oliva D (2023) A robust parameter-free thresholding method for image segmentation. Multimed Tools Appl 86(26):40701–40743
8. Montero RS, Bribiesca E (2009) State of the art of compactness and circularity measures. Int Math Forum 4(27):1305–1335
9. Zhang DS, Lu GJ (2004) Review of shape representation and description techniques. Pattern Recognit Lett 37(1):1–19
10. El-Ghazal A, Basir O, Belkasim S (2009) Farthest point distance: a new shape signature for Fourier descriptors. Signal Process Image Commun 24(7):572–586
11. Musso E, Nicolodi L (2009) Invariant signatures of closed planar curves. J Math Imaging Vision 35(1):68–85
12. Mokhtarian F (1995) Silhouette-based isolated object recognition through curvature scale space. IEEE Trans Pattern Anal Mach Intell 17(5):539–544
13. Mori G, Belongie S, Malik J (2005) Efficient shape matching using shape contexts. IEEE Trans Pattern Anal Mach Intell 27(11):1832–1837

14. Alajlan N, Kamel MS, Freeman G (2008) Geometry-based image retrieval in binary image databases. IEEE Trans Pattern Anal Mach Intell 30(6):1003–1013
15. Ranzato M, Taylor PE, House JM, Flagan RC, LeCun Y, Perona P (2007) Automatic recognition of biological particles in microscopic images. Pattern Recognit Lett 28(1):31–39
16. Kim HK, Kim JD (2000) Region based shape descriptor invariant to rotation, scale and translation. Signal Process Image Commun 16(1):87–93
17. Ricard J, Coeurjolly D, Baskurt A (2005) Generalizations of angular radial transform for 2D and 3D shape retrieval. Pattern Recognit Lett 26(14):2174–2186
18. Zhang DS, Lu GJ (2002) Shape-based image retrieval using generic Fourier descriptor. Signal Process Image Commun 17(10):825–848
19. Chen YW, Xu CL (2009) Rolling penetrate descriptor for shape-based image retrieval and object recognition. Pattern Recognit Lett 30:799–804

MIX
Papier aus verantwortungsvollen Quellen
Paper from responsible sources
FSC® C105338

If you have any concerns about our products,
you can contact us on
ProductSafety@springernature.com

In case Publisher is established outside the EU,
the EU authorized representative is:
Springer Nature Customer Service Center GmbH
Europaplatz 3, 69115 Heidelberg, Germany

Printed by Libri Plureos GmbH
in Hamburg, Germany